Strategy and Tactics of the Mexican Revolution, 1910-1915

Joe Lee Janssens

PREFACE

My objective when setting out to write the *Maneuver and Battle in the Mexican Revolution* trilogy was to create the definitive military history of the revolution. I am sympathetic, however, to those who have an interest in the Mexican Revolution but can do without all the minutiae. Therefore, I have pulled together this simple book, endeavoring to keep it as brief as possible. Consequently, the narrative and analysis have been limited to basic strategy and tactics, eschewing expositions on logistics, field medical practices, military culture and militarism, and esoteric insights into grand strategy (psychological and economic warfare, propaganda, and diplomacy). No attention has been given to the why's and wherefore's of the revolution's many shifts in alliances, nor to debunking tired conspiracy theories about undue influence and covert operations allegedly perpetrated by the United States. To do that would necessitate thousands of pages, which has already been done in the aforementioned trilogy. Additionally, guerrilla warfare, which was ancillary to the outcome of the revolution, will not be covered.

Finally, I realize that the enormous cast of characters may overwhelm some readers. Consequently, and keeping with the spirit of simplification, I recommend focusing attention on the following: Francisco Madero and Venustiano Carranza; Generals Victoriano Huerta, Pascual Orozco, Álvaro Obregón, Pablo González, Jacinto Treviño, Gabriel Gavira, and, of course, Francisco "Pancho" Villa. It might also help to read the Introduction, Conclusion, the body text, and then the Conclusion again, in that order.

ISBN 13: 978-0-9964789-5-3

Revolution Publishing, Houston, TX

To the Mexican soldier

Contents

FIGURES AND MAPS

INTRODUCTION

At the dawn of the twentieth century a tremendous angst gripped the armies of the world. Europe had experienced almost a century of relative peace since the devastating Napoleonic wars, but tensions on the continent were high. Technology had made the battlefield more lethal and many wondered if the lessons learned (Clausewitz, Jomini) from the Napoleonic wars were still applicable, and if current tactical doctrine sufficiently incorporated innovations to deal with the latest developments. It appeared that the answer to both questions was "no."

WEAPON SYSTEMS AND THE FIN DE SIÈCLE BATTLEFIELD

By 1910, ammunition had undergone significant improvements, most especially with the advent of smokeless powder that removed a large amount of haze and virtually eliminated any "signature" produced by gunfire. Still, other factors could obscure visibility in combat, namely fog, rain, darkness, and, particularly in northern Mexico, dust. But the modern battlefield would have a visibility not seen since the Middle Ages, especially given the application of artificial light sources. Additionally, bullets had become smaller and therefore lighter, meaning that soldiers could carry more into battle and once fired the projectiles followed a flatter trajectory, remaining in the kill zone longer and farther. Rapid-fire crew-served weapons also filled the air with more lead, but at the same time required numerous men to operate, soldiers who otherwise would have been available to serve on the line of fire. And the issue of how to maneuver machine guns during assaults—both mounted and on foot—still had not been entirely resolved. Taking all these factors together, the dilemma that emerged was: how to mass soldiers for offensive firepower but remain dispersed to attenuate the devastating effects of return fire.

Most military theorists agree that it is innovation, not technology, per se, that drives changes in tactics. Innovation is at the top of the pyramid, or tip of the spear, if you will, formed by Innovation (how to make use of the available technology), Technology (applied science), and Science (knowledge about how the material and physical universe functions). At the time of the Mexican Revolution the two most recent wars from which to draw lessons were the Boer War and, more importantly, the Russo-Japanese War. Indeed, the Federal Army General Staff had Martynov's book on the latter conflict translated into Spanish.

Matynov accurately described the *fin de siècle* battlefield and provided tactical prescriptions for future combat. Accordingly, the solution for infantry assaults inclined toward the dispersion of troops and bounding assaults, which created new concerns. Men charging from one covered position to another might get comfortable in one location and stop advancing without noncommissioned officers, in their traditional role, being able to maintain order and keep the men moving forward. Officers, too, worried about the effects on command and control of dispersed troops that—in the words of Martynov—gave the battlefield "an almost completely empty look." Units scattered in such a manner could rarely orchestrate volleys of massed fire, making it harder to gain fire superiority. As a result, a certain stasis developed in battles that caused them to go on for days and even weeks. Adaptations to these conditions included improved communications such as heliographs, telegraphs (wired and wireless), telephones (field and civilian), and a new style of leadership that emphasized positive motivation over the negative reinforcement that had been so common to eighteenth century armies. The school of the soldier also went through a transformation as soldiers were expected to embrace professionalism, meaning rudimentary knowledge of tactical doctrine and competency in marksmanship, among other combat skills. Still, much of this was beyond the patience of most officers and the abilities of the raw material available as recruits. Therefore, many European armies simply crafted a cult of irrational battlefield morale that they hoped would drive offensive operations to a successful conclusion.

A small minority of officers, however, came to realize that although the offensive may have been deemed superior in the past—because it had salutary effects on morale and afforded freedom of action and initiative—the advent of mass armies composed of citizen-soldiers, attended by a shortage of professionally trained staff officers, now made maneuver difficult. These new personnel limitations led to the favored tactic of choosing one point to assault on the opponent's line, hoping to break the weakest link in the chain. Yet this practice held no guarantees for success and in the event of failure could produce enormous casualties. Superior firepower and the defensive were gaining in attractiveness, with the preferred combination being a strategic offense and tactical defense.

Best practices for infantry on the defense were much easier to glean and implement than for offensive maneuvers. A battalion's two machine guns emplaced on each flank could direct grazing fire in quasi-enfilade against the opposite wings of the enemy. Machine guns could also be employed in indirect fire missions, to clear the reverse slope of a hill or railroad embankment, for example. Wire strung outside the perimeter served to keep opposing forces out of grenade range or, in the case of the Mexican Revolution, from hurling dynamite either by hand or with a sling.

For artillery, the new operating environment called for grouping cannons in hidden positions several kilometers behind the lines and laying down indirect fire, as directed by forward observers from concealed and/or covered vantage points. The idea was to saturate extensive zones occupied by enemy infantry with shrapnel and to inhibit the effectiveness of runners (communications). Direct fire (lining up the objective in sights) became a rarity as gunners calculated fire missions to hit unseen targets according to complex formulas, to include the coefficient of gravity, which for most of Mexico is a little less than ten. The days of artillery duels were long gone with only a minority of guns assigned to counterbattery fire missions.

Gone, too, for the most part, were cavalry charges and use of the white arm— saber and lance—for shock value. The cavalry had its mission sharply curtailed to its traditional ancillary roles: scouting and reconnaissance, foraging, screening the movements of the main body, and protecting the lines of communication.

FEDERAL ARMY DOCTRINE

Federal Army Doctrine in theory, if not always in practice, conformed rather well to these new combat realities.

Beginning with the revised 1887 regulations the infantry adopted the ternary system, which consisted of an echeloned approach to combat broken down into a firing line ("chain of shooters"), support, and reserve. Each echelon performed a distinct role during the various phases of combat, which were designated as: 1. Reconnaissance of the position to be attacked or, alternatively, occupied and defended, 2. Preparation of the Attack or, on the defense, skirmishers and scouts harassing and maintaining contact with the enemy during his approach, 3. Combat, and 4. Pursuit or Retreat. The commanding officer paid the utmost attention to formally identifying THE KEY to the position to be attacked or defended to his subordinates—an acknowledgement of the practice of locating the most vulnerable point on the line to assault.

To facilitate application of the ternary system, the Federal Army reorganized its infantry on a base of three's. The new company had 125 men composed of three sections, which in turn had three 13-man platoons, with associated staff and leadership at every level. The new doctrine confirmed the primacy of the company as the unit of maneuver in the new open order of combat and fixed the minimum distance between soldiers at 1.5 meters. Each battalion consisted of about 550 men, including a staff and four companies, which could deploy for combat with two companies (separated into left and right wings) forming the firing line and the other two as support and reserve.

On a micro level, the ternary system made the cannon fodder of yore obsolete. A soldier had to know how to judge distances and the advantages of terrain, to shoot his rifle accurately, and to understand how the ternary system worked as a whole during the various phases and types of combat, as well as his own role under the corresponding scenarios. The army expected soldiers to apply themselves in learning and instructed officers to give orders to their charges during training with respect and in a clear and concise manner.

In a country founded on Iberian culture, army regulations paid particular attention to the defense of plazas. The Federals realized it could be difficult maneuvering troops between streets and houses and delivering messages to all units in preparation for a retreat or counterattack. Therefore, doctrine recommended detaching a force outside the plaza in a concealed location to be able to execute a counterattack around the plaza, that is, on the flanks, in combination with a push forward by the firing line. In the case of defending a plaza "at all costs," protocol entailed successive lines of defense and the use of a sturdy structure, such as a bank, barracks, church, or customs house, as a rally point and redoubt—a castle keep of sorts—even though some military theorists discount the effectiveness of successive lines of defense because of its impact on morale; those troops on the outer perimeter may see themselves as sacrificial lambs for the protection of the inner perimeters.

The defensive could also have deleterious effects on morale because it left the defenders in a constant state of anticipation, which made "it very inferior to the offensive." Alternatively, the Federals recognized that the defensive provided soldiers with the best use of terrain and maximized firepower although "paralyzing to a certain point their means of action." Ergo, the Federal Army did not revel in the idea of preparing complex field works, such as later became common in World War I, but rather focused on hasty defensive positions that made use of existing fences, walls, and buildings.

In open terrain against cavalry, the sole prescription for the infantry was to avoid complex maneuvers, such as forming a square from a column of march that might unsettle the troops and waste time, and to simply trust in the overwhelming firepower of their rifles (again, use of the white weapon, the infantry's bayonet, was vitually unheard of during the Revolution). Contrary to popular impressions, doctrine did not dictate forming a square—which exposed the laterals to enfilading fire—but instead recommended a triangle (although never when enemy artillery accompanied the cavalry) or a circle, if the commander felt compelled to arrange his men in a predetermined geometric formation. And, these formations were intended only as transitory, not for a prolonged defense.

In sum, Federal Army doctrine for infantry was ahead of many armies in that it put faith in firepower over irrational battle courage and shock assaults, eschewed infantry squares instead favoring triangles and circles, but only for a short time and under limited conditions, and it paid close attention to the defense of plazas, particularly addressing communications and counterattacks. The prevailing culture, however, discouraged the use of fortifications except in a transitory manner, which would have dire consequences in certain future battles.

Cavalry tactics and units had also been appropriately reordered. During an advance, the cavalry was to make contact and secure and hold the favorable ground (with support from horse artillery) until the main body of infantry could arrive. During combat—the third phase of battle—the cavalry would fall back to the rear to secure supply lines, attack vulnerable enemy artillery, and possibly execute rearguard attacks delivered over long distances with speed and surprise. In the culminating fourth phase, the cavalry would come up to cover the retreat of the infantry and artillery or, in the event of victory, attack with weight to press home the assault and carry out the pursuit.

Regulations confirmed the unsuitability of cavalry for protracted dismounted engagements precisely because such fighting automatically resulted in a twenty-five percent reduction in combat strength, since one in four troopers—the worst marksman—had to hold the horses of the other three. Conversely, the mobility of mounted troops could act as a force multiplier, enabling them to appear in multiple locations over a short period of time. Still, doctrine explicitly confessed the supremacy of firepower over the shock value of the white arm and reaffirmed the cavalry in its traditional auxiliary roles such as scouting and foraging.

The cavalry regiment at peacetime strength only numbered about 425 men including staff and four squadrons, which in turn consisted of three sections made up of two 16-man platoons. Since revolutionaries in the North (Sonora excepted) consisted almost entirely of dragoons and adopted the Federal Army table of organization, it was possible to have a "minimum" (only two regiments) brigade with slightly fewer than one thousand men. Also common were "corps" of about three hundred mounted men, somewhat modeled on the organization of the Rural

Police Corps ("*Rurales*"). These tactical corps were distinguished from an "army corps" formed of two or more divisions.

During a movement to contact, doctrine called for the horse artillery to accompany the cavalry in the vanguard to locate the enemy. Employing superior firepower and maneuver, the two combat arms would then work together to gain the favorable ground for the follow-on infantry, or retreat and inform the commanding officer of the location, composition, and numbers of the enemy.

The 9:1 ratio of shrapnel to high explosive shells as required by regulations indicated that in battle the artillery would focus its attentions on enemy infantry, attempting to break its morale and keep it from maneuvering while providing friendly forces with the freedom of action to conquer terrain. In order to fulfill that mission, and by rule, field pieces were to be employed in groups, and as individual batteries only as an exception, such as in the vanguard or in mountain fighting. The detaching of individual sections and direct-fire missions were expressly prohibited, and batteries were to be sent to forward positions only if their egress could be assured. The use of surprise was encouraged and achieved by keeping cannons hidden or by speedily and unexpectedly advancing, going into battery, and shooting. Objectives of opportunity included those clearly observable, of extensive size, and within three kilometers—a range appropriate to the Federal artillery's mainstays: the 75-mm. Saint Chamond-Mondragón and Schneider Canet.

Although artillery duels were understood to be a thing of the past, the artillery was expected to assign a priority to destroying the enemy's artillery on par with his infantry when determining fire missions, of which regulations mentioned three types: precision, grazing, and progressive.

Precision worked best against a single target, such as to destroy objects and cover, while grazing was to be employed against an extensive target such as a line of deployed infantry. Finally, progressive fire functioned best in the case of an enemy moving along a single axis perpendicular to the front, such as advancing or retreating down a road.

In the pursuit phase of battle, the artillery would employ its mobility to quickly displace and accompany the infantry or simply adjust the range of fire, if practicable. In defeat officers were to make every effort to save the artillery assets, most especially the pieces, although no dishonor attached to those officers who lost guns in the performance of their duty.

Because coordination between guns, batteries, forward observers, and the other combat arms was so critical to its mission, the artillery assumed responsibility for battlefield communications. Forward observers received training in various modes of communication such as hand signals and, most of all, field phones. Indeed, responsibility for setting up the field phone system in battle fell to the artillery, and battery personnel had to know the telegraphic alphabet.

The Federal Army's main weakness in terms of artillery resulted from the makeup of its arsenal, which consisted of guns, a handful of mortars, and no howitzers. The preponderance of guns significantly reduced the artillermen's ability to apply indirect fire. And the largest guns in the arsenal were 80-mm., too small to destroy any works except those lightly constructed for cover.

Finally, the Federals did not resolve the mobility issue of how to employ machine guns in attacks, and instead treated them like artillery to be deployed in batteries and in defilade, a worst-in-class practice.

WAYS OF WAR

Every historian worth his salt knows that it is not enough to discuss the state of the art, or professed military doctrine. Only what actually took place matters. In the case of the Mexican Revolution, any discussion of the tactics practiced requires a reference to Where and When, which in turn speaks to Who and points to How, since there were at least six different Mexican Ways of War in play during the critical five years of maneuver warfare: Chihuahuan, Lagunero, Sonoran, Northeasterner, Federal Regular, and Suriano. Hence, as maneuver warfare in the revolution progressed from its beginnings in Chihuahua to neighboring states and then spread to the heartland and eventually encompassed the entirety of the nation, it pulled in new sections, each with its own martial traditions, or ways of war. In other words, as the geography of the conflict changed, so too the actors and their tactics, and with the increase in territory and population joining the fight, over time, battles and campaigns grew in scope, scale, sophistication, and speed of development. Therefore, discussing strategy and tactics in the Mexican Revolution requires distinguishing both time and space. To aid in this effort we have recourse to Jean Bodin's methodology for historiography: a call for plain language instead of purple prose; a focus on geography and climate, which powerfully influences cultural behaviors and practices; and an attention to chronology, with a view to reducing errors.

What follows are six chapters arranged according to the main stages of maneuver warfare: Madero's Rebellion in Chihuahua (November 1910 to May 1911); Orozco's Insurgency throughout the North (February 1912 to October 1912); the Constitutionalist Army's founding in response to General Victoriano Huerta's palace coup (February 1913 to January 1914); the Constitutionalist Army's strategy and campaigns that defeated the Federal Army and conquered Mexico City (January to August 1914); the Civil War (September 1914 to April 1915), which witnessed the creation and high point of the Conventionist Army and campaigns from Baja California to the Yucatán; and the final strategy and operations (June to December 1915) that resulted in the Constitutionalist victory.

Each chapter contains a narrative of the battles and campaigns as related by the combatants themselves, interwoven with tactical analysis and discussions of the strategies involved. Occasionally there will be commentaries on the received history as perpetuated by the Mexicanists (those of the professoriate who study Mexican history and civilization), which are corrected in virtually every respect through the lens of the military historian or contemporary observers. The conclusion recapitulates the strategies in each phase of the revolutions and the tactics as practiced between 1910 and 1915 according to the various traditions, or ways of war, by combat arm.

Analyses of campaigns and pivotal battles, the author hopes, will finally dispel the common misperception of the Mexican Revolution as a principally guerrilla—or "small"—war.

1. MADERO'S REBELLION, NOVEMBER 1910 – MAY 1911

FIGHTING FORCES

lthough the movement directed by Francisco I. Madero to topple the regime of Porfirio Díaz was national in scope, its success counted entirely on the efforts of Madero's adherents in Chihuahua. The leader of the Maderistas in that state, Abraham González, put together an alliance of successful, struggling, and failed businessmen (especially from the mountainous, western part of the state) who enjoyed the loyalties of their customers and employees and resented the economic and political stranglehold of the Terrazas-Creel oligarchy. Additionally, politically-repressed citizens joined local liberal clubs sympathetic to Madero, which in turn provided a network conducive to subversive activities and paramilitary organization, augmented by a second network of malcontent Protestants with their own societies, temperance leagues, and press. These Maderistas tended to be family men who brought their friends, neighbors, and sons in train. They were joined by career outlaws, such as Pancho Villa, who believed that the closed legal and economic system had forced them into a life of crime. They knew the terrain and how to ride and shoot and had their own underground networks, all this in a state where civilians had taken a leading role in defending their homelands against Indian incursions, the last of which had taken place only about twenty-five years earlier.

The Federal Army, meanwhile, had largely been engaged in fighting Indians in Sonora and the Yucatán and putting down periodic flash rebellions. In 1910 the army contained as few as 14,000 men and as many as 30,000, depending on sources, but most likely just under 25,000. In the 2nd Military Zone, which covered the states of Chihuahua and Durango where the Maderista movement posed the greatest threat, the Federals had two generals, 13 *jefes* ("field officers"), 69 line officers, 1340 soldiers, 416 horses, and 77 mules at the outset of hostilities.

Even at that late date the Federal Army largely retained the institutional culture of an eighteenth-century army, such as the one made infamous by Santa Anna, in spite of the highly educated, skilled, and motivated soldiers required by Federal Army doctrine. Its soldiery consisted mostly of political dissidents, criminals (sometimes impressed wholesale from jails), vagrants, and hapless Indians dragooned into service. Brutish noncommissioned officers, whose main function was to administer corporal punishment, kept malcontents under control. Line officers were mostly political appointees unexceptional in every regard, especially intellect, whose overriding concern was to keep their charges from misbehaving or escaping—in sum, nothing more than glorified jailers. At the apex of the organization stood field and general officers who were often excessively aged to the point of being physically incapable of performing their duties; corrupt; ignorant of recent developments in tactics, strategy, the operational arts, and how to properly employ modern weaponry; and skilled only in the services of the garrison. There were pockets of excellence, mainly among the Technicals of the artillery, engineers, and the general staff, but they formed a minute minority in an army where egregious incompetence and criminality obtained.

Professional education was nonexistent for noncommissioned officers and minimal for officers after commissioning. Field and general officers, who could have benefitted from large unit exercises, were too distracted with administrative responsibilities such that maneuvers on that scale never took place. Aside from simple instruction at the battalion and regiment level, no schools or programs for line officers existed except the annual "academies," where participants tested in the same rudimentary material year after year, which they learned to master. When evaluated in unfamiliar settings reflecting real world scenarios they failed miserably. Ignorant in their profession, most officers simply trusted in their ability to muster enough courage to obtain victories in future combat.

INITIAL OPERATIONS, 1910

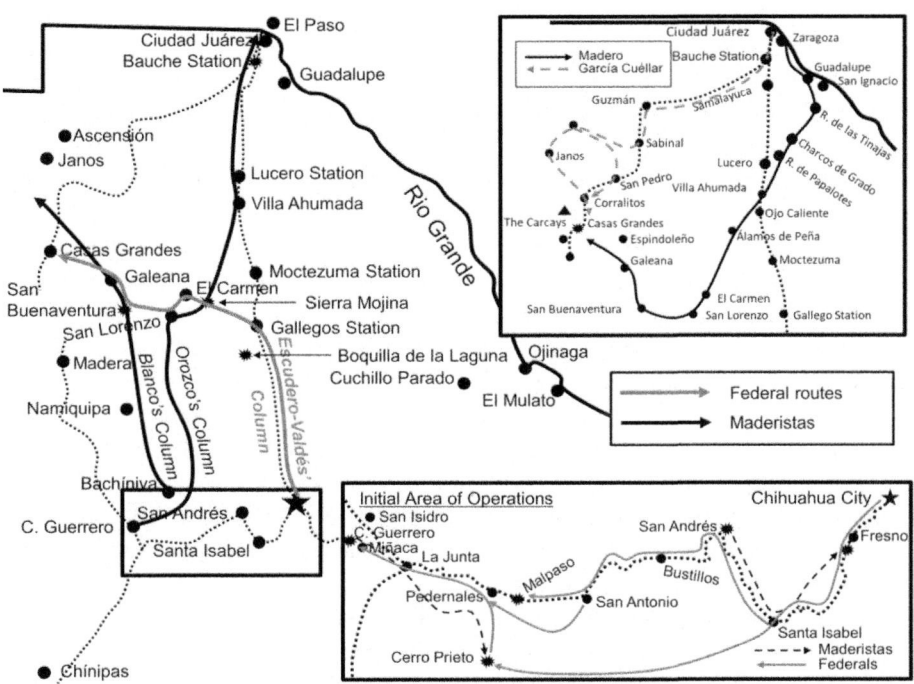

Map 1 Chihuahua, 1910 – 1911

Francisco I. Madero established November 20, 1910, as the official date for his rebellion to begin. Nevertheless, Toribio Ortega rebelled in Cuchillo Parado with sixty men on November 14 upon learning that he was about to be arrested for subversive activities. He and his men chased off the local *jefe político* and fled into the sierra where they were soon joined by Don Abraham González, Chihuahua's Maderista leader. To quell the disorder, the Federals sent Colonel Alberto Dorantes with a cavalry regiment and two machine guns (just over 400 men) from Chihuahua City to reinforce the garrison at Ojinaga threatened by Ortega.

Ortega waited for Dorantes at Rancho de Vanegas where his men pulled telegraph wires half-way to the ground, lit campfires inside the ranch, and retired to the hills to the west. On a pitch-black night, the advancing Federals caught sight of the fires and supposing the rebels to be gathered there charged ahead into the ambush. The rebels opened fire from the hills and the frightened Federals fled the field along the banks of the Rio Grande where they ran into the telegraph wire and were unhorsed. The majority of the Federals crossed over into American territory and returned machine-gun fire back into Mexico from the banks. Don Abraham ordered his men to break contact and retreat since they had limited ammunition. The Federals continued firing into the darkness as the rebels moved to El Mulato on the border to begin stockpiling ammunition.

On December 18, 1910, Dorantes attacked Ortega at El Mulato, but once again Ortega's rebels repulsed the Federals. Abraham González crossed over into Texas and went to El Paso to meet with Francisco Madero, who was rumored to be in that city.

In western Chihuahua the citizens of San Isidro, chief among them Pascual Orozco Jr., also had to take up arms early because of threats from the local authorities. On November 19 Orozco and his men marched off to Miñaca, an important rail and commercial center, where they obtained arms, provisions, and additional adherents. The men then returned to subdue their hometown, gained more war materiel and men, and proceeded to lay siege to the Federal garrison in Ciudad Guerrero beginning on November 21.

In the center of the state, the Maderistas from around Chihuahua City had gathered in Sierra Azul under Cástulo Herrera in preparation for the attack on the state capital. Rebels inside the city planned to dynamite the Federal barracks, cut the telephones, telegraphs, and power and then apprehend the local authorities on November 20 as Herrera's men came down from the sierra and assaulted the town. But events did not go according to plan.

On the appointed date, Herrera's men left the sierra and entered San Andrés, taking the town without any resistance. Shortly thereafter a passenger train approached on its way from Chihuahua City to Ciudad Guerrero with two companies of Federals aboard to crush Orozco. Herrera's Maderistas shot up the train, killing the Federal commanding officer, as it hastened on its way westward to escape the ambush. Herrera spent three days in the area, recruiting additional men before again moving toward Chihuahua City.

Citizens continued to flock to the rebel standards in Ciudad Guerrero, but on November 24, Orozco had to divert his attention from the siege to the Federal relief force that Herrera had shot up in San Andrés. Taking about fifty men from his growing command, including his chief lieutenant Marcelo Caraveo, Orozco slipped off to Pedernales and attacked the two companies of Federals commanded by Captain Manuel Sánchez Pazos. Those Federals hailed from Aguascalientes and, unaccustomed to the cold North, had taken refuge from a blizzard in a *bodega* without establishing proper security. The rebels surrounded the building and in the early morning of November 28 surprised the Federals during roll call. Captain Sánchez Pazos was killed as he tried to climb to the rooftop of a house to get a commanding view of the situation, and the remaining terrified Federals fled into the sierra. The Maderistas collected arms and ammunition and even recruited a few of the dispersed Federals. On December 3 a mere fraction of the original 160-man force arrived in Chihuahua City.

Second Captain Joaquín Castillo, who took over after the death of Sánchez Pazos, blamed the defeat on the energy of the rebel attack force, which he estimated to outnumber the Federals two to one, the initial demoralization from being surprised that caused some of his men to scatter, and a lack of ammunition.

Meanwhile, after learning about the ambush of the train transporting the relief force at San Andrés, 2nd Zone command had prepared a combined-arms column of 650 men under General Juan J. Navarro to leave Chihuahua in the morning of November 27 to provide additional relief to Ciudad Guerrero. The Federals were headed on a collision course for Herrera, who had occupied Santa Isabel, recruited, organized his 250 men, and departed on November 26 to attack Chihuahua City.

On the morning of November 27, Herrera was waiting on reinforcements from Ciudad Guerrero when he ordered his men to break up into teams of about fifteen men to reconnoiter the capital. For undisclosed reasons, Navarro had orders to send one hundred troopers back to Chihuahua City after he had traveled a distance of sixteen kilometers. As this returning cavalry detachment passed by Tecolote Hill, it began to take fire from the Maderistas on the heights—including from one of the teams commanded by Francisco "Pancho" Villa—who had previously allowed the larger column to pass by undisturbed. The main body under Navarro soon returned to give aid to the cavalry and in the ensuing battle dealt a drubbing to the woefully outmanned and outgunned rebels. Thereafter, the Federal column returned to Chihuahua City. Villa blamed himself for the loss of fifteen killed and three wounded out of his team of twenty-three, citing his "nothing knowledge in the art of war and the impetuous blood of boys."

By December 3, General Navarro's combined arms column was once again underway to rescue Ciudad Guerrero but had only reached as far as Santa Isabel. The next day the Ciudad Guerrero garrison surrendered after the rebels informed the Federal commander that they were tunneling underneath the barracks and would soon blow it up with dynamite. The Federals realized the futility of holding out any longer and surrendered all their war materiel in exchange for their lives. The officers changed into civilian clothes to escape reprisals and arrived at Chihuahua City by train on December 7. In the meantime, Cástulo Herrera's Maderistas (once again back in Sierra Azul) left to join Orozco in Ciudad Guerrero to present a united front against the advancing Navarro.

At Santa Isabel, Navarro decided to abandon the rails and continue cross-country for three reasons: to avoid being lured into another fight against Herrera that would detour him from his mission to attack Orozco; to screen his movements from rebel sympathizers, especially railroad employees; and to surprise the rebels with a rearguard attack. Initially his plan seemed to be succeeding, but he lost the element of surprise on December 8 when the Maderistas received reports that Navarro was on the plains of San Juan Bautista.

During his movements Navarro had been reinforced by another three hundred men coming from Chihuahua City. Additionally, the War Department had ordered another four hundred Federals and Sonoran National Guardsmen to move up to Chínipas from the southwest in order to cut off any rebels trying to flee the approach of Navarro's column.

Far from running away, the Maderista who had been placed in charge of a group of rebels to attack Navarro, Francisco Salido, ordered his forces to Cerro Prieto and sent scouts to ascertain the exact location of the Federal column. Informed that Navarro was close and approaching, on December 11 the Maderistas took up position behind stone walls and rocks on the heights to the east of Cerro Prieto and waited. Soon, the Federal column passed by in a column of march parallel to the heights, which ran in a general north-south direction. Distracted by the burdens of their personal gear and impedimenta, they were oblivious to the presence of the rebels.

Once the Federal column had advanced far enough to cover the extent of the entire rebel front, shooting rang out on the south end, preventing the Federals from reversing course without rolling up their own column. In the initial pandemonium, the Federals panicked, some firing wildly, others hiding, and still more dropping to feign death. Eventually, however, the Federal officers imposed order and organized a proper disposition. The artillery moved in an oblique direction from the line of march to take up position behind the infantry, went into battery, and opened fire. The infantry then began to advance up the hills, threatening to envelop individual groups of Maderistas on the right. Salido ordered a retreat, and coming out of his covered position was hit by an artillery shell that ripped open his chest. When the rest of the Maderistas observed what was happening on their right, it was their turn to panic and flee. Navarro sent his cavalry around the Federal right to cut off the retreating Maderistas.

Some Maderistas continued to move and shoot, in textbook fashion, as they retreated. Mounted rebels stopped to grab others and slung them up on the backs of their horses to help them escape. A group from Namiquipa rode up to the Federal machine guns and attempted to lasso them—a tactic that persisted throughout the revolution that sometimes succeeded in damaging the weapon, but almost always resulted in the death of the cowboy.

Upon passing through the town pursued by Federal cavalry, the Maderistas noticed Pascual Orozco and his contingent arriving on the battlefield to reinforce them.

Orozco's men came up from the southwest and received a lethal volley from the Federals who had taken up position behind the low stone walls of the town's cemetery. The discharge instantaneously decimated Orozco's force and his remaining men immediately abandoned their horses, looked for any available cover, and began returning fire from the prone position. Fighting continued as some Maderistas took refuge in various houses until the Federals burned them out. As night fell, Orozco and the remainder of his troops, virtually out of ammunition, slipped away into the darkness. Navarro had the prisoners shot except the badly wounded, whom he ordered bayoneted in the cemetery. Thus began the blood feud between the rebels of Guerrero District and the Federal general.

After the Battle of Cerro Prieto, Navarro remained in the town for four days before moving to Pedernales on the rail line. There he waited for a reinforcing column of Federals commanded by Colonel Martín Luis Guzmán coming from Chihuahua City. The Maderistas regrouped and occupied a place called Ojo de Polanco about two kilometers to the west of Navarro to block the Federals' possible advance toward Ciudad Guerrero. Another group of rebels occupied the heights of Malpaso to the east of Navarro, a treacherous pass through which he would need to egress and, alternatively, any trains bringing reinforcements to his Operations Brigade would have to traverse.

On December 16, Navarro sent Colonel Fernando Trucy Aubert with a combined column of cavalry and infantry to secure Malpaso for the arrival of Guzmán's column, but the Maderistas bitterly contested the terrain as Trucy Aubert tried to enter the canyon from the west. They shot up his Federals from the heights and then extended the attack back toward Navarro's camp at Pedernales. Navarro dictated his dispositions for defense of the bivouac and then personally went to join Trucy Aubert. After about two hours of combat, Navarro managed to repulse the rebel attack at Malpaso but had to pull Trucy Aubert's column back to the main camp where the attack had grown in intensity. After close to six hours of fighting, the Maderistas finally broke contact and concentrated at Malpaso.

Guzmán's battalion consisted mostly of recruits from Querétaro, many of whom had not been issued coats or even footwear, in spite of the freezing cold. The departure of that relieving force had been delayed until December 17 by rebel threats to kill any railroad personnel who might operate the trains serving Guzmán's column. Thereafter, Guzmán found the going slowed by the need to continually reconnoiter danger zones and respond to harassing fire from rebels along the way.

By December 18, Guzmán's relief train had reached Malpaso Canyon. Apparently the colonel knew that rebels occupied the pass, but he discounted their martial abilities and superior tactical positions. He sent ninety dragoons with two companies of infantry two hundred meters ahead of the trains as a vanguard. As soon as the train reached Malpaso Station inside the canyon, the Maderistas on their flanks and to their front shot the Federals to pieces. The Maderistas employed the stratagem of placing hats on top of bushes to make their numbers look greater than actual and to draw the fire of Federals, while they assumed covered positions a few meters away.

The Federals in Pedernales could hear the battle taking place to the east so Navarro dispatched a column under Colonel Emilio López to assist the trains trying to force their way through the canyon. But shells from Guzmán's mountain guns aimed at the Maderistas to his front sailed high over their targets and landed to the west, in the direction of López. López stopped his advance and took up position to the west of the canyon.

Finally, a mortally wounded Guzmán ordered the retreat, and the Federals withdrew as best they could. Since Maderista José de la Luz Blanco had failed to seal Guzmán's path of egress, the Federal relief column managed to limp back to Bustillos. The highest ranking Federal of each arm in the column—infantry, cavalry, artillery, and engineers—was wounded in the engagement. Guzmán died in the Salas Hospital of Chihuahua City a few days later. Reflecting on the battle, Caraveo remarked that Guzmán, "with more pride than tactical sense, attacked in the open, the greater part of his troops being annihilated."

Stuck on the plains of Pedernales, Navarro's Operations Brigade remained under a virtual state of siege. He was down to seventy rounds per man and needed food and medicine. He could not attempt to pass through Malpaso Canyon by rail and his impedimenta and the many wounded overruled any attempt at an overland route. Moreover, the Maderistas captured and killed any runners Navarro sent to get word to zone headquarters in Chihuahua City about his condition, leaving him incommunicado.

Zone headquarters did not wait to hear from the Operations Brigade and immediately sent two more relief columns, one after the other, that joined the remnants of Guzmán's column in Bustillos. This combined force had pack mules, supplies, and a wireless. Reaching San Antonio de los Arenales on Christmas Day, the relief column crossed overland—bypassing Malpaso Canyon—to link up with Navarro's Operations Brigade on December 28, and thus re-established communications with Chihuahua City. Navarro finally reached Ciudad Guerrero on January 7 to great fanfare but little effect. The rebels sought refuge in the hills and plotted their next moves.

Analysis

The commission subsequently formed by the Federal Army to evaluate its performance in the field during Madero's rebellion divided its observations into five broad categories: manpower, communications, supply and service support, security and intelligence, and command and control.

Because of the lack of funds, a proper espionage service could not be recruited when such a service was imperative given the populace's sympathy for the rebels. Commanders aggravated the situation with poor reconnaissance, scouting, and security practices. Incredibly, the Federals frequently failed to send messages in code so the telegraphers passed the communications on to the rebels.

The Federals also lacked the requisite numbers. Given the small size of the Federal Army and its duty to cover the entirety of the nation, it would have been impossible to garrison all the important population centers and guard the means of production and distribution in the State of Chihuahua even in peacetime. Nor should the Federals have tried to perform a constabulary role while the Maderistas were massing for operations. Moreover, with the poor condition of the state's arteries of communication, it was impossible to respond to all the uprisings. Organizing flying columns against concentrations of Maderistas would have been a better solution, although the capabilities and eighteenth-century culture of the Federal Army would not permit such a mission requiring mobility and initiative on a sufficient scale. Therefore, reinforcements sent to Chihuahua were assigned to join with the one main column, under General Navarro, which was more in tune with the Federal tradition of keeping troops tightly controlled and under close scrutiny to prevent desertion.

However, given the woeful logistical system and practices of the Federals, such a large concentration as at Pedernales resulted in the predictable shortages of provisions, fodder, ammunition, medical supplies, payroll, and so on, which impeded the advance of the Operations Brigade. The lack of proper equipment such as tools for rail repairs, entrenching tools, communications apparatuses, pack mules, and horses required relief columns to convey those resources, putting the smaller supply columns bringing that materiel, namely Guzmán's, at risk of annihilation.

Also at fault was the government's refusal to admit to the true seriousness of the uprising and declare martial law, which would have united the security services of the ministries of War and Interior operating in the state under one single command: "The lack of centralization of all those armed elements in the hands of the General who commanded the zone could not but gravely harm operations." Instead, under the prevailing system local jefe políticos, 2nd Zone Headquarters, the Secretary of War, and even the president of the republic competed for resources and issued contradictory orders that introduced chaos to "the detriment of the orderly march of operations, with delay, waste of expense, and diminution of the initiative of subordinates who did not think themselves authorized to act without detailed and constant instructions." As a result, detachment and column commanders remained tied to the communications umbilical in expectation of detailed instructions with the result that "the rebels in that region were the ones who frequently took the initiative to attack."

The deeper problem, one that plagued the Federal Army until its dissolution in 1914, was a failure to develop a strategy, to appoint "one jefe with wide-ranging faculties and general instructions for repression of the revolutionary movement, so that with complete initiative and sufficient means of action he could rapidly and safely achieve such objective." The government initially viewed the rebellion as a "simple uprising local in character constituted of a few foci that had to be easily extinguished," and then never corrected those initial impressions as it became clear that the problem was much more substantial.

As a result of the reactive nature of Federal operations, units were detached to garrison duty then reassembled for short-term operations after which they were detached again such that units became atomized; columns of operations, when formed, comprised a pastiche of major units: "It is true that often in times of such a war one has to resort to that dislocation of units, but at the beginning of the war and when the subdivision of forces is not so necessary, such dismemberment should not be done to the extreme that it occurred, to the detriment of administration and *esprit de corps*." Over the course of the revolution the condition only worsened because the Federals refused to recognize the true nature of the conflict. It was never a "small" or "guerrilla" war, as conceived by some contemporaries and portrayed by historians. Rather, as the preeminent chronicler of the Maderista rebellion in Chihuahua, Tomás Serrano, wrote, "the conduct of the rebels, who did not refuse combat and knew how to measure their weapons against a disciplined army waging formal battle, and not just in guerrilla warfare as some imagine, is worthy of eternal remembrance."

MADERO ENTERS CHIHUAHUA

After yielding the Chihuahua City-Ciudad Guerrero line to Navarro's Federal Operations Brigade, Pascual Orozco Jr. held a council of war in which it was agreed to accept Abraham González's invitation to head toward the border and attack Ciudad Juárez, a potentially important entrepôt for war materiel.

The Federals had already begun moving assets into the northwestern part of Chihuahua around Casas Grandes, considered the second city of the state, after forces loyal to the Partido Liberal Mexicano (PLM) attacked Janos. Federal Colonel Antonio Rábago left Chihuahua City on January 4 with about 350 men traveling by rail north to Ciudad Juárez and then turned south with the ultimate destination being Casas Grandes. Around this same time, Orozco left Guerrero District and headed north taking a route farther to the east with the intention of cutting the rail line between Ciudad Juárez and Chihuahua City. A second rebel column commanded by José de la Luz Blanco and others left Bachíniva taking a course for San Buenaventura and Galeana.

Blanco's column bypassed San Buenaventura on January 18 after exchanging shots with the Federals inside the city and the next day destroyed the Federal garrison at Galeana. The two Federal captains had decided to leave the plaza and attack the rebels, who caught the two section-size groups in a cross fire and defeated the two commands in detail. Colonel Rábago, on his way to render aid to San Buenaventura, passed through Galeana on January 20, a day too late to save the garrison. The 2nd Military Zone headquarters immediately began pulling together another column to leave Chihuahua City and reach the Galeana-San Buenaventura area from the east. Around this same time, Orozco's column of rebels had also entered Galeana District and encamped at San Lorenzo, on line with San Buenaventura and about a third of the way between that latter plaza and Gallegos Station.

The Federal combined arms column under Colonels Antonio M. Escudero and Agustín Valdés (about 650 men from the three arms with a section of mountain guns and two machine guns) left Chihuahua City to reach Gallegos Station on January 26. The Federals detrained and began moving westward on a collision course with Orozco, who was moving toward the rails.

The next day, January 27, the Federals spotted Orozco's Maderistas and sent a squadron of cavalry to try to detain the rebel group. Orozco responded in kind, sending a mounted company to confront the Federal squadron while the rest of his men occupied favorable ground on a mountain called Chorreadura de la Mojina, which had an arroyo passing along the base. There in the arroyo, among the mesquite, and farther up the hill, the rebels prepared to receive the Federals.

The Federal pursuit squadron abandoned its horses to the care of a second, skeleton, squadron that served as the reserve, and advancing across the open ground took up position on what would become the Federal right and opened fire. The third squadron arrived and likewise dismounted and took up position in the center in a line of shooters. Advancing under the protection of a machine gun, they forced the Maderistas out of a large stretch of the arroyo and back up into the mountain. After about an hour of combat, Valdés showed up with the infantry and artillery.

The Federal section went into battery and fired on the crest of the mountain, as two companies of infantry deployed in echelon by platoons, with other small units providing covering fire for the artillery, guarding the impedimenta, and forming the reserve. The cavalry then cleared from the front and the infantry advanced and opened fire, coming to within one hundred meters of the rebels. The Maderistas returned fire and eliminated an entire platoon in one volley, causing the

Federal infantry to falter. The Federal artillery and machine guns continued to pour in copious fire.

Reinforced on the left, the Federal infantry reached the arroyo and rolled up the Maderista line. The rebels fell back to a second gully, which served as a trench, and continued to resist as the Federal infantrymen assumed the prone position surrounded by the chaparral and attempted to advance. At this point Escudero began to sense that the Maderistas had been reinforced by those Maderista companies that had already reached Hacienda El Carmen and now rejoined the main body. He hadn't seen them arrive, but the number of rebels on the crest and the intensity of fire increased. His left became compromised and he had to send a squadron of cavalry and more infantry to reinforce that section of the line. With a flanking movement the Federals managed to push back the rebels, but the Maderistas had an extensive line, held the high ground, and the Federals could not determine how to reach their positions. The Federals also attempted a flanking movement on the right, but the Maderistas repulsed that effort. Then the Federals resorted to a stratagem as it was growing dark. They sent a party under white flag to parley and when the Maderistas approached the Federals opened fire, killing a member of Orozco's staff. After night had fallen, the Federal cavalry provided security as the infantry and artillery marched off to Hacienda El Carmen to regroup and tend to their wounded. Escudero claimed that the dark and a lack of food and water had compelled him to break contact. He fully intended to resume the battle the next day, but by then Orozco had continued on his way northward. The Federals listed about twenty wounded, five missing, and thirty killed, but these numbers were probably understated.

The news of the Battle of Sierra Mojina revitalized the Maderista movement. Just as at Cerro Prieto, the rebels assumed the high ground and fought a battle against a Federal Army combined-arms column. This time, however, they refused to yield and won the battle. It was also the first indication that shock was the Federal artillery's only tactical value, and as the revolution wore on and troops became more inured to its effects, its impact on victory diminished to the point of virtually zero.

Also on the night of January 27, Blanco's rebels circled back to attack San Buenaventura after Rábago had returned to Casas Grandes. The Federal garrison of some one hundred men held out against Blanco and rebel sympathizers inside the town until the next morning when they ran out of ammunition and dispersed. Rábago pleaded with Colonel Escudero to hasten his march to join him in Casas Grandes. Escudero, in turn, ran into Blanco's rebels as he tried to pass through San Buenaventura, placed the town under siege, and waited for Rábago, who now was supposed to be on his way to reinforce him according to orders from 2nd Zone headquarters.

In the meantime, Orozco's column continued toward Ciudad Juárez, throwing the entire border region into a panic. In response, the Federals recalled General Navarro and his Operations Brigade, which had spent the month of January repairing the rails and telegraph as far as Madera Station in western Chihuahua, to Chihuahua City. To give Navarro time to arrive, the Federal commander of the Ciudad Juárez garrison, Colonel Manuel Tamborrel, sent a detachment of about seventy-five men to dynamite the bridges about thirty kilometers south of the city in order to slow the advance of Orozco. The detachment ran into the advance party of Maderistas, engaged in a brief nighttime firefight in which it lost its dynamite, and returned to Ciudad Juárez without accomplishing its mission.

The failure of Tamborrel's mission forced a reshuffling of Federal troops. Since Orozco continued his advance, Colonel Rábago had to abandon Casas Grandes and march northward to reinforce Ciudad Juárez while Colonel Escudero, who had grown tired of waiting for Rábago to reinforce him and was just about to assault San Buenaventura, received orders on February 4 to bypass the town and go directly to Casas Grandes. Orozco reached Bauche Station and was waiting for Blanco's column of Maderistas to arrive when he received news that Rábago was on his way. The rebels loosened the railroad tracks and prepared an ambush.

On the morning of February 4, Rábago's train hit the loosened tracks and skidded to a stop as the Maderistas opened fire. The fighting continued into the next day at which point the Maderistas decided to allow the Federals to escape—they had already spent two days without food or water for the horses as they waited to spring the trap and they were desperate for both. They planned to resume the battle with Rábago's Federals, who detrained and continued on foot, but the Maderistas never managed to catch up to the column, which arrived in Ciudad Juárez to reinforce the garrison on the evening of February 7.

Orozco remained in the Ciudad Juárez area for about ten days before deciding to leave and return south. The revolutionary junta had tried to place his men under the command of the notoriously incompetent José de la Luz Soto. Moreover, with Ciudad Juárez reinforced by Rábago and with Navarro on the way, Orozco did not think he could carry the city, so he left to return to Guerrero District while José de la Luz Blanco remained in Galeana District. Meanwhile, Francisco I. Madero and his coterie of foreign advisors finally crossed the Rio Grande into Chihuahua on February 14 to take charge of the rebellion. The Mexican government had pushed him into action by trying to have him arrested in Texas for violating U.S. neutrality laws.

Navarro's column arrived in Ciudad Juárez the next day and sought out Madero around Guadalupe, to the southeast, without success. Soon more rebels poured across the border to the southwest of Ciudad Juárez and once united the group of Maderistas headed southward.

Around this time, President Díaz became alarmed by the insecurity of Casas Grandes, the state's second city and a key center for command and control. At the time, communications to the border went from Chihuahua City by telegraph to Madera Station, from that location to Casas Grandes by telephone, and from Casas Grandes again by telegraph to Ciudad Juárez. Accordingly, Díaz ordered Colonels Manuel Gordillo Escudero (who probably had arrived in Ciudad Juárez with Rábago back in early January) and Antonio Escudero to return to Casas Grandes, assuming they were not already there and that these orders did not conflict with the successful ongoing pursuit of rebels. The two were to cooperate in operations and use the town as a "strategic pivot" from which to send out flying columns. Said columns were not to travel more than three days' distance from the city. The two Federal commanders did as instructed, trying to locate Blanco (who had left the area to invade Sonora) and Orozco.

February 23 turned out to be a critical day. After six days of pursuing Madero, Navarro's column returned to Ciudad Juárez with numerous sick soldiers, some from the freezing cold and the worst suffering from an outbreak of typhus, which Navarro blamed on the inattention to hygiene caused by the numerous marches and work details. In Casas Grandes, the Porfirian authorities from Ascensión arrived to report that Blanco had sacked their town and forced them to flee. The Maderistas controlled all the land north of Casas Grandes to the border, south at Hacienda San Diego, to the east around San Buenaventura, and the sierra to the west; their movements were constant, and more parties seemed to arrive to the district daily.

Colonel Agustín Valdés, who commanded the Casas Grandes garrison, immediately requested reinforcements from zone headquarters and redoubled the number of pickets and patrols. A combined-arms column of about five hundred Federals under Rábago left Ciudad Juárez headed for Villa Ahumada on the Central Railway in pursuit of Madero, but it had to countermarch to Lucero Station at some point in order to get payroll, food, and fodder. In the meantime, Madero left Villa Ahumada on February 23, continuing on his way.

Four days later, on February 27, zone headquarters unadvisedly and without reason ordered the Escudero-Gordillo column that had been operating in the Galeana District to return to Chihuahua City just as Madero was entering the district. It was expected that Rábago would handle the pursuit of Madero, and zone headquarters informed the general that he could count on whatever resources he might need at any of the numerous haciendas owned by the Terrazas that he might happen upon. Rábago would have no more excuses to suspend the pursuit for logistical reasons.

Also on February 27 a combined-arms column commanded by Colonel Samuel García Cuéllar with about five hundred men and two 80-mm. mortars departed Ciudad Juárez heading for Ascensión to clear José de la Luz Blanco from the area. Once on the scene the colonel determined that Blanco had never been there, that Ascensión actually had been sacked by José Orozco, and that only small parties like José Inés Salazar with ten men remained in the area. Then, he received reports that a party of rebels had passed through the Carcay Hills headed for the south of Casas Grandes, and that Madero had passed through San Lorenzo and San Buenaventura. García Cuéllar concluded that the two groups intended to gather south of Casas Grandes and then attack the plaza, so maintaining strict communications silence, he began moving in that direction. Indeed, Madero had continued to elude Rábago, entering San Buenaventura to cheers where another two hundred rebels rallied to his cause. He sent word to Pascual Orozco and other rebels to join him, but he did not wait for them to arrive before deciding to attack Casas Grandes.

Valdés had been assigned to Casas Grandes and had immediately begun preparing the town's defenses, arranging for a small interior defensive line that the garrison's troops could occupy, and an expanded exterior line that could accommodate both the garrison and any reinforcements, which he had requested on multiple occasions, that might arrive. The west side had naturally clear fields of fire, so it was only lightly manned. The Federals dug a trench on the north side, established a covered outpost constructed of adobe to the south, and on the east side selected choice houses on the periphery of the town that could provide interlocking fire and function as forts. They also strung barbed wire, established observation points, and assigned men to the various sectors. Between Federals, Rurales, Auxiliaries, and local volunteers, about five hundred men defended the plaza. On March 1 the phone lines to Madera Station had gone dead, leaving them incommunicado with the outside world.

In the afternoon of March 5, the Federals inside Casas Grandes observed the Maderistas pass through Chocolate Pass, arrive at Rancho del Refugio about three kilometers away, and detach their outposts about eight hundred meters to the south of the Federal positions. Later that night the magistrate of neighboring Nuevo Casas Grandes phoned to announce the impending arrival of García Cuéllar's column. The news spread like wild fire and came as very welcome news since Rábago had merely reported Madero gone from the area around Moctezuma Station and then returned with his Federals to Ciudad Juárez at the beginning of March. Just before midnight Valdés and García Cuéllar held a telephone conference in which the two discussed the Federals' tactical situation in general terms and the probable Maderista plan of attack. Valdés noted that there was no water in the main irrigation ditch to the east of town. He suspected that the rebels had cut the water to dry it out so that it might be used as a trench in the battle, which indicated an attack from that direction. Therefore, the two Federals agreed that García Cuéllar would remain in Nuevo Casas Grades, about six kilometers to the northeast, until after the Maderistas had initiated the assault. At that moment, Valdés would telephone García Cuéllar, who would quietly exit Nuevo Casas Grandes to hit the Maderista flank and roll up their positions. During the night and early morning hours the Maderistas moved into positions on the north, east, and south side of the plaza, occupying and barricading themselves in some houses outside the Federal interior perimeter.

In the morning of March 6, with Orozco and other reinforcements only a short distance away, an overly-confident Madero attacked Casas Grandes with his rebel column that had grown to as many as eight hundred men, but may have been only slightly more than four hundred. The battle was a disaster for the Maderistas.

The Maderista assault began on the south and northeast sides at 5 a.m. After two solid pushes over the course of one and a half hours of combat, the rebels were repulsed. The shooting attenuated with only isolated shots coming from rebels occupying the houses outside the Federal perimeter. A short time later, the Maderistas made another forceful assault from the main irrigation ditch to the southeast of the plaza, in an effort to reach the smaller irrigation ditch closer to the town, and also used the main ditch to move around to the north. The southeastern attack fizzled, and from that time the fighting was generalized on the east side. The Federals' sole machine gun hammered away at the Maderistas from its location atop the church inside the plaza. No Maderistas attacked the west side, and there were only a few to the northwest.

At 6 a.m., Colonel García Cuéllar began his movement on Casas Grandes, crossed the river by the same name, and using the alamedas to screen his approach surprised the Maderistas with an attack from the rear. His Federals rolled up the Maderistas on the north and east side of the plaza and prepared to push southward while his impedimenta and its infantry support remained on the right margin of the river with instructions to enter the plaza at an opportune moment. The two 80-mm. mortars went into battery and began laying down fire to the south and southeast to prepare for an infantry assault, while a detachment of 150 dragoons rode out to cut off the Maderista retreat and capture the rebels' impedimenta, if possible. By 8 a.m., the Federal infantry was pushing hard to the south along the ditch. The Maderistas kept up their fire for about three more hours, trying to pull back toward the south. Then García Cuéllar threw in his reserves and put the rebels to headlong flight. It was said that José de la Luz Soto was the first to break and run, sowing panic in the line. The fleeing Maderistas ran into the Federal cavalry and infantry that had been sent to cut off their escape. Isolated groups of Maderistas were subdued, but most escaped with their lives, abandoning valuable materiel, which the Federal cavalry recovered. García Cuéllar's impedimenta train safely entered the plaza at 10:20 a.m.

Isolated fighting continued around the plaza until the late afternoon, as the Federals cleared Maderistas holed up in buildings around the perimeter, mainly on the north and south sides. The Federal garrison incurred thirteen killed and twenty-three wounded, while García Cuéllar's column had about thirty wounded and twenty-five killed. The Maderistas suffered fifty-eight killed, forty-one captured, many wounded (including Madero), countless dispersed, and the loss of their baggage train.

Madero and his men retreated to Hacienda San Diego and remained there for days, unmolested. García Cuéllar requested cooperation with other zone forces to "give a decisive blow" to Madero, but Valdés was more interested in holding Casas Grandes, citing its strategic worth to the rebels as a key point on the pipeline to the United States for contraband arms and ammunition. He recommended reinforcing the city and holding it at all costs.

Madero left for Galeana where he continued to gather in adherents, including Orozco, and test fired a homemade bronze cannon before moving toward Chihuahua City intending to attack it. But the Federals divined his plans and had several columns placed just north of the state capital, so Madero decided to move into the western districts and unite his followers there. He reached Hacienda Bustillos at the end of March and continued to gather in all his supporters, including Pancho Villa, who had been marauding in the southern part of the state after Cerro Prieto and arrived with five hundred men on April 2. The news of so many Maderistas massing so close could not help but cause concern inside the state capital, and the Federals began concentrating their forces in the center of the state. García Cuéllar left for Ciudad Juárez with his wounded and prisoners from the Battle of Casas Grandes, and Colonel Valdés left with his column for Chihuahua City. In spite of Valdés' warnings to the contrary, the government had abandoned Casas Grandes.

Analysis

The Battle of Casas Grandes was the first Federal defense of a plaza. Colonel Valdés properly prepared a perimeter defense appropriate to the size of the garrison with an expanded perimeter that could be used in the event of being reinforced. Trenches and barbed wire also increased the efficacy of the defense, but he employed his one machine gun in an elevated and central position atop the church. This was the first manifestation of a common practice by Federals to emplace machine guns—almost exclusively—in defilade to lay down indirect fire from an elevated position instead of on the flanks for grazing direct fire in enfilade. The battle was also notable for the employment of two mortars, almost unheard of in the Federal arsenal, and the unusual cooperation of two Federal commanders in developing effective operations in the face of incomplete intelligence, mainly owing to the professionalism and initiative of Colonel García Cuéllar, a personal favorite of President Díaz for obvious reasons. Still, many believed that none of the Maderistas should have escaped that battle to fight again and blamed García Cuéllar for letting them go. It was a missed opportunity.

The rebels, meanwhile, failed to make a proper reconnaissance before initiating combat, did not have a detailed battle plan, and do not appear to have maintained any reserves, which Madero might have had if he had waited for Orozco to arrive with reinforcements. The entire rebel effort was so amateurish that Valdés later stated that he could have repulsed the Maderista attack even without García Cuéllar's column. Because the Maderistas did not coordinate their assaults, Valdés was able to blunt them in succession and toward the end of the battle they appeared to run out of ammunition without any discernable means to resupply the line.

On a strategic level, the Maderistas continued to control the pace and direction of operations, turning north in January (the deadliest month of the uprising for the Federals) with two columns that confounded Federal efforts to locate and trap them. The Federals continued to bemoan the detailed orders coming out of Mexico City—those of the Secretary of War and the President sometimes conflicting—that overruled those of commanders in the field and zone headquarters. Additionally, the Federal flying column missions were frequently of short duration and distance, often because of a need to return to base for fodder and provisions, and therefore produced little effect. The larger columns lacked coordination and sometimes resources, suffered from a dearth of intelligence about the rebels, and demonstrated little initiative by commanders, except in the case of Colonel García Cuéllar's movement to reinforce Casas Grandes. But even he later faced criticism for not following up on that victory and obliterating Madero as the rebels regrouped at Hacienda San Diego and then later in Galeana. The Federals blamed the inability to pursue defeated rebels, in general, on the woeful state of their own cavalry. Therefore, rebels could pick and choose areas and districts to occupy, increase their numbers, avail themselves of resources, and extend their radius of activity.

Finally, the Federals blamed the Americans for allowing the Maderistas to cross the border from Texas at will and organize inside Mexican territory under the watchful eyes of the Federals. But in this, the Federals were equally to blame.

REVOLUTIONARIES WHEEL NORTH

It may have been Villa who convinced Madero to take another chance at conquering Casas Grandes, or possibly Ciudad Juárez, as the Federals began concentrating about Chihuahua City. No matter, the Maderistas left Bustillos on April 7, passing through Ciudad Guerrero to Madera where they collected two guns built in the shops of the Northwest Railroad. To support their movement northward, the rebels sent a second column of three hundred men commanded by Colonel Agustín Estrada headed in the same direction but farther to the east, just as they had done back in January, to Galeana District. The Maderistas intended for Estrada to block the passage of any Federals coming from the Central Railroad toward Casas Grandes. They did not know that the Federals were moving in a direction opposite to Estrada's. Colonel Valdés' mixed column of about eight hundred men was headed for Chihuahua City along with government employees from Galeana District, prisoners, and volunteers from the earlier battle at Casas Grandes.

On April 9 Estrada's Maderistas encountered and attacked Valdés' extreme rearguard near Galeana. Then, racing ahead it set an ambush at a place called Boquilla de la Laguna. Because of the heat and out of special consideration for conserving water, Valdés had decided to travel at night. At 3:30 a.m. the Federals passed through the kill zone of Estrada's ambush. The fighting lasted into the daylight hours when Estrada had to break contact and continue his march to the Central Railroad. Once there he tore up track and moved toward Ciudad Juárez to participate in the attack on that town.

Valdés reported recovering the bodies of forty-two Maderistas while suffering a measly three killed, ten wounded, and twelve dispersed. These numbers should be viewed with a high degree of skepticism, both in the unbelievably high number of rebels killed and the low numbers of his own casualties. After the engagement, Valdés also made his way to the Central Railroad where he linked up with Brigadier Antonio Rábago and continuing southward the two entered Chihuahua City on April 17.

Zone headquarters knew within days of the rebel departure from Bustillos that Madero had turned north because as early as April 11 the jefe político of Ciudad Juárez began reporting on rebel movements around Casas Grandes and that many more Maderistas were waiting to cross over from El Paso to attack the border town. But for whatever reason, the War Ministry was more concerned about Ojinaga, which had been under a light siege since the beginning of February. Brigadier Rábago was told to organize a column of almost 1,100 men to march to the aid of the garrison. The relief column left on April 18, but the next day President Díaz cancelled the mission and ordered Rábago to return to Chihuahua City. His column would be augmented by another four hundred men and sent to Casas Grandes in forced marches to cut off the rebels who might be retreating from Ciudad Juárez in the event of a failed attack on that plaza.

By April 15, Madero's vanguard under the command of Lieutenant Colonel José Orozco had reached Bauche Station, only fifteen kilometers from Ciudad Juárez. Federal General Juan Navarro sent about one hundred troopers to attack Orozco that morning, but after a few hours of combat the Federals had been unable to put the

150 rebels to flight, so Navarro sent a follow-on force of 100 infantry to join the action. When three wounded infantrymen returned to the city saying that the fight was still going on, Navarro committed his two 80-mm. mortars and about fifty artillerymen to the battle. The rebels also received reinforcements, some three hundred men commanded by the provisional President's brother, Raúl Madero, and Pancho Villa. As night began to encroach the Federals disengaged and returned to Ciudad Juárez, which the Maderistas then invested.

Navarro immediately sent urgent requests for ammunition and President Díaz responded by ordering a repair train with a significant armed escort to leave Chihuahua City for the border, to repair the rails of the Central Railroad and bring urgently needed provisions, ammunition, and water. Díaz apparently ignored the warning from the jefe político of Ciudad Juárez that the only way to get materiel past the rebels was via the United States. The War Department sent detailed instructions for the relief column including Díaz's order to make it known publicly that the train was headed elsewhere to deceive any rebel spies inside the state capital.

For days on end Madero negotiated with the Federals inside Ciudad Juárez, even reaching an armistice that among other provisions called for the freezing of Federal troop movements in northern Chihuahua, including Rábago's column. These negotiations dragged on until May 7, a day after the armistice had expired. At that point, Madero announced to the press that his forces, which had grown to an estimated 2,500 to 3,500 men, would not be attacking the border town out of patriotism, since a battle might cause damage to neighboring El Paso and provoke an armed U.S. intervention in Mexico. Instead, the rebels would be turning south to march on Mexico City. This did not sit well with Madero's supporters, especially those from Guerrero District who wanted Navarro's head as revenge for the murder of prisoners at Cerro Prieto. Others suspected that Madero might be trying to reach an accommodation with Díaz's government. Accordingly, on May 8, Orozco and Villa had their men initiate the battle.

Shooting began on the west side of the city, closest to the river. Madero was in the act of planning the march through the center of the country when the battle began and he tried to stop the fighting, but the battle escalated beyond his control, with his forces attacking from every direction except the east side. According to rather incomplete records, the Federals had close to nine hundred men, including auxiliary policemen, to defend the plaza. Almost two hundred of these, however, were ill because of an outbreak of typhus and various other maladies. Moreover, although Navarro had overall command, the army had not been set on a war footing, meaning the lines of authority between the War Ministry, Interior Ministry, and zone headquarters, on the one hand, and the Operations Brigade, civilian paramilitaries, and garrison, on the other, were not so well defined. Navarro commanded the Operations Brigade, which was mainly located on the southwest side, while the garrison troops were on east side around their barracks and headquarters, and the jefe político guarded his offices as well as the jail, on the north side. The largest concentrations of government forces were in the barracks and the trenches on the west side of town, but the trenches ended some 150 to 200 meters from the river. Worse still, an irrigation ditch with levies high enough to conceal a mounted soldier from view separated the terminus of the trenches from the river, and Navarro had assigned a mere ten auxiliary policemen to cover this area with only natural terrain to serve for defensive works. The Maderistas easily punched through this area, rolled up the trenches with enfilading fire, and then attacked the plaza from the north side, forcing the Federals to shoot back at them in the direction of the United States.

After two days of combat, the Maderistas had forced the Federals to abandon several key landmark defensive positions and closed in on the few remaining. The rebels burrowed through the walls of buildings to advance without risking the barren streets. Alternatively, the Federals had no communications trenches and trying to resupply the east side with ammunition by running through uncovered streets was becoming nearly impossible. Thus, Navarro decided to concentrate his forces on the southwest side of the city around the Federal Barracks, where the ammunition was stored. That meant, however, that the garrison forces on the east side had to run a gauntlet of some six hundred to one thousand meters through the streets, effectively taking fire from the rooftops and at the numerous intersections, subject to copious flanking fire. The Federals, nevertheless, succeeded in executing the precarious concentration of forces and continued to fight, but not for long. No provisions had been made to keep the men fed, and after more than forty-eight hours of combat through hot days they were exhausted. Moreover, the Maderistas had cut the water supply almost from the start of the battle, which meant the Federals were dehydrated and lacked the ability to clean their wounds. Some men were so enervated they could barely work the action of their rifle. Most of the livestock had been killed or captured (along with the fodder), which made any attempt to break out of the city and across the desert with their wounded, ammunition, and artillery out of the question. The rebels finally overpowered the Federals, who surrendered.

Afterward, Navarro confirmed his losses of thirty-five killed and as many wounded, but those were only the numbers he could ascertain and he personally believed them to be much higher. He put the rebel casualties at an unbelievably high four hundred. Many observers put the figure at two hundred casualties for both sides.

Unable to swallow defeat at the hands of armed civilians, Federal Army apologists offered up a litany of excuses: they were prohibited from shooting toward the U.S. (north) side of the plaza; a fifth column of more than four hundred sympathizers inside the town opened fire on the Federals from the early moments of the battle; the Federals respected the cease fire that Madero tried to arrange in the opening moments of combat, but the rebels took advantage of the decreased fire to roll up the Federal trenches on the west side; the Federals were outnumbered by the Maderistas, whom some Federals put at an unbelievable 6,000 men; and, of course, the Federal government's questionable orders to evacuate Casas Grandes.

Yet in many respects the Federals had been their own worst enemy. In spite of several months under threat by various Maderista groups and with plenty of available construction material, the Federals inside Ciudad Juárez had prepared only the lightest of defenses and had not stockpiled any water or food beyond what zone headquarters had sent—which still sat inside a boxcar in the railyard, unloaded. The trenches on the west side were of rudimentary construction without interlocking fire, nor had any communications trenches whatsoever been dug or erected, meaning that problems with the resupply of ammunition could have been easily foreseen. And the Federals had advantages that the rebels did not enjoy. The Federals had machine guns and artillery, which the rebels did not (their homemade cannons inspired more ridicule than fear), and given the 3:1 (some say as high as 10:1) numerical advantage of men required to successfully assault a fixed fortification, the Federals should have had enough men to repulse the rebel assault.

Analysis

Although the topic of logistics is outside the scope of this book, in at least one way it had direct tactical implications. Under the Federal Army system, soldiers received "*haberes*," a combination of wages and per diem. With this pay they were expected to feed themselves, whether in garrison or on the march, thus freeing up officers from attending to this duty. It was essentially a continuation of the Spanish "*rancho*" system. The problem arose when the army went into sparsely-populated areas where food vendors could not be found, and during battles in defense of a plaza, such as that of Ciudad Juárez, when the restaurants and shops closed until the end of combat. Under those circumstances the soldiers could not obtain the food that they needed, especially since in this case they had not been paid. Moreover, General Navarro later stated that he did not want his soldiers to be distracted with feeding themselves so he purposefully did not concern himself with arranging food for them. This belief in the distracting force of eating seems to have gained some currency among Federal Army officers, and in at least one instance was redeemed when Federal soldiers abandoned the firing line during the September 1913 Battle of Second Torreón for lunch. Thus, culture combined with the logistical system resulted in Federal soldiers frequently going hungry during combat, which could last for days, and contributed to defeat, such as at Ciudad Juárez. Other Federal tactical deficiencies that would become chronic were the lack of competently prepared defensive positions (Casas Grandes being an exception to the rule), and the misapplication of machine guns in defilade and in indirect fire missions.

The Chihuahuans, conversely, constantly outmaneuvered the Federals. They began threatening Chihuahua City along the Guerrero-Chihuahua line in November, drawing Federal Army resources into the southwestern part of the state. After putting up a tough fight, the Maderistas moved northward along two axes, pulling the main Federal maneuver forces out of the southwestern part of the state and into the northern section, while Pancho Villa wreaked havoc in the south. Madero then entered the state and after a failed attempt to take Casas Grandes, the main rebel groups began converging around the state capital. The Federals panicked and began ordering columns back to Chihuahua City, abandoning northwestern Chihuahua, whereupon the Maderistas wheeled north and attacked Ciudad Juárez, precipitating the fall of Díaz's government.

They also employed frontier tactics, such as ambushes and roping machine guns, and improvised others, like burrowing through buildings and using dynamite for warfare.

Yet they also exhibited professional behavior, especially in the case of Cerro Prieto, where the Maderista leaders sent scouts to locate the Federal Operations Brigade in a movement to contact, chose favorable terrain for combat, properly placed their men in covered and concealed positions on the military crest, and then during the retreat continued to move and shoot instead of just showing their backs to the enemy. In the words of one Federal major at Cerro Prieto who had fought in Sonora and the Yucatán: "The Yaquis and Mayas are brave, but there is no word for these serranos when it comes to combat." From the very beginning, the Mexican Revolution was a conventional war, not a guerrilla, or "small" war, and it was about to get much bigger.

2. OROZCO AND ZAPATA, FEBRUARY – OCTOBER 1912

O n November 28, 1911, Suriano General Emiliano Zapata and his followers proclaimed the Plan de Ayala out of fear that President Madero would fail to follow through with his campaign promises. The Plan de Ayala named Pascual Orozco as the military commander of the new Liberating Army, even though Orozco was not as yet in rebellion. That changed on March 1, 1912, when Orozco resigned his commission and pronounced himself in rebellion.

The Federal Army had been withdrawn from Chihuahua as a condition of the Treaty of Ciudad Juárez, and General Orozco commanded the Rural Police Corps that had been assigned to provide security in the state. Many of these Rurales followed Orozco into rebellion, and he received support from the rebels of the PLM, who captured Ciudad Juárez on February 27, and the followers of Emilio Vázquez Gómez.

Alarmed by these events, the Federal Army commander in Torreón sent about 175 soldiers under Major Adolfo Ramírez into the state in February—before Orozco had officially rebelled—to intercept any rebels pushing southward from Chihuahua City, and placed them under the command of Comandante José de la Luz Soto, commander of the 40th Rural Police Corps at Parral. Additionally, Governor Abraham González reactivated former Maderista insurgents as Auxiliaries, such as Toribio Ortega and Pancho Villa, mainly to protect the railroads from being destroyed.

FIGHTING FORCES

Orozco's insurgency marked a sharp escalation in the revolution. Whereas Madero's supporters in 1910 consisted of more established heads of households, many young Chihuahuans from various walks of life followed Orozco into war, further democratizing the revolution. The ensuing battles increased in size as both sides fielded division-sized elements that fought set-piece, open-field engagements. Campaigns became more sophisticated, with multiple operations carried out simultaneously and, just as important, the scope of those operations expanded, bringing other regions into play and putting them on a war footing, most especially the state of Sonora and the region known as La Laguna.

Sonora was unique to Mexico in its economic development and military traditions. A distant region with no discernable economic promise in the sixteenth century, the Hapsburgs had tapped the Jesuits to bring it under Spanish authority. The good fathers settled missions and organized the Yaquis and Mayos under a centralized military hierarchy to be able to repel marauding Indians of other tribes. Later, the regular army established presidios to check the threatened encroachment of foreign European powers. Together, the missions and presidios, along with subsidies paid by the crown to the Apaches, did a fairly good job of maintaining the peace. There was no need for the farmer-soldier construct so common throughout the rest of northern Mexico.

By the eighteenth century the presence of mineral wealth had been established in Sonora, attracting Spaniards of the hidalgo class to the area, and the Enlightenment arrived as the Bourbons kicked the Jesuits out of Sonora and claimed their mission properties. The Yaquis naturally believed those lands had always belonged to them and struggled to regain them.

At independence, Sonora presented an unusual aspect with a largely militarized and sedentary Indian population that resisted miscegenation, an urbane and worldly class of Spaniards who favored regulars for defense over grassroots militia, and a decreased Church presence.

During the nineteenth century another tradition was born: politicians enlisting the aid of the Yaquis and Mayos in the many conflicts between liberals and conservatives. In the Mexican revolution, the two tribes would again be called upon to join the fight and the Sonoran state forces would become the best practitioners of Federal Army doctrine.

The region known as La Laguna, which straddles the Durango-Coahuila border, was likewise atypical of Mexico in that until the middle of the nineteenth century it had a sparse population consisting of a few indigenous people and large landholders. After the French Intervention those *hacendados* who had supported the French lost their properties, and in 1880 the Madero family, which had grown wealthy from dealing in contraband Confederate cotton during the U.S. Civil War, branched out into the area to found a cotton kingdom of their own. With irrigation, they managed to harness the erratic waters of the Nazas River to grow cotton in the fertile alluvial plain. The traffic in cotton attracted railroads, which lowered transportation costs and made the region's moribund mining sector once again profitable. Smelters and factories that used cotton-seed to manufacture soap and dynamite soon followed, and guayule, a plant indigenous to the area, became valuable on the world market for its use in making tires for automobiles.

With all this activity industrial and agricultural workers flooded into the area. The work in the cotton fields demand a transient labor force that could react quickly to the weather and harvest conditions, and soon wars over water rights erupted between upriver and downriver haciendas. The atmosphere had all the makings of Wild West boomtowns and range wars and the rootless itinerants it attracted provided perfect stock for a mobile rebel army. Many of the region's workers travelled to work other areas—including the United States—according to seasons and cycles and had been exposed to foreign notions of employment relations and economics and had been radicalized. They made willing recruits for Pascual Orozco.

THE ROAD TO RELLANO

On March 1, at a celebratory dance, Orozco became aware of Pancho Villa's presence on the outskirts of Chihuahua City and immediately mobilized his followers to exit the city and present battle. Villa and his men, however, had orders from Abraham González to avoid combat and promptly fell back, retreating southward. At the same time, PLM Generals Emilio Campa and José Inés Salazar came down from Ciudad Juárez and Orozco's soldiers, variously referred to as "Orozquistas" and "Colorados" without distinction, began pushing southward to link up with the Zapatistas and conquer Mexico City.

Orozco and Zapata, February – October 1912

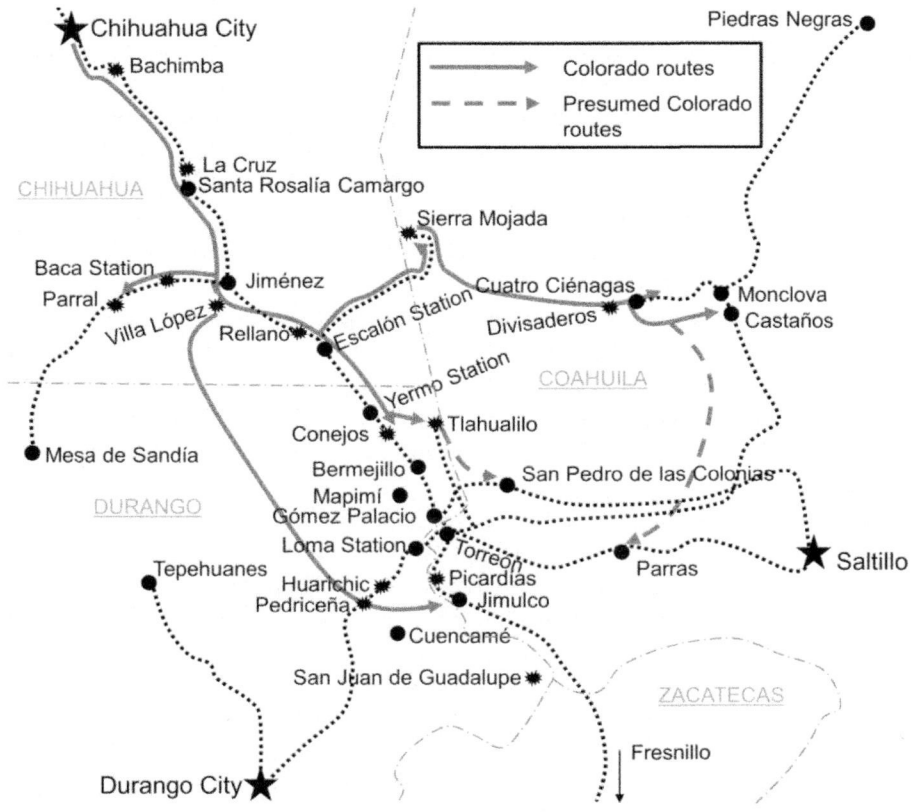

Map 2 Multi-State Maneuver Warfare

As the Orozquista juggernaut barreled down the Central Railroad, Federal Major Ramírez blew up a few bridges and retreated toward Jiménez. Villa also headed southward, only through the center of the state in the direction of Parral. Comandante Soto then ordered Ramírez to join him in Parral, but Emilio Campa's Orozquistas attacked and defeated the major at Baca Station.

Meantime, a Federal division commanded by General José González Salas left Mexico City headed north to organize at Torreón and then march into Chihuahua to face Orozco's army. Having completed its preparations, on March 19, the Federal Division of the North left Torreón, which was under a loose siege by Colorados from the area. González Salas with his infantry and one small cavalry brigade under Brigadier Joaquín Téllez moved northward along the rails while General Fernando Trucy Aubert took the other cavalry brigade and in a pincer movement marched farther afield to the west intending to reach the Central Railroad at Jiménez.

On morning of March 24, the Federal division encountered Orozco's troops commanded by General Emilio Campa at Rellano, Chihuahua. The Colorados began the battle by launching a *máquina loca* (a runaway locomotive loaded with dynamite) at the first train in the Federal convoy. Because of a lapse in security procedures, due precautions had not been taken and the máquina loca slammed at maximum speed into the lead gondolas carrying troops of the 6th Battalion, sending men, machine guns, and metal into the air and causing the gondolas to telescope, thereby saving the locomotive and the railcars behind it.

The rest of the Federals immediately detrained and began marching north to engage the Orozquistas, who occupied the high ground, along a saddle to the northwest of the train station.

The Federals deployed in a classic formation with the infantry in the center according to the ternary system with a chain of shooters backed up by a support and reserve. Most of the cavalry took up position on the flanks while smaller detachments provided security for the trains, a section of machine guns, and the field artillery located to the rear. A second machine-gun section was on the right flank.

General Aurelio Blanquet on the right ordered his infantry to make a flanking movement against the Orozquista left, but González Salas' chief of staff, a general staff officer, overruled the maneuver and ordered an oblique movement against the Orozquista left. The countermand had the effect of turning a flanking movement into a direct assault against the high ground. At this point a reporter accused the 20th Infantry Battalion of firing on the 29th Battalion. Treachery was alleged. By early afternoon the chief of staff had been killed and the Federals were losing many officers and men, including a wounded Blanquet, when Colorado General José Inés Salazar arrived by train with reinforcements for Campa. The Colorados began coming down from the hills and started a flanking movement against the Federal left, forcing the entire line to begin pulling back.

González Salas lost his nerve. He ordered General Téllez to take charge of the retreat to Escalón Station as Federals began tossing their weapons and abandoning their horses to climb onto the trains that had begun slowly pulling away. The cavalry was left to cover the retreat of the artillery that did its best to fend for itself. Using small eminences to screen its movements, the cavalry and artillery bounded in retreat to the entrance of a canyon where it endured a brutal assault. Looking to his right, Téllez could make out a battery of Schneider Canets and a group of infantry; it was the Federal general reserve commanded by Colonel Salvador Mercado. Continuing to fight and retreat, hill by hill, and gun by gun, Téllez managed to reach open ground at Asúnsolo Station, where the battle had opened with the crash of the máquina loca. Assuming a column of march with the infantry on the flanks, the artillery in the middle, and the cavalry covering the rear, the Federals marched over open ground followed by the Colorados who turned around a short while later. Téllez finally managed to reach Escalón, where he found the trains that González Salas had left for the men now under his temporary command. He waited there another day to collect any stragglers and then the Federals continued the retreat to Bermejillo at a rather dilatory pace because of the condition of the mules and horses.

At Yermo Station, González Salas nervously waited for word about the condition of his command; finally, he could endure the uncertainty no longer, and at five o'clock in the morning on the day after the battle he went to the water closet in his railcar and shot himself in the head.

Villa eventually reached Parral, Chihuahua, while the Battle of Rellano was raging and disarmed Soto and his Rurales on suspicions that they might join the Orozquista rebellion. Many of those Rurales, including Maclovio Herrera, chose to join Villa.

The Colorados had to pull back to Jiménez from the area around Rellano and then continue to the southwest in order to confront General Trucy Aubert's Federal cavalry brigade, which they trounced handily at Hacienda San Pedro and Villa López over three days beginning on March 26. Trucy Aubert reached Mapimí on April 1, after losing his entire impedimenta, machine guns, and a battery of mountain guns.

Next, the Colorados had to turn their attentions to Villa, who at Parral sat dangerously close to their supply lines. Colorado General Emilio Campa attacked Villa at Parral but was repulsed. The Colorados returned with reinforcements and on April 5 finally defeated Villa, who retreated through northern Durango to Mapimí and Gómez Palacio. Reaching La Laguna with nine hundred men, Villa joined the Federal Division of the North, now under the command of General Victoriano Huerta.

Analysis

There were rumors that one Federal captain had ordered his men to shoot at other Federal units during the Battle of Rellano, a charge that he denied vigorously. More likely, the friendly fire, if indeed it occurred, was either due to the ternary system itself (which required skillful coordination among the three echelons) or because of the flanking movement that the chief of staff unwisely countermanded. That changed order had forced Blanquet's men to oblique back toward the center where they may have come under friendly fire. The artillery and cavalry performed admirably in battle and later expertly retreated under fire and—in spite of persistent rumors of utter obliteration—lost only one field piece. However, the Federals had once again failed to employ effective scouting patrols, allowing the vanguard to be surprised. Still, it is precisely the vanguard's role to make initial contact, and in direct contradiction to conventional wisdom, the initial casualties from the máquina loca do not appear to have been that great. Additionally, overall casualties did not exceed acceptable limits for such a hard-fought battle, and González Salas never threw his reserve battalion under Mercado into the fight. The decision not to employ all the forces at his disposal led some to believe that the Federals might have won the battle under a more experienced general. The choice of González Salas to lead the Federal division had been a political one, made by Madero. Once again the President had been responsible for the humiliating defeat of a Federal command, overwhelmingly blamed on a failure in leadership and the defective skills of the commanding general and his hand-picked chief of staff. (In the Federal Army it was customary for the commanding general to choose his chief of staff; the position was not assigned by the Secretary of War).

On the level of strategy, some historians blamed Trucy Aubert for swinging out too far on a "flanking" movement. However, as his official report makes clear, Trucy Aubert's mission was a pincer not a flank attack, and he blamed his defeat at Villa López on González Salas for splitting up the Federal division, especially given the fact that the Orozquistas had plenty of Mausers and artillery. By dividing his forces, González Salas allowed his Federals to be beaten in detail—a rookie error. And if, as General Jacinto B. Treviño contended, Trucy Aubert had been sent southwest of Jiménez because General Salazar had been seen there, then the Colorados had successfully used Jiménez as a pivot point to operate on interior lines against two separate threats in succession.

Additionally, the Orozquistas achieved a strategic offense with a tactical defense, regarded as a superior combination by many military theorists. But Trucy Aubert's presence near Jiménez and Villa's at Parral threatened the Orozquista supply lines and did what González Salas had been unable to do at Rellano—halt the advance of the Colorados. One Colorado later blamed Orozco specifically for not giving General Emilio Campa enough men to clear Villa out of Parral on the first attempt. The resulting delay in operations would prove the movement's undoing.

MULTI-STATE MANEUVER WARFARE

Federal General Victoriano Huerta arrived at Torreón to assume command of the Division of the North on April 15, 1912. One of his first acts was to build defensive works around the city against a possible attack by Colorados from the surrounding area. Huerta spent the next few weeks recruiting men and preparing for the upcoming campaign, assembling firepower and the best and brightest officers that the Federal Army had to offer. He also promoted Pancho Villa to the rank of brevet general.

Meanwhile, the Colorados put their campaign into action. On April 5, General José Inés Salazar left with a column of perhaps 2,000 men to drive into Coahuila via Sierra Mojada, Cuatro Ciénagas, and Monclova, in the center of the state. There is some disagreement over whether his objective was strategic—to occupy the state and its valuable coal fields in the northern part while recruiting more men—or operational, to attack Cuatro Ciénagas and then drive southward to cut the rails and telegraph lines at Parras in order to isolate Torreón and Gómez Palacio from Saltillo to the east. However, General José "Cheché" Campos also had orders to cut communications to the east of Torreón, taking his brigade on a course from Tlahualilo through San Pedro de las Colonias, which would weaken the operational hypothesis.

Simultaneous to Salazar's mission, Colorado General Emilio Campa was to take 1,500 men on a route to the west, essentially the same path that General Trucy Aubert had taken earlier only in reverse, and link up with General Benjamín Argumedo, who had already rebelled in the Laguna region. Together, the two commands would cut the rail and telegraph lines to the south of Torreón at Pedriceña and Jimulco stations, thus severing communications and preventing Federal reinforcements from arriving in La Laguna from Durango or Zacatecas. Finally, Orozco would take the main force comprised of his serranos and march southward headed directly for Torreón and the Federal Division of the North.

Salazar's column departed first, but it got into trouble almost immediately. Before this campaign, "General" Salazar had not commanded more than a handful of men, and his commitment to anarchism augured ineffective command and control. He got bogged down trying to cross the difficult terrain, and only subdued the Coahuila state Auxiliaries at Sierra Mojada on April 17, losing precious time and allowing the commander of the fort to alert his commander, Lieutenant Colonel Pablo González, to the invasion. González took up position to the west of Cuatro Ciénagas at Divisaderos with about four hundred men and battled Salazar and his Colorados for fourteen hours, a full two weeks after Sierra Mojada. Under the pressure of overwhelming numbers of Colorados armed with Mausers, while the state forces had short-range .30-30 carbines, González had to retreat through Cuatro Ciénagas and farther to the east until he came to El Carmen Pass. At that chokepoint, González stopped and took up position to block the Colorado advance. Salazar entered Cuatro Ciénagas where he engaged in some heated disagreements with his key subordinates. In those moments, a column of Federals commanded by Trucy Aubert left Torreón by rail to reinforce González. Salazar sent a smaller column of four hundred men to cut the rails at Castaños Station, to the south of Monclova, but that detachment failed in its mission and Trucy Aubert reached Monclova without incident.

The battle between Salazar and González was joined on May 6, when General Trucy Aubert arrived on the scene from Monclova with his Federals, more regional forces, and artillery. The combined forces of the government easily overpowered Salazar and forced him to give up on the invasion of Coahuila. To compound his woes, Salazar abandoned stragglers and did not bother to attend to matters of logistics such that many of his men and horses died of thirst and hunger during the long march back to Chihuahua. The result was that Salazar's column virtually disintegrated during its retreat, and key subordinates such as José Flores Alatorre and Lázaro Alanís wanted to kill Salazar. Trucy Aubert and his Federals returned to Torreón to participate in the invasion of Chihuahua.

As all this was transpiring, General Pascual Orozco remained in Chihuahua City, inexplicably inactive—some said he was waiting on ammunition, while others claimed mere complacency—and providing Huerta and his Federals the precious time they required to organize and field a competent and complete division.

By May 8 the Federal Division of the North moved north to join Pancho Villa's brigade, which had been performing scouting duties around Hacienda Santa Clara, near Bermejillo. That afternoon, in a state of inebriation, Huerta told Villa to attack Cheché Campos' Colorados at Tlahualilo. Villa asked for permission to wait until dark to attack since the dust kicked up by his horses would announce his approach, but Huerta refused the request, accusing Villa of insubordination. It was a bad start to the campaign. Villa repeated his appeal for a night attack to another Federal general who got Huerta to change his mind; the commanding general even gave Villa an additional 150 troopers from the 7th Cavalry for the mission but admonished Villa that he had better be engaged in combat by the morning, or else...

Villa's men were able to slip in between Campos's pickets and the main Colorado force and initiate contact. As the sun came up Villa was still fighting when the Federal artillery and reinforcements arrived on the scene at 11 a.m., and with four cannon blasts precipitated a retreat by Campos toward the north. Because of this action, we will never know if Campos' mission had indeed been to cut Torreón's communications to the east because he never made it past Tlahualilo.

To the south of Torreón, General Emilio P. Campa's campaign initially enjoyed some success but ultimately failed. The Federals first detected Campa's presence when he entered the Nazas District, Durango. A combined column of some five hundred Rurales, Federals, and Auxiliaries left Durango City for Pedriceña to protect the rail lines, but Campa and his Colorados—joined by Argumedo with another 800 to 1,000 Colorados during the battle—drubbed the smaller column on May 14 and forced it back southward to Cuencamé.

Colonel Ricardo Peña commanded a second Federal column—composed mostly of Auxiliaries—that left Torreón to repair the rail lines as far Pedriceña and thus prepare the way for the Federal relief column that was on the way. The combined forces of Campa and Argumedo now gave Peña their undivided attention. The repair crew had just finished their work at kilometer 678 on the line when, once again underway, Peña's convoy came under heavy fire from Colorados arrayed along a high rail bank as the lead train rounded a hill at the entrance of Huarichic Station. The

Colorado artillery landed shells to the left and right of the repair train and the Federals detrained and deployed for action.

Soon the Colorados had Peña's command under a virtual state of siege, but could not force it to capitulate. Not one of their artillery shells found their mark during the battle. Peña later opined that with a cavalry charge he could easily have routed the Colorados and captured their guns since after each assault they fell back in complete disorder. Unfortunately for him, the train carrying his horses, equipment, fodder, and water tanks had retreated back toward Torreón leaving only the repair train. At one point, the Colorados resorted to stampeding their horses and mules and following behind them in an assault. The Federals simply rounded up the animals and hurled them back at the Colorados, allegedly with better results. The Colorados made one final assault under an artillery barrage at around 7:30 in the evening before retiring to a point on the Nazas River about nine kilometers northwest of Peña's position. Peña pulled back to Loma Station in order to get water for the repair train. He remained stuck there between Campa to the north and Argumedo to the south.

On May 19 General Blanquet arrived at Torreón from Mexico City—where he had been recuperating from wounds suffered at Rellano—with his own 29th Infantry Battalion, one section of mountain guns, and another section of field artillery. The *jefe de las armas* in Torreón informed Blanquet that Campa was around Loma advancing toward Torreón from the southwest while Argumedo was around Jimulco, threatening from the southeast. Blanquet divided his command, sending about 250 men under Lieutenant Colonel Teodoro Jiménez Riveroll to face Argumedo. Upon arriving at Picardías Canyon (site of a rather high bridge), Jiménez Riveroll found a captain who had been battling approximately nine hundred Colorados under Argumedo since midnight but had finally driven them back. The lieutenant colonel continued southward and at 8:30 entered into combat with the Colorados who were returning to the canyon. The Colorados fired their mountain guns at the Federals and attempted several flanking and rear attacks from the heights in a running battle of more than eight kilometers that lasted for about five hours and ended about four kilometers from Jimulco. Obviously, Argumedo had been trying to blow those bridges according to the campaign plan. After neutralizing that threat, Jiménez Riveroll returned to Torreón and, upon learning that Blanquet was engaged in combat with Campa, requested permission to reinforce his general. The jefe de las armas of Torreón, worried about an assault on the plaza, denied the request.

Blanquet had taken about four hundred Federal regular cavalry and infantry, two field pieces and a section of machine guns and began moving to the southwest along the railroad when he ran into Campa's Colorados on May 21. That very day, Campa had also begun an attack on Peña at Loma Station. Blanquet managed to repulse the Colorados but he could not pursue because of the condition of the railroad. Repairing the railroad as he went, Blanquet hoped to catch Campa in a pincer with Peña. At eight o'clock in the morning of May 23 Blanquet again caught up to Campa at the Las Sabinas Bridge over the Nazas, four kilometers northeast of Loma Station. Blanquet waited for Peña to arrive and attack Campa's rearguard, but Peña was no longer in the area. Only two hours earlier Peña had entered into Picardías Canyon on the Central Railroad, twenty-five kilometers to the southeast.

In effect, Peña had been cut off from Torreón ever since the May 17 battle at Huarichic Station. He had put together two trains that he sent to Torreón in order to retrieve reinforcements, but neither one had been heard from again. When he tried to move northward from the station toward Hacienda Loma he encountered Campa's Colorados occupying the hacienda and nearby high ground, with their artillery to the east. Peña's Federals constantly engaged Campa and waited to hear the report of Blanquet's guns to join in formal battle, to no avail. They did hear large booms on May 21, but mistook the explosions for rebels blowing the bridge at Sabinas, or Campa's artillery. By the night of May 22, running low on stores and water for the trains, his condition had become untenable, so just before midnight he abandoned his rolling stock, which included another five engines and about one hundred cars, and cut over to the Central Railroad. In the dark and confusion of the early morning hours, he got into a firefight of little importance with friendly forces before the two commands recognized the error.

At first the jefe de las armas in Torreón thought that Peña's forces were Argumedo's Colorados returning to the fight and sent Colonel Jiménez Riveroll to reinforce Picardías. When the colonel reported that the supposed Colorados were actually Peña's Federals, it became apparent that Blanquet was alone in facing the forces of Campa. The jefe de las armas finally gave Jiménez Riveroll permission to reinforce Blanquet at Sabinas. Campa had employed his overwhelming superiority in numbers to shift troops back to the north of Blanquet in an attempt to cut him off from Torreón. By the time Jiménez Riveroll arrived on the scene at 5 p.m., Blanquet had been forced to pull back to Avilés. Two hours later the battle ended when Campa's Colorados abruptly quit the fight.

Three days later, Blanquet again defeated Campa at Avilés by employing a stratagem. He had his men pass in review and then prepare to board trains in order to retreat. Entraining in the face of an enemy was a precarious operation and sensing the vulnerability of the Federals, the Colorados rushed to the attack. When they were about four hundred meters from the plaza, Blanquet ordered his mounted troops to charge with the support of about one hundred deployed infantry. His cannons and machine guns also came alive and together broke the Colorado assault. That ended the Colorado campaign to cut off Torreón from the south. Campa retreated back to Chihuahua taking the same course to the west that had brought him to La Laguna, avoiding the Central Railroad that was now dominated by Huerta's Division of the North.

Also on May 26, General Argumedo tried to take the strategic plaza of San Juan de Guadalupe in eastern Durango, defended by the Veracruzan Comandante Cándido Aguilar. The Federals sent a column of about one hundred men from Aguascalientes to reinforce Aguilar, and it arrived in the middle of the Colorado attack on the town. The Federal relief force managed to force its way into the plaza and once joined to Aguilar's Rurales succeeded in repulsing the attack. Aguilar gave pursuit and later defeated Argumedo's Colorados again at Fresnillo, Zacatecas.

While Blanquet and Peña took care of neutralizing Campa and Argumedo, Huerta finally got underway with his Division of the North to conquer Chihuahua. It would take almost two months. The first major battle took place at a small train station called Conejos, which was situated just to the west of large mountain called Banderas.

As the Federals moved northward, they had their artillery moving in a column of batteries flanked by mounted Maderista irregulars in the vanguard. In the extreme vanguard, the army gendarmes and a squadron of "guides" marched in a line of foragers. All rode at a gallop to raise enough dust to convince any unfriendly forces that the vanguard also had infantry, although the main body of the division actually followed behind in trains. As they got closer to Conejos, Federal Colonel Guillermo Rubio Navarrete told Villa that he should send some troopers to feel out the area to the right, around Banderas Mountain. When Villa's men began taking fire from the hills, the Federal column came to a halt. Huerta continued to assemble his forces, went into a night defensive position, and waited for the next morning, May 12, to begin the battle.

The Orozquistas were dug in at various points on the west side of Banderas, which was situated to the east of the railroad with a general north-south orientation, meaning they had effectively assumed an echeloned disposition. They had one mountain gun, one field gun, and numerous mortar tubes. Huerta positioned his mounted troops under Trucy Aubert and Pancho Villa on the right, directly opposite Banderas, and to the left of them some infantry and machine guns, followed by most of his artillery and its infantry support in the center. On the west side of the railroad tracks he had another battery of artillery, more infantry, and on the left flank mounted units under the command of General Emilio Madero, the president's brother. Because of the disposition of forces and the terrain, it could be considered that Huerta's division was in an oblique order of battle in which his suspended (left) wing functioned as the reserve.

Huerta had about 5,000 men, including four batteries of 75-mm. field guns, two batteries of 70-mm. mountain guns, and eight machine guns. The Orozquistas had a similar number of men.

At eight o'clock in the morning, Villa's troops on the far right started to advance against the hills as the Federal artillery landed rounds to their front. Originally, Villa thought the artillery was poorly aimed and endangered his own men. He sent a runner to tell Colonel Rubio to check his fire, but received a response that the artillery was properly zeroed and to proceed with his attack. As his men resumed the advance so did the artillery shells, walking into the Orozquistas' lines ahead of Villa's brigade.

As the government forces attacked, the Orozquistas returned small arms and artillery fire and mounted several counterattacks. It did not take long, however, for Rubio's guns to silence the opposing artillery, and the Colorados took cover in their positions. At that point the entire Federal line moved up in echelon, closing to within 1,500 meters. Huerta then ordered Villa and Trucy Aubert to attack and turn the Colorados' left. The Colorados put up a stiff resistance and Huerta had to throw First Captain Manuel García Santibáñez's battery of mountain guns into action.

On the Federal right, the opposing forces charged ahead, increasing the rate of fire as the two lines closed to within three hundred meters and finally crashed into one another. In close quarters combat, the government forces and Colorados alike had recourse to their pistols. At this critical juncture it was reported that Huerta's chief of staff, Lieutenant Colonel Carlos García Hidalgo, "moved" Villa's brigade, which most likely meant that it was drifting toward the center. As the Colorados tried to return to their positions, Villa's men outflanked them. Then Trucy Aubert's brigade, which was to the left of Villa's, dislodged the Colorados to his front on Banderas Mountain. Being dislodged from their main positions, some Colorados attempted another counterattack, and then their entire disposition came undone as they fled in disorder to the safety of their awaiting trains. General José Inés Salazar's brigade tried to cover the retreat but the trains departed before General Luis Fernández's brigade could arrive and his men had to retreat cross-country. The entire battle lasted some ten hours.

The Federals incurred forty killed and seventy wounded, although these figures seem quite understated. Huerta estimated that he had killed four hundred Colorados (including the eighty prisoners captured by Villa's brigade that Huerta had executed) and wounded another two hundred. The booty was vast, including 107 railcars with clothes and equipment, three locomotives, three artillery pieces, seven mortars, and 550 horses, most likely abandoned by Colorados jumping aboard the retreating trains. According to some accounts, Orozco personally directed the battle for the Colorados, but Huerta's intelligence indicated that Orozco had left for his headquarters at Yermo Station on the eve of battle, placing General Salazar in command of the forces at Conejos Station. After this latest loss, some of the Colorados were growing disaffected by Orozco and Salazar's faulty generalship.

Some of Huerta's officers also started to grumble about their commanding general's handling of the campaign. Captain Jacinto Treviño even went so far as to call Conejos a sham battle designed to defeat but not annihilate the Colorados. He specifically blamed Huerta for the idleness during the entire battle of the three thousand men who made up the left wing. However, the left wing actually consisted of fewer than one thousand men and, as previously explained, in this oblique order of battle served as the reserve. Moreover, as the Colorados fled from the mountain to their trains, their general movement was away from the Federal right toward the left wing. In such a fluid situation, Huerta most likely did not want to commit the totality of his forces. Huerta also received criticism for his slow, general advance so that he could keep his name in newspaper headlines for as long as possible, some speculated. Huerta, however, did not want to wind up like González Salas, and therefore he advanced very slowly and deliberately so as "to guarantee success of the operations," as he stated.

The advance continued slowly and cautiously because of the numerous burned bridges and mines that the Colorados had placed under the tracks. The Federals suspected that the Colorados had taken up position at Rellano, site of González Salas' earlier ignominious defeat, and on May 22 initiated a movement to contact. At 5 a.m., a squadron of mounted "guides" commanded by Trinidad Rodríguez left Escalón Station followed an hour later by a squadron of army gendarmes, two companies of regular infantry, a company of railroaders, and a battery of machine guns and Rexers. Leaving a half hour after this second group were two batteries of machine guns, four batteries of field guns, and a squadron from the 7th Regiment. The field batteries marched in a double column each carrying four carts of artillery shells and one cart of infantry ammunition. Colonel Rubio commanded the vanguard.

After clearing several mines, the vanguard reached Asúnsolo Station at eleven o'clock in the morning and waited for the main body to arrive. In the meantime they observed the Colorado patrols, then pushed the Colorados back to the hills north of Rellano Station and reconnoitered the area. At 1 p.m. Huerta arrived at Asúnsolo, and after receiving several reports told Colonel Rubio, "The enemy seems to be concentrating in Rellano. With all the forces at your command you will make an offensive reconnaissance in the direction of Rellano endeavoring to [get a] fix on the enemy and protect the deployment of the Division, which will start arriving on the field at four in the afternoon. For this operation, besides the forces of your command, you will have at your disposal the 'Madero' and 'Villa' Irregular Forces and all the forces in the immediate vicinity until the arrival of the main body of the Division. The position of the General in Chief to receive reports will be in Asúnsolo."

The Colorados commanded a front of about seven kilometers in length, running from a hill to the east of Rellano Station to the west, through the station, the hills to the west, along the Sauz Arroyo, past the dam, and ending at the hills west of the dam. They were well-entrenched and supported by one Schneider Canet, two mountain guns, two 80-mm. Mondragón mortars, five homemade cannons, and several mortar tubes positioned along the hills. They also had four Colt and two Hotchkiss machine guns.

Colonel Rubio sent the railroaders and regular infantry to reconnoiter in the center and ordered various mounted units to contest the farthest extents of the Colorado lines. Under the protective fire of Rubio's cannons and machine guns, the mounted units all conquered their assigned positions, forcing the Colorados to a much-reduced front of only four kilometers running from the hills north of the station to the dam. Rubio could not take the dam because any force occupying that position would have been exposed to enfilading fire from Colorados in the hills.

After that initial encounter, the Federal disposition was perpendicular to the railroad, and in an oblique against the Colorado left wing, with the Federal left suspended and quite a distance away from the Colorado right at the dam. Rubio's guns kept the Colorados from attempting to swoop down from the hills in the center. An hour later, at four o'clock in the afternoon, the main body of infantry began arriving on the scene, as did the commanding general. Huerta ordered General Antonio M. Rábago, who had taken over Trucy Aubert's brigade while he convalesced in Mexico City from a wound suffered at Conejos, to extend the Federal left and then attack the Colorado right. Rubio and Villa both panned this maneuver, since Rábago should have been sent against the Colorado left, which was weak and about to be turned by the Federals.

Rábago's maneuver was a disaster. The Colorados brought their artillery into action and repulsed the Federal assault, perhaps because Rábago had not waited for Emilio Madero's irregulars before beginning the attack and had not thrown the full weight of his brigade into the fray. For whatever reason, by 7 p.m. the Federal left was fleeing in disorder and only Rubio's artillery in the center and Villa holding fast on the right prevented a complete collapse of the line. Encouraged by that success, the Colorados pressed the attack against the Federal right with such vigor that in less than two hours the Mariano Escobedo Battalion had been decimated and was out of ammunition. But the Federal line held and during the night there was only sporadic combat as Rubio's guns rained metal down upon the Colorado trenches, firing at objectives that had been identified during the daylight hours.

Most of the damage inflicted on the Colorados came from Santibáñez's battery near Rancho de Sauz, opposite the Colorado right, where it was able to provide enfilading fire against the Colorado center and left. Therefore, under cover of darkness the Colorados attacked the battery and its infantry support provided by the 15th Battalion, killing a second captain. The Colorados also moved back into Rellano Station, extending their left. At three in the morning, the O'Horán Brigade finally arrived and Huerta sent it to reinforce Rábago on the left. There still was no word from Téllez's Brigade, which functioned as the rearguard. As the sun came up, the Federals were surprised to see that the Colorados remained in their positions in spite of the night's terrific cannonade.

Just before six o'clock in the morning, the Federal guns again came alive, dislodging the Colorados from their positions at the dam. There is some disagreement about what happened next, but apparently the Colorado's initiated a flanking movement against the weak Federal right flank preparatory to retreating. At that point Huerta pulled Miguel O'Horán's brigade from the left and sent it to the right where it came under fire from Colorados in the hills to the east of the Federal line. The Federal flanking guard then occupied a hill to the east to extend and anchor the right, but that small group was soon surrounded by Colorados and subject to intense fire. Had it been pushed off the hill, the Colorados could have rolled up the Federal right and perhaps the entire line. Huerta had to throw the Xico Battalion and a battery of machine guns into the line to support the right flank, and Colonel Rubio ordered Santibáñez's battery to fire on the Colorados rushing to reinforce points opposite the Xico Battalion. That was enough to turn the tide and "the rebel column, if not completely undone, was completely dispersed," in the words of Huerta.

After the flanking movement had been repulsed, Rábago delivered an assault in depth against the hills in the center of the Colorado line that was supported by Rubio's guns and Madero's Rurales on his left. The government forces put the Colorados on the hills to flight at a little past noon. Rábago then turned attention to his left and rolled up the Colorados at the dam as the artillery moved forward. The Saint Chamonds continued to fire as the shorter range Schneider Canets displaced. O'Horán's battalion moved up to occupy the dam, but soldiers of the Railroader and 15th Battalion had already secured the point.

Huerta reported to the Secretary of War that the last and most important Colorado positions had just been taken at 12:10 p.m., and he put their casualties at eight hundred to one thousand killed and wounded and another fifteen hundred dispersed. General and historian Miguel Sánchez Lamego estimated the Colorado losses at closer to 650, and put Huerta's at 140 men, which seems extremely low for the government's forces. The Federals also captured three cannons, six mortar tubes, and numerous horses. To cover their retreat the Colorados blew up sections of the railway and set several bridges on fire.

Federal officers present at the Battle of Second Rellano once again had plenty of unflattering things to say about the effort. Captain Treviño bemoaned the lack of any maneuver, saying that the Federals resorted to crass frontal assaults, and then followed up the victory with a weak pursuit. Rubio blamed the latter on General Téllez, who showed up only at the conclusion of the battle and then his men "began to pour into our left without having taken part in the combat and in the position opposite [that] necessary to impede the enemy's retreat." Regarding the lack of maneuver, it seems quite possible that with fire superiority and a division that consisted mainly of irregulars and raw recruits, Huerta did not want to risk maneuvering under fire, which would have been difficult for even the most seasoned troops.

The Colorados also made several crucial mistakes. First, by digging in and allowing the Federals to bring up their artillery and go into battery, the Colorados were unable to execute their planned tactics of getting close to, and mixing with, the government forces in hand-to-hand combat in order to neutralize the effectiveness of the Federal artillery as much as possible. The time to execute that tactic would have been during the movement to contact when all the Federal batteries were in the vanguard protected by around eight hundred mounted and dismounted soldiers. Second, they allowed the government forces to force their disposition into a reduced front of a mere four kilometers, thus cramming them into a target-rich kill zone for the Federal guns. Indeed, this battle was one of the few in the course of the entire revolution where artillery appears to have played a defining role. Still, ample evidence suggested that their trenches "made by technicians" (in the opinion of Huerta) provided fairly adequate protection against direct fire guns that hurled shrapnel *at* the Colorados rather than detonating it *over* them, as mortars or howitzers might have done. The effectiveness of trenches against guns helps explain why the Federals were so surprised that the Colorados were still in their positions the morning of May 23.

The Battle of Second Rellano also supplied a perfect example of why artillery in the revolution provided an advantage in signals. The artillery set up the "rapid telegraphic and telephonic installation [that] intimately connected the Commanding General with Commanding General of the Artillery, the firing line, and the military trains, and permitted the transmission of successive telegrams to the President of the Republic during the battle right from the field of action," boasted the division's chronicler, José Juan Tablada.

As the Division of the North moved into Chihuahua, it relieved pressure on the counterinsurgency elements commanded by Colonel Toribio Ortega and others who had been operating out of Ojinaga under the direction of Federal Brigadier Agustín Sanginés. Ortega had provided a crucial service to Madero's government by denying Ojinaga to the Colorados as an entrepôt for contraband and his presence there had kept the plazas of Ciudad Juárez and Chihuahua in check; now he moved down to Santa Rosalía Camargo ("Camargo") to join with the division. At the same time, the Federals began assembling a blocking force in northeastern Sonora to be commanded by Sanginés, who left to take charge of his new assignment. Sanginés would act as an anvil against which Huerta's hammer would smash the Colorados.

Continuing northward, General Rábago, in command of the Division of the North's vanguard, fought a two-day action against the Colorados at La Cruz starting on June 16.

Afterward, the Colorados withdrew to Bachimba, a railroad station at the entrance of a canyon formed by two large mountains where Orozco's men dug in and, once again, awaited the approach of the Division of the North. The destruction of several large bridges as well as the mining of the railroads delayed the arrival of the Federals, and during this long march the commanding general decreased the depth of the various columns and had the elements of the vanguard march in a "column of battalions" formation in order to be better prepared to assume a combat disposition. The artillery and impedimenta followed to the rear.

Finally, on July 2 the Federal division arrived to within about four or five kilometers of Bachimba. In days prior they had sent about four hundred regular cavalry to test the Colorado right, but those troopers were turned back. They had also reconnoitered the Colorados' positions at Bachimba. Here the railroad entered Nogales Pass formed by the Ocotillo and Bachimba mountains where six thousand Colorados had entrenched along a concave line of about 3.5 kilometers in length. Orozco had one Schneider Canet, one Mondragón mountain gun, and ten "breach-loading American cannons" on Bachimba Mountain. At the foot of Bachimba Mountain and anchoring the Colorado left was Hacienda Bachimba. Because of the natural curve of the terrain, the Colorado right was in enfilade to the general direction of the Federal approach, where Orozco also had some artillery, probably mortars tubes. In contradiction to the prevailing narrative among Mexicanists, a journalist with the Colorados reported being shown large stockpiles of ammunition and said the troops' morale was high. They chalked up previous defeats to fighting on foreign soil, but now they were on familiar terrain and that would redound to their favor.

On July 3, the Federals began their evolutions in perfect view of the Colorados. Huerta placed his five batteries of field and mountain guns under Colonel Rubio in the center of the line, supported by the Xico and Railroaders Battalions and the Gendarmes squadron. The Manzano Brigade assaulted Hacienda Bachimba directly while the Rábago Brigade made a flanking movement against Bachimba Mountain to the rear of the hacienda. On the Federal left, the Téllez Brigade attacked along the railroad headed directly for Bachimba Station and Nogales Pass beyond. On the far left, the O'Horán Brigade supported by machine guns rolled up the Colorado right. The Federal artillery fired in support of the maneuvers and the Colorado artillery returned fire, for a short time. Toribio Ortega was with O'Horán on the left, and he made a complaint similar to Villa's at Conejos about Federal artillery landing dangerously close to his men. In fact, Ortega's private secretary later asserted that the Federals were purposefully trying to kill the Maderista irregulars, but more probably the gunners walked in the shells ahead of the ground assault as planned. On the Colorado left, at the hacienda and along the creek at the foot of Bachimba Mountain, the resistance put up by Cheché Campos' men was impressive. At one point Orozco's chief of staff even mounted a counterattack, which was turned back when Huerta threw one of his reserve battalions into the fray. The main action was over by 2 p.m., although sporadic fighting continued into the evening, as the Colorados began a full retreat. The Federals did not pursue, perhaps with visions of Malpaso in their heads.

Analysis

There are two narratives frequently parroted by historians as to why Huerta defeated Orozco at Bachimba: Orozco lacked ammunition, which he blamed on the Americans, and Huerta had overwhelming firepower. These facile explanations are self-serving for all concerned: the Federals—especially Rubio—were able to boast of the superiority of their arms, while the Orozquistas had a convenient excuse for their defeat, and Mexicanists had proof of "imperialist" American influence in the outcome of the revolution and could easily sidestep having to stoop to writing anything resembling military history. But is the prevailing history correct?

For the rest of his life, Colonel Rubio never stopped crowing about the effectiveness of his batteries during the campaign and especially at Bachimba, the coup de grace of Orozco's army. But Edwin Emerson, an American reporter who had accompanied the Division of the North on its campaign, was unimpressed with both Rubio and the Federal artillery, saying that it did not cause much harm to the opposing forces ("the Federal artillery there accomplished little beyond frightening an enemy who had no artillery.") This opinion was born out by reporters who accompanied the Colorados and said that Orozco's men experienced few casualties in general, and as a result of the shelling in particular. Correspondent A. Ruiz Sandoval said, "The revolutionary artillery fired its cannons three times, their shots being answered by two cannonades from the Federal batteries, but none of them hit the mark." Another member of the press, José V. Soriano, said that "The shells of the Government's troops caused little damage in the revolutionary camp. At nightfall few casualties had been reported." The reasons for the ineffectiveness of Federal Army artillery can be explained technically: it only possessed direct fire guns and employed them against an entrenched enemy. The lack of skill by Federal gunners may also have played a role, as Emerson wrote, "General [Rubio] Navarrete enjoys a high reputation as an artillerist among the Mexicans, but I consider him much over rated. He was in command of Huerta's artillery at Bachimba, where I saw him directing the work of twenty guns in a very exposed position on the flat from a poorly chosen observing station."

As regards the Colorados' shortage of ammunition, according to testimony given to the U.S. Senate Subcommittee on Foreign Affairs by Orozco's representative in the United States, Manuel Luján, the Orozquistas had five million rounds of ammunition at the beginning of the campaign but after the ten-day Battle of First Rellano they had run through all their reserves. This assertion can be rejected out of hand: the Battle of First Rellano did not last ten days and even if it had, if we take as a guide the pivotal seventy-two-day 1915 Battle of El Ébano, where the Constitutionalists expended almost six million rounds, then we can see that this testimony was not reasonable. Even Orozco's right-hand man, General Marcelo Caraveo, stated that the Colorados only really began to feel the effects of the U.S. arms embargo after La Cruz, or rather, in mid-June. Yet the war correspondent Emilio Valenzuela said: "As regards what the *capitolino* [Mexico City] press has said, that ammunition is scarce and that it is probable that the rebels will surrender because of it, is a lie. When I was in Bachimba, several revolutionary leaders took me to the places where they have the ammunition and I was in admiration of its enormous quantity." Thus, we can surmise that claims of the Colorados' shortage of ammunition were either overwrought, not true until late in the campaign, or totally false. In the throes of the campaign, the narrative of an impending Colorado surrender due to the lack of ammunition served the government to boost the confidence of the public in a quick victory and was pure propaganda.

And there are at least five additional, indisputable factors, aside from the reasons already expressed, to doubt the responsibility of the U.S. embargo for Orozco's defeat. First, the United States enacted an embargo of all arms and ammunition into Mexico on March 14, 1912. But on April 22, 1912, President Madero applied for and received a resumption of the exportation of war materiel into Mexico (mainly because it was hindering the progress of industry and making the counterinsurgency operations of General Sanginés in northeastern Chihuahua difficult). Interpretation: nothing had really changed from 1911, with the encumbrance to the Orozquistas being nothing more than neutrality laws. Second, it was perfectly legal for Orozco's agents to purchase arms and ammunition in the United States. Third, Marcelo Caraveo admitted that the Orozquistas had plenty of money to buy war materiel. Fourth, because of the particulars of neutrality laws and American jurisprudence, the smugglers had to be caught at the border in flagrante delicto. That meant agents for the Mexican government inside the United States had to track gunrunners and then corral U.S. authorities and smugglers at the border, a virtually impossible scenario. Finally, the Orozquistas chose places to cross the border with contraband where "the oversight of the [American] authorities [was] deficient and the personnel very scarce," according to one Mexican consul.

Indeed, word on the street was *not* that the Orozquistas lost because of a deficiency in war materiel, or because of the Federal Army's massive firepower, but rather because of a lack of unity of command (an unfortunate nature of the beast among anarcho-syndicalists like the Colorados), Orozco's untalented subordinate generals, and, most importantly, that Orozco delayed almost two months after First Rellano before resuming the drive southward to Mexico City, giving the government plenty of time to organize an armed response.

Moreover, if Orozco were truly concerned about his ability to get more war materiel, there was plenty awaiting him in Torreón if he had pressed onward after First Rellano.

In spite of all the headwinds blowing against an Orozquista victory, plenty of theoretical and anecdotal evidence exists to suggest that the Orozquistas could have won nevertheless. First, as consistently stated, artillery played an indecisive role throughout the revolution as Orozco himself had already experienced at the Battle of Sierra de la Mojina in 1911 when in a similar circumstance to Bachimba, albeit on a reduced scale, he prevailed against a combined-arms Federal column with artillery. Second, Federal Army doctrine made it potentially vulnerable during movements to contact because on the one hand their doctrine required that cavalry and artillery be detached to the vanguard in order to secure the best terrain, but on the other hand Federal cavalry was so woefully inept ("The Federal cavalry is laughed at by the northern rebel horsemen," according to one U.S. Army report) that Huerta had to rely overwhelmingly on mounted gendarmes and irregulars. If, as mentioned previously, the Colorados had executed their tactic to neutralize the effectiveness of the Federal artillery against the vanguard during the movement to contact, and not in pitched battles, they might have prevailed, as evidenced by outcomes in other engagements during the revolution. For instance, just before the Battle of Culiacán in 1913, the Federals found one of General Álvaro Obregón's batteries undefended and removed the breechblocks, so such bold moves were possible. In another example, in 1915 at the Battle of El Cuatro, disciplined Constitutionalist infantry without any artillery attacked Villistas holding the high ground with fourteen field pieces and emerged victorious. The problem for the Orozquistas was not Federal artillery, it was a lack of discipline to execute chosen tactics and a force composition almost totally comprised of dragoons, which foreshadowed Villa's defeat in 1915.

On a strategic level, Salazar's, Orozco's, and Campa's multi-state operations came off uncoordinated such that the Federals could use interior lines to race Trucy Aubert to reinforce González, defeat Salazar, and return to participate in the campaign against Orozco while minor Federal Army commands did the same using Torreón as a pivot to attend to Campa and Argumedo.

OROZCO'S SMALL WAR

If there was any sham battle during Huerta's drive north, then it was at Bachimba, and the perpetrator was not Huerta but Orozco, who had no intention to be drawn into a decisive battle and started withdrawing his troops in an orderly fashion almost from the outset of combat. Days before the battle Orozco had already revealed that if Huerta attacked him at Bachimba, he would divide "the seven thousand men that currently form the revolutionary army...into three columns of two thousand men and use a system of guerrillas" to frustrate the plans of a regular army that had proven itself incapable of successfully waging such a war. Specifically, Orozco cited "Zapata's men fighting in just one state [with]...three thousand men...only a few kilometers from the capital of the Republic." He calculated that with the number of men backing his movement, the Federal Army would need two hundred thousand men to quash his insurgency, which was fragmenting.

During this time there appeared to be a split among the Colorados, with the hardline PLM members under Emilio Campa, José Inés Salazar, and Luis Fernández staging an invasion of Sonora from western Chihuahua, and Orozco's closest collaborators following him to Ciudad Juárez, where he established his new capital. It did not last long.

On August 16, Orozco and his followers abandoned Ciudad Juárez without a fight, leaving it to the Federal Army column moving up the Northwest Railroad under Generals Téllez and Rábago. Escaping down the Rio Grande, the Colorados defeated Colonel Toribio Ortega at Cuchillo Parado and then took the border town of Ojinaga on the night of September 11. Three days later Federal Colonel Manuel Landa and Maderista Colonel Ortega recaptured Ojinaga. Orozco crossed into the United States, and Colorado General Marcelo Caraveo conducted an invasion of the neighboring state of Coahuila. Landa sent Ortega and other irregulars in pursuit because his own Federal Army troopers were too exhausted, but the Maderista colonel had to turn back at Punta del Agua.

As Caraveo penetrated into Coahuila, Federal General Blanquet moved up from Durango with a combined arms column that rode the rails to Piedras Negras. The Federals considered entering Texas by rail to Marathon with plans to detrain and march overland back into Coahuila, but Blanquet ultimately decided against it. His presence on the rails between Piedras Negras and Monclova kept Caraveo in the hinterlands where irregular Colonel Luis Alberto Guajardo's state troops defeated the Colorados on numerous occasions in the many canyons, valleys, and passes, forcing the Colorados to flee on foot into the hills, which explains why the Coahuila state troops reported recovering so many horses after each encounter.

The campaign for Sonora envisioned a three-pronged invasion of the northern, central, and southern parts of the state. The Colorados planned to plunder the state of its resources and had high hopes since they predicted that the Federal Army would not be able to move its artillery assets over the mountains and follow them. In this assumption they were correct, but Salazar faced an early defeat at Ojitos, just inside the Chihuahua state border, at the hands of Federal General Sanginés, where his commander of infantry, Lieutenant Colonel Álvaro Obregón, received a baptism by fire. Sanginés and his irregulars penetrated farther into Chihuahua to take part in the only successful counterinsurgency operations in western Chihuahua before winding up in Ciudad Juárez. Afterward, Sanginés returned with his troops to Sonora via the United States and Obregón again defeated Salazar at Hacienda de San Joaquín, a battle that marked the beginning of a very long and successful career that would ultimately land him in the presidency. The other invasion columns commanded by Campa and Fernández were also defeated by the Rurales, National Guard, and Guerrillas (posse-like units) working hand-in-glove with the clumsy Federal regulars. By October 1912 all the Colorados were back on home soil, in Chihuahua and La Laguna, where they continued to elude the Federals.

Analysis

Orozco's small war gave proof once again of the prevailing narrative about the incapability of Federals to range far and wide in independent operations. The performance of the regular cavalry under General Manuel Landa dispatched on August 30 from Agua Nueva, where it was guarding repairs to the rails, to reinforce Ortega proved risibly incompetent. His movement was so slow that he did not reach Ojinaga in time to keep it from falling to Orozco on September 11. Then, after retaking the border town on September 14, his Federals were too worn out to carry out the pursuit, having to leave that job to Maderista irregulars even though the regulars had not been subjected to any hardships in excess of those endured by the irregulars. In western Chihuahua, Sanginés' irregulars were the only ones who developed effective mop-up operations because Huerta's regulars did not depart from the Northwest Railroad. Meanwhile, in Coahuila, Blanquet did nothing to separate from railroads, leaving Coahuila state troops to carry out the difficult and effective search-and-destroy operations.

THE CITADEL

In October 1912, President Madero faced a new challenge to his government when Félix Díaz, the nephew of Don Porfirio, rebelled in the port city of Veracruz. The young general had hoped that the officers of the Federal Army would fall into line behind him and carry him to the presidency. Instead, the Madero administration began moving forces toward Veracruz and placed them under the command of Federal General Joaquín Beltrán.

In the culminating battle, Beltrán's artillery bombarded the port from the west as the bulk of his forces assaulted from the north while two smaller commands pressed the rebels from the south and southwest to block their exit. Beltrán had intended the attack force to the north to make a frontal assault supported by a flanking movement, but the commanders in the field decided to make a slight adjustment and instead of a flanking movement initiated a *martillazo* ("hammer strike"), which consisted of an exaggerated swing out of forces on the right to attack the Felicistas along a line perpendicular to the line of attack. An offensive hammer required strict fire discipline along designated fire lanes by the attacking force in order to avoid friendly fire casualties. Beltrán's soldiers carried out the maneuver effectively, forced the Felicistas back toward the plaza, and pressed onward to take the main barracks and command center. The capture of that installation forestalled what might have been a protracted battle had the rebels occupied that sturdy redoubt. As it happened, General Díaz and his supporters took refuge in the town hall, unsuitable for a defense, and had to surrender.

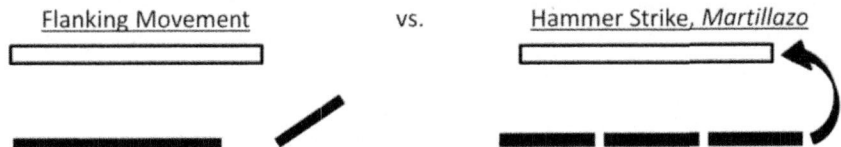

| Flanking Movement | vs. | Hammer Strike, *Martillazo* |

Figure 1 The Martillazo

Díaz was tried, convicted, and sentenced to death for rebellion. However, the magnanimous President Madero commuted the sentence and had Díaz transferred to the penitentiary in Mexico City. At the time there were many conspirators in the nation's capital scheming to overthrow Madero. The most viable effort was headed by Generals Bernardo Reyes, in prison for an earlier abortive rebellion, and Manuel Mondragón (designer of the army's cannons), among others. In the early morning hours of Sunday February 9 the rebels hatched their plot, supported by most of the artillery forces in and around the capital, the 1st and 2nd Cavalry Regiments, the students of the *Escuela de Aspirantes*, the mounted gendarmes and security battalions, part of the 20th and 24th Infantry Battalions, other Federal Army units, and numerous civilians. Federal General Lauro Villar, Military Commandant of the Capital, commanded the loyal troops of the army. After successfully defending the national palace against the initial rebel attack, during which he had been wounded and Reyes killed, Villar handed over command of the government forces to Victoriano Huerta, who conveniently appeared on the scene to offer his services.

Unlike at Veracruz, where Beltrán's Federals had kept Díaz's troops from securing the eminently defensible barracks and command center, during the Mexico City *cuartelazo* ("barracks uprising") the rebels managed to secure the *Ciudadela* ("Citadel"), a key armory with the construction of a bunker, complete with a roof and walls ninety-five centimeters thick. So critical was the defense of this site to the ultimate success of the rebellion that it became known as the "*cuartelazo de la Ciudadela.*" Over the next ten days, referred to as the *Decena Trágica* ("Tragic Ten Days"), government and rebel forces battled each other in the nation's capital. The artillery from both sides spewed metal promiscuously and indiscriminately in unsuccessful attempts to gain fire superiority. As the citizenry and foreign diplomatic corps tired of what can only be described as medieval conditions, President Madero issued an ultimatum to Huerta: take the Citadel by storm or perish in the attempt. Given the virtual impossibility of a military success Huerta chose a political way out of the predicament and initiated a palace coup with the backing of General Blanquet and others. He then negotiated an arrangement with the remaining leaders of the cuartelazo, Félix Diaz and Manuel Mondragón. Days later, Madero and his vice president were murdered during their transfer to prison.

Numerous tactical conditions frustrated any successful government operation to subdue the Citadel, starting with the artillery. Many expected the Federal artillery to simply pound the redoubt into submission, yet as Colonel Rubio pointed out the Federal arsenal consisted solely of the relatively small 75 and 80 millimeter guns. These were ill suited for the "arched and vertical" trajectory necessary to destroy rebel ground forces and too small to destroy any structures except "cover of little resistance," and certainly not a fortress such as the Citadel. Moreover, most of the shells were anti-personnel shrapnel—not high explosive—and did not have reduced-charges necessary for this type of combat. Similarly, the guns needed to be firing in concert at a precise sector, which was never designated, directed by telephone communications and observations points, which were lacking, and in coordination with a major ground assault, which never occurred. So many soldiers and Rurales were participating in mop-up operations against the Orozquistas in the North, working to subdue the Zapatistas in the South, or trying to maintain civil obedience around the country that there simply were not enough bodies available to tip the balance in the government's favor. And the government fell.

3. THE CONSTITUTIONALIST REVOLUTION, FEBRUARY 1913 – JANUARY 1914

The response to the overthrow of constitutional government was not long in coming from the northern states. The main opposition came from Governor Venustiano Carranza of Coahuila, who refused to grant recognition to Victoriano Huerta's new government. On March 26, 1913, Carranza drafted his Plan de Guadalupe, which formed the Constitutionalist Army, and appointed himself to the position of First Chief. The sole purpose of the plan, a document signed by state militia officers that contained no socio-economic agenda, was the overthrow of Huerta and the restitution of constitutional governance.

GEOGRAPHY AND ARMED RESPONSE

Although the state forces of Coahuila stood ready to oppose Huerta and the Federal Army, Carranza's soldiers suffered many disadvantages. First and foremost, the group was quite small, amounting to possibly as few as five hundred men. When we think of the Mexican Revolution, our minds conjure up images of Zapatista guerrillas attacking trains or mounted Norteños fighting heads-up battles against Federals. Yet those revolutionaries from the Northeast were significantly weaker than their Norteño brethren from Sonora and Chihuahua/Durango, although that had not always been the case.

The state of Nuevo León had been colonized beginning in the sixteenth century with the establishment of communal *mercedes* (large land grants). Ownership of these communities was vested in the hands of *accionistas* ("stockholders") from extended families who exercised rights and obligations more typical of citizens than subjects of the crown. They had a duty to bear arms in protection of the land and elected their own municipal leaders and created militias. It was the earliest form of the *campesino*-soldier construct designed to protect the heartland from invading forces. They had fought long, hard, and frequently against Indians, in the Mexican-American War, when they were known as "rancheros," and in the War of the French Intervention, when they provided one of the best armies, called the Army of the North, for Benito Juárez. By that time the states of the Northeast—Nuevo León, Tamaulipas, and Coahuila—frequently functioned as one, sometimes officially as when Juárez granted Governor Santiago Vidaurri permission to join the states of Nuevo León and Coahuila politically and gave him military command of Tamaulipas. After the French Intervention, one of the most famous leaders of the Army of the North, Mariano Escobedo, settled San Pedro de las Colonias in Coahuila with some of his soldiers, perpetuating the marriage of the agricultural to the military. Yet of all the sections of the North, the Northeast benefited most immediately from Pax Porfiriana, as the Indian wars there finished earliest permitting industry and commerce to blossom. The first steel mill in Latin America was founded in Monterrey, the happy coincidence of coal fields in neighboring Coahuila, iron-ore in Durango, and the transportation grid, finance capital, and educated workforce of Nuevo León. Similarly, the petroleum industry in the Northeast had been well established while the *hacendados* of the region remained relatively weak compared to the rest of the nation. By the time of the revolution the campesino-soldier tradition had faded and the populace was largely prosperous with numerous options for employment and therefore not so interested in revolution. Thus, the Northeast provided the weakest contingent of the Constitutionalist Army.

To bolster his meager forces, Carranza gave all Federal officers thirty days from April 20 to join the Constitutionalist Army, except those who had participated in the Citadel *cuartelazo* and subsequent palace coup. After that term expired, Benito Juárez's Law of January 25, 1862, would be reinstated, subjecting all Federal officers to summary execution upon capture.

Second, since the Northeast contained some of the nation's most vital economic and advanced transportation infrastructure, the Federals assigned top priority to the region and had numerous axes of advance to exploit in developing military operations (Torreón-Saltillo-Monterrey, San Luis Potosí-Saltillo-Piedras Negras, Monterrey-Nuevo Laredo, Monterrey-Ciudad Victoria-Tampico, San Luis Potosí-Tampico, and Matamoros-Nuevo Laredo). In contrast, Chihuahua entailed only one major axis running from Ciudad Juárez southward through the middle of the state, and Sonora had a similar operating environment with one axis from the south up through Empalme (Guaymas) and Hermosillo to Nogales, with a dog-leg to the east that ended at Agua Prieta. Additionally, the Federal Army's base of operations, Mexico City, was connected by numerous modes and routes to the Northeast.

Third, and related to the previous point, what would become the Constitutionalist Army Corps of the Northeast had to operate over three states (four if one includes San Luis Potosí) and contend with three Federal divisions (the Division of the Nazas, the Bravo, and the Center) while those Constitutionalists from Chihuahua and Durango only had to deal with one each, the Federal Division of the North and the Division of the Nazas, respectively. With such a vast territory to control, the future commander of the Army Corps of the Northeast, Pablo González, would have to deal with numerous far-flung regional *caudillos* who resisted his authority and complicated operations.

The situation in Sonora, meanwhile, was most advantageous for the revolutionaries, since the Sonorans only had to manage one state and handle one division, the Federal Division of the Yaqui. And the state was relatively isolated from the Center, requiring troops and supplies from Mexico City to travel by rail and then overland and/or by sea to reach the state. (After June, 1913, Chihuahua and Durango were also inaccessible to Mexico City via Zacatecas and Federals intending to reach those two states had to travel through Saltillo, which became a transportation hub right in the midst of the Northeast, one more factor retarding the development of military operations for the Constitutionalists in that section.)

Most importantly, the Sonoran government had obtained backing from the central government to pay for its state militia, which was quite large, and had gained some rudimentary training during the Orozco Insurrection. And the Sonorans moved swiftly and decisively.

NORTHERN SONORA CAMPAIGN

On February 26, Governor José María Maytorena of Sonora requested a six-month leave of absence and abandoned the country rather than choose sides. Meanwhile, the highest-ranking officer in the militia, Colonel Álvaro Obregón, had begun concentrating the state's forces in Hermosillo, which was located in the center of the state and ideally situated for subsequently conducting operations on interior lines of communication. Under interim Governor Ignacio Pesqueira the state legislature voted to renounce "Señor General Huerta as Interim President of Mexico" on March 4. Two days later, Obregón entrained his 4th Irregular Battalion and began heading north along the rails toward Nogales.

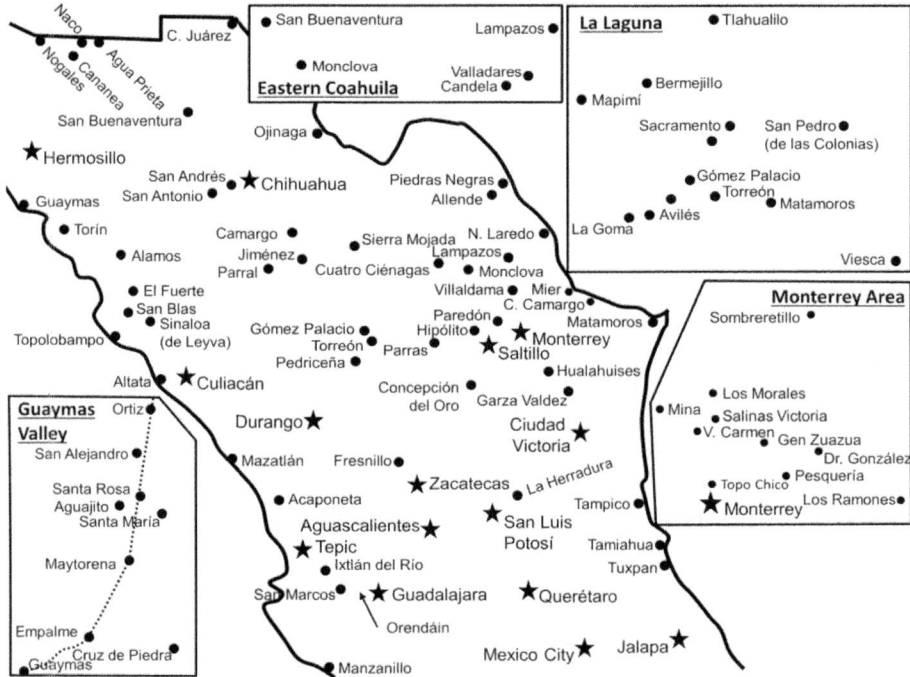

Map 3 The Constitutionalist Army Area of Operations

The Federal zone commander, General Miguel Gil, had about six hundred men in Torín, in the south, to manage the unruly Yaqui Indians, and had begun forming a force of about 1,200 men in Empalme to strike at Hermosillo, but he had given no definitive signs that he planned to begin moving against the state capital any time soon. Gil also had approximately 1,400 soldiers in northern Sonora, mainly in Nogales, Cananea, Naco, and Agua Prieta, where the highest ranking officer,

General Pedro Ojeda, had his headquarters. Federal Colonel Emilio Kosterlitzky commanded fewer than three hundred Federal soldiers and customs officials for the defense of Nogales against an estimated nine hundred men approaching under Obregón. Kosterlitzky sent several frantic telegrams to Ojeda requesting reinforcements, but Ojeda downplayed the threat. In the subsequent battle for Nogales, Kosterlitzky tried to defend the plaza but the course of the perimeter as required by the terrain was too large for the number of men under his command and he soon began to run out of ammunition after only a few hours of combat. The Sonorans easily captured the border town on March 13 and with it a crucial entrepôt for war materiel. Tactically, the battle was of little note, but strategically it was of vital importance. Obregón gave more credit for the victory to the Federals, meaning General Ojeda for not promptly reinforcing the plaza, than he took for himself.

Ojeda finally departed Agua Prieta to reinforce Nogales on March 12. He arrived at Naco the next day to learn to that Nogales had fallen. Originally Ojeda had about 550 men, but he had to leave behind a group of thirty to garrison Agua Prieta and then another seventy deserted, so by the time he reached Naco his force had been pared down to about 450, of whom only about thirty were mounted. The conditions of the rails and Constitutionalists in the area prevented the Federals from advancing any farther.

Obregón did not leave Nogales until March 19 and by the time he reached Naco he had collected enough state forces along the way that his command now totaled about three thousand men. Because Naco was a border town like Nogales and Obregón wished to avoid an international incident that might arise from damage to American property during a battle, he first tried to get Ojeda to agree to fight in open territory. The Federal wisely refused. Then, Obregón tried a feint, attacking with only two hundred men to make Ojeda think that he was weaker than in fact he was. Ojeda did not fall for the ruse, so Obregón decided to wheel south and attack the Federal garrison in Cananea, thus forcing Ojeda to abandon Naco in order to come to the rescue of that Federal force. This made better sense anyway, since the force of about six hundred Federals to his rear constituted a minor threat for Obregón and besides, Cananea had wealth and the promise of more recruits since the town had always been revolutionary in sentiment. Moreover, by now Colonel Plutarco Elías Calles's Sonorans had already captured Agua Prieta, meaning that the revolutionaries had access to an entrepôt in northeastern Sonora so there was no urgency to subdue Naco.

In Cananea, Federal Colonel José Ramón Moreno had scarcely prepared any defensive works until the night before the attack when his men began stacking stones and filling sand bags. He mostly intended to use the Federal barracks as a redoubt where he placed all his men, except for a section of troops with one machine gun at the small lunette fort about one thousand meters to the west of the barracks and about another one hundred men at the meat packing facility just outside town. The latter group fell back to the barracks almost immediately upon the approach of the Sonorans.

In order to carry off the attack, Obregón divided his force into five groups: one to watch over the rails to Naco and guard against a possible attack by Ojeda, and four more groups to assault from each cardinal direction. He apportioned his four machine guns with two assigned to the east group, and one each to the south and north side assault forces. In the early morning hours of March 24 the battle began.

After two days of fighting Moreno's men had had enough. The Federals had not stockpiled any ammunition, food, or water and Obregón's men had cut the water supply to the barracks. As it happened, Ojeda had left Naco that same March 26 to come to the aid of his fellow Federals. Along the way Ojeda learned that the garrison had surrendered. He suspected that Obregón intended to turn and face him while the state forces of Colonel Calles hit his rearguard. He was correct, but Obregón's men were too exhausted after two days of combat to catch up to Ojeda, who left railroad bridges destroyed in the wake of his retreat. At San José on March 28, Ojeda easily pushed Calles aside and reentered Naco where he remained ensconced.

Obregón sent Moreno and his officers to the penitentiary at Hermosillo. He reported recovering five hundred Mausers, 30,000 rounds, horses, mules, and other equipment, and he put the Federal casualties at about thirty killed and sixty wounded while he only lamented six killed and about twenty wounded, although his official numbers seem too light to be taken literally.

Obregón's initial plan to conquer Naco was to once again try to embolden Ojeda to attack, that is, to get him to assume the offensive. Therefore, he spread the rumor that he was leaving with the bulk of his troops for Hermosillo to attack Federal General Gil and sent his trains back to Nogales—empty. Meanwhile, the few troops remaining on the firing line kept up a feeble pace of fire. That stratagem failed.

Ojeda, for his part, was confused by the scheme. He could not figure out why the revolutionaries maintained harassing fire for eight days, unless it was to fatigue the Federals or get them to waste their ammunition. He used the time to improve his defenses, erecting barricades made of stones and sandbags at the street entrances into town. He planned to use windows in the buildings on the perimeter as loopholes and placed machine guns on top of one of these buildings on the south side, that is, elevated and in defilade. He placed more men and machine guns on the top of key buildings protected by sandbags, and located his two 80-mm. mortars in the bullring with infantry support. Finally, he established two outposts and created a pile of boards and railroad crossties that he planned to light to provide illumination in the event of a night assault.

The terrain around the town afforded the Federals clear fields of fire, so Obregón decided to try sending a máquina loca, dubbed the "emissary of peace," to detonate inside the plaza and sow death and mayhem. At that point the revolutionaries would assault from the east and west. It was too dangerous to attack from the south because it might cause bullets to land on the American side, even though converging on the town simultaneously exposed Obregón's men to possible friendly fire.

After several failed attempts to get the máquina loca to explode, the Constitutionalists decided to make an all-out assault in the early morning hours of April 13. The Federals had begun to run low on ammunition of every variety and in the midst of the vigorous assault Ojeda ordered the woodpiles ignited, but instead of throwing light on the situation the resulting smoke obscured the defenders' vision. Soon the revolutionaries were mixing with the Federals in the trenches and penetrating the town's perimeter, making a special effort to advance down the street along the international dividing line to cut off any escape. In no time the Federals were reduced to the bullring, a few key buildings adjacent to the border, and the cemetery. Federals often manned cemeteries because they were typically situated on the outskirts of town, on high ground, and contained solid construction that could be used for cover. They could not be left for an attacking enemy to occupy and use to his advantage, but at the same time extending the perimeter to occupy these landmarks had the adverse effect of stretching their available manpower.

Obregón's men continued to close in, and there were reports of treachery, with Yaquis in federal service deserting to join their revolutionary brethren. But most especially, the Federals began to feel the dearth of ammunition—blamed on the Americans for not allowing the passage of the resupply sent by Mexico City—and ultimately Ojeda had to quit the fight and order his remaining men to cross over the border into the safety of the United States.

By noon the fighting had ended and the Federals lost about eighty killed, an equal number captured, and twenty-three wounded. Obregón reported only about twenty killed and forty wounded, again, numbers too low to be considered accurate. The Sonorans recovered the two mortars, 30,000 cartridges, about 100 Mausers, horses, mules, and other equipment. They controlled the entire northern part of the state.

SOUTHERN SONORA CAMPAIGN

Colonel Benjamín Hill, Obregón's cousin, commanded the Sonoran state forces in the south. Hill left Hermosillo at the beginning of March headed for La Colorada and La Dura, gathering in minor commands that had been recruited by local leaders. The Federal commander in the area, Colonel Jesús P. Díaz, sent a small force of about two hundred Federals (he only had about six hundred total) from his base at Torín to stop Hill and the revolutionaries from taking the important mining center of La Dura, but Hill's men managed to turn that force back in a fight at El Realito, just south of La Dura. The reasons for the Federal mission failure sounded the familiar refrain of poor economy of force: too few men, too little ammunition (the men only had one hundred rounds each with no way to resupply them), and no food; the soldiers carried no provisions and had eaten only a small breakfast two days earlier before they had to turn around. The commonplace scarcity of food in adequate quantities for Federals on campaign earned them the moniker *bocas de palo*, "for having lost the custom of eating."

After that encounter, Colonel Hill continued driving southward toward Alamos, essentially taking the same route that the Colorado invaders had taken the year before. His main objectives were to take control of the garbanzo harvest, deny that wealth to the Federals, and keep the Federals in the southern part of the state occupied so they could not join General Gil's main force in Empalme. Colonel Hill managed to conquer Alamos, which had been defended mostly by Huertista volunteers. As such, he levied stiff fines on those locals and then paroled them. With the money, he was able to feed and supply his men, finance an expeditionary force to invade neighboring Sinaloa, and still have money left over for repairing the railroad. He then began moving back to the north along the Southern Pacific Railroad reaching as far Cócorit.

As Hill cleared the south of Federals and gathered revolutionaries to his command, Major Ramón V. Sosa moved down the Southern Pacific Railroad from Hermosillo to the Guaymas Valley with the mission of recruiting revolutionaries, tearing up railroad track between Empalme and Ortiz Station, and acting as a blocking force against the Federals in Empalme and Torín who might try to advance on the state capital. Sosa experienced enormous success in increasing the size of his force from about 150 men to around 800 and had reached as far as Batamotal Station, where he established his headquarters. Perhaps emboldened by Obregón's success in the north, Hill's victory at Alamos the day before, and his own achievements, he decided to attack Gil at San José de Guaymas on April 18—the same day that representatives from the states of Chihuahua, Sonora, and Coahuila signed the Monclova Accords to officially form the Constitutionalist Army. Yet the Federals managed to turn back Sosa's men, marking a rare setback for the Constitutionalists in Sonora.

After Sosa's failed attempt to take Guaymas, and reflecting upon his own shortage of ammunition, Obregón realized that he could not take Guaymas, so he sent about three hundred riders to link up with Hill and attack the Federal zone headquarters at Torín. Those men reached the advance Constitutionalist forces at Bácum on April 23, bringing the total under Hill's command to about seven hundred men, against Díaz's estimated six hundred in Torín. In the ensuing Battle of Lencho Station, however, the Federals managed to prevail against Hill and remained in the area until May 22, a constant threat to Hill yet virtually incommunicado with the main Federal command in Guaymas.

SOUTHERN CHIHUAHUA CAMPAIGN

One of the first to stand up to the new Huertista regime was Colonel Manuel Chao, who rebelled on February 18, in Rosario, Durango, and crossed over into neighboring Chihuahua with three hundred men and defeated the garrison in Santa Bárbara. Chao's success encouraged Captain José Rodríguez in Guerrero and Colonel Maclovio Herrera, who was engaged in mop-up operations against Colorados in San Buenaventura, to join Chao in the south. Major Rosalío Hernández in Santa Rosalía Camargo also joined the Constitutionalists, setting out on February 21 for Coahuila to put himself under Governor Carranza's command on orders from Governor Abraham González. When the commander of his rearguard accepted amnesty from the Federals, however, Hernández had to return quickly to southeast Chihuahua where he nevertheless remained active.

By March 5, Chao's force had swelled to six hundred men and he felt sufficiently strong to attack Parral, which was defended by Federal Brigadier Salvador R. Mercado. Mercado placed his men on the hills surrounding the town and inside the town on the high buildings, such as the church, from where they hoped to cover the gaps in the perimeter with machine-gun fire. The Federals only had hasty defensive positions and Mercado kept his reserve inside buildings in the city's center. Almost from the start of the battle a group of his volunteers defected to the revolutionaries and then citizens inside the town began rioting, sacking commercial houses, and making a play for Mercado's headquarters. The fighting was heated and possession of the hills changed several times such that by the next morning the Federals had been reduced to one position on La Cruz, the highest hill. Mercado considered surrendering but the recently-amnestied Colorado General Luis

Fernández convinced him to hold out. When a 500-man Federal relief column reached Parral the next day, Chao had to break contact and retreat.

The Constitutionalist effort in Chihuahua got off to a rapid but uncoordinated start, mainly because on March 7 Federal Army officers murdered the Maderista governor, González, and Francisco "Pancho" Villa only reentered the country the next day, after having escaped from a Mexico City prison to the United States in December.

Villa and his fellow revolutionaries in Chihuahua faced a three-fold challenge arising from the recent Orozco insurgency: an economy devastated by two full years of revolution; numerous and experienced Colorados who accepted amnesty and joined the Federal Army as irregulars; and a full Federal division occupying the state with strong garrisons in the major cities and towns such as Chihuahua City, Casas Grandes, Ciudad Juárez, Ciudad Guerrero, Santa Rosalía Camargo, Jiménez, and Parral with the most powerful contingent of artillery in the army to include six field artillery batteries, a few mountain gun batteries, and about twenty machine guns. Moreover, Villa only had eight companions when he crossed the border into Mexico, so unlike those paramilitary commanders who simply led their men into rebellion, Villa had to raise a brigade from the ground up.

Word soon spread that Villa had returned and several of his previous lieutenants sought him out for instructions. He sent them to make the rounds of various ranches, haciendas, and towns to recruit men while he did the same all over south-central Chihuahua. Returning back toward the north after recruiting about four hundred men, Villa battled and defeated a Federal column sent from Chihuahua City to destroy his infant brigade at San Antonio de los Arenales.

While Villa recruited, Colonels Chao, Hernández, and Herrera battled the Federals in southern Chihuahua at Rancho de Charcos, Camargo, Santa Bárbara, Conchos Station, Hacienda del Corraleño, and San Pablo de Meoqui. Then, on April 10, Villa attacked a train near Santa Isabel and had the good fortune of recovering 122 silver bars that he used to finance his new brigade. Next he moved into northwest Chihuahua and began recruiting among the military colonies there. In the course of his bandit days during the Porfiriato, whenever "things got hot" Villa had often hidden out in the military colonies of western Chihuahua, where the local populace simply knew him as "el Güero." He had made many friends there and some of those had begun the work of gathering men for Villa ahead of his arrival.

Chihuahua's Military Colonies

The military colonies provided fertile ground for raising a brigade. They had their origins in a measure initiated by the Spanish crown at the end of the eighteenth century to study the problem of frontier defense. Situated along the eastern watershed of the Sierra Madre Occidental, the first five military colonies—Namiquipa, Buenaventura (Galeana), Casas Grandes, Janos, and Las Cruces—occupied points along the warpath used by Apaches to invade the heartland and were intended to disrupt those incursions and at the same time provide supplies to area timber and mining operations. Similar to the communities of stockholders in Nuevo León, the colonies were granted commons to be used for grazing, water, and timber, while the individual colonists received "possessions" to be held in perpetuity. Any colonist interested in giving up his possession could only sell it to fellow colonists. In exchange for these land grants, the colonists accepted the obligation, not just the right, to bear arms in defense of the realm. Unlike mere subjects of the crown, these soldier/farmers or armed campesinos were *vecinos*, with rights and duties closer to those of a citizen. In the national period, the Mexican government established more military colonies, including Cuchillo Parado, hometown of General Toribio Ortega, after the French Intervention and the Mexican-American War.

The military colonists continued to fight the Apaches until the mid-1880's when they encountered another, more formidable, foe—the Terrazas-Creel oligarchy. Through a series of laws the oligarchy sought to dispossess the colonists of their lands purchased over decades with blood and sweat. The effect of these assaults on the rights of colonists combined with their near-recent participation in the Indian wars meant that these campesinos-soldiers were able and willing to fight for the revolution.

While Villa recruited in western Chihuahua, one of the most momentous actions to date occurred on April 13. Constitutionalist Colonel Tomás Urbina was forced to abandon Mapimí, Durango, by the fierce Colorado Brigadier Jesús José "Cheché" Campos and, setting his sights to the north, Urbina attacked Jiménez with his six hundred Constitutionalists. The Constitutionalists came at the town from every direction but pulled back when a train with about one hundred Federals coming up from Escalón to the south reinforced the plaza. Afterward, Urbina renewed the assault and then a *pistolero* inside the town shot the Federal commander dead as he passed a tailor shop on Hidalgo Street. Panic soon spread among the Federals who dispersed and made for the nearest large towns to the north, west, and south, leaving Jiménez in Urbina's hands. The possession of that town effectively cut communications between the Federals in Torreón, Chihuahua City, and Parral.

General Mercado, still in Parral, then sent a train with about four hundred men and an 80-mm. gun commanded by First Captain Eudaldo Traconis to go to the aid of the Jiménez garrison. Repairing the rails and battling Constitutionalists along the way, on April 15 that column reached Jiménez, which Urbina abandoned after three hours of combat in order to spare the townsfolk the unpleasantness of a cannonade. Those Federals who had been dispersed in the earlier fighting made their way back to the plaza.

A week later, Colonels Chao and Herrera attacked Camargo, which was defended by Colonel Manuel Pueblita, the son of a famous general of the French Intervention, and his 350 or so Federals. Pueblita had constructed an outer perimeter anchored on nine key positions, each manned by eight soldiers, and inside the plaza his men also occupied the barracks, a school, a flour mill, and the church, upon which he had placed his two Madsen machine guns. At 4:30 a.m. on April 23, the Constitutionalists attacked from the north, east, and southwest. Pueblita reinforced the points on his perimeter, since they were significantly undermanned, at the beginning of combat. Fighting continued all day and by the next morning the fighting was inside the plaza at close quarters. Similar to the battle for Jiménez, the commanding officer Pueblita was killed in the streets and terror gripped the remaining Federals who made a last stand in the hospital on the outskirts of town. Colonel Hernández then arrived with his Constitutionalists and delivered the final assault of the battle, against the hospital. A few Federals escaped to Chihuahua City, but most were captured or killed. The Federal 3rd Cavalry Regiment had been wiped off the face of the earth.

Pueblita committed several errors common to Federal commanders: trying to man a perimeter too large for the size of his force, placing his machine guns up high and in the center instead of on the flanks to deliver grazing fire, and an inattention to security, since his officers were drunk from attending a party the night before the battle and thus ill-prepared for combat.

Captain Traconis had tried to make his way north from Jiménez to reinforce Camargo, but once he learned that the plaza had fallen he travelled all the way south to Torreón, leaving Jiménez to the Constitutionalists. The loss of those two towns left General Mercado vulnerable, so he abandoned Parral on May 6 for Chihuahua City. It had been a smart move because he did not have enough men to successfully prosecute a defense of the town, and he was short on money, provisions, and ammunition. Traveling by road, he crossed paths (according to some reports) with Pancho Villa, who was now returning to southern Chihuahua from the western part of the state.

Upon reaching Camargo, Villa was greeted by the populace as a conquering hero. Also in Camargo, representatives sent by the First Chief interviewed Villa, who accepted a brigadier's commission in the Constitutionalist Army on May 26. That effectively made him commander of all the forces in Chihuahua since he was the only general officer in the state; this appointment did not sit well with Colonels Herrera and Chao. They, along with Hernández, had done all the fighting, reducing the Federals to a single axis of operations in the state stretching from Ciudad Juárez to Chihuahua City, since Toribio Ortega's revolutionaries had run the Federals out of

Ojinaga almost two months earlier. Colonels Ortega and Hernández, on the other hand, became eager and committed Villistas. Still, it did not escape the attention of anyone that Villa had done virtually nothing to this point for the Constitutionalist cause.

At Conchos Station on May 30, Villa directed his first battle as a Constitutionalist against a 1,300-man Federal column sent from Chihuahua City to try to reopen the lines of communication to Torreón. With an estimated two thousand men under his command, Villa succeeded in turning back the Federals, who first withdrew to Saucillo and later all the way to Chihuahua City. There Mercado received command of the Federal Division of the North from General Rábago, who had been fired for being too weak, and for having ties to Félix Díaz. Villa left Chao, Herrera, and Hernández to hold the southern part of the state and on June 12 he ordered Colonel Toribio Ortega to begin mobilizing from his base of operations at Ojinaga toward Ciudad Juárez while he returned to the military colonies and Ascensión in the northwest part of the state. Villa intended to recruit, organize, train, and equip his own brigade with help from Señor Carranza and the jefes of Sonora and then attack Ciudad Juárez. He would be on the border for over two months.

THE GUAYMAS VALLEY CAMPAIGN

On May 1, a five-ship Federal Navy flotilla reached Guaymas bearing Brigadier Luis Medina Barrón and 1,500,000 7-mm. cartridges, five artillery pieces, two machine guns, and approximately 1,200 soldiers, mostly raw recruits. The Huertista chronicler of the 1913 campaign in Sonora, Rómulo Velasco Ceballos, marveled in bewilderment at the physical specimens before him:

> In the main from the jails of México [City], showing traces of inebriation from pulque on their countenance and in the whole of their organism a lamentable lack of vigor. Used to the delightful climate of the Valley of Mexico, were such men going to have the spirit and courage to fight under the furnace of these fields, where the Fahrenheit thermometer reaches the 119 degree mark? Habituated to the asphalt streets of the Metropolis, would it be easy for them to climb steep and uneven hills, carrying on their back their weapon, clothes, ammunition, their supplies, and all this [while] suffering a frightful thirst because the burning deserts of Sonora refuse even one drop of water? On what would they

establish the enthusiasm to transform their weakness into strength? On their political opinions? ... If they do not have them! And if they do, they are reduced to the simple cry of: ¡Viva Madero! or ¡Viva Zapata!

Alternatively, the Sonoran Constitutionalists had been fighting Indians for thirty years, were robust and inured to the variations in temperature of the desert, and could count on "the very important assistance of the Yaqui, a truly military type, a warrior race [known] for its bravery, obedience, and somberness," as Ceballos lamented. Still, the Constitutionalists observed the unloading of copious numbers of men and an abundance of materiel with alarm. Obregón correctly divined their objective to be the capital, Hermosillo, but he mistakenly estimated the reinforcements alone at 3,000. In reality, Federal General Gil had 2,100 men in the Guaymas Valley, yet he felt sufficiently strong enough that he did not need to call in Colonel Díaz's six hundred Federals from down south at Torín, at least not yet.

General Gil was a reluctant warrior, most probably because he realized the impossibility of his position. Mexico City expected him to conquer Hermosillo, yet no one in his right military mind would have traded the state capital for Guaymas as a base of operations, meaning that he had to hold both, which entailed splitting up his forces, thus enabling the Constitutionalists to divide and conquer. Moreover, he had to worry about keeping the invading column fed and hydrated through hundreds of kilometers of Sonoran desert. And although he had artillery, which the Constitutionalists did not, his opponents had him outmanned both in numbers and quality. Consequently, he began the campaign for the conquest of Hermosillo with a naval cannonade of Empalme in order to push back Obregón's advance forces and cut Obregón's communications with Hill in the south. Then, leaving three hundred men behind to hold the port he started moving forward with the rest of his command. Obregón's intentions were to permit Gil to advance, thus using up his provisions and stretching his lines of communication and forcing him to have to detach security details and reduce the number of effectives available for combat. At the right time the Constitutionalists would attack. However, Gil barely penetrated the Guaymas Valley and then spent several days traveling in circles until the Secretary of War fired him and told him to turn over temporary command to General Medina Barrón. General Pedro Ojeda, after being forced into U.S. territory upon his defeat at Naco, was making his way to Guaymas and would assume command of the Division of the Yaqui once he arrived.

Santa Rosa

General Medina Barrón was not one to wait for Ojeda, however, and began pushing forward after reinforcing his vanguard. Upon arriving at Santa Rosa, the lead Federals stopped and set up security, the 28th Infantry to the north and west of the railroad toward the sierra, which ran parallel to the railroad, and the 53rd Infantry Battalion off to the right of the rails. By this time Obregón had fallen back as far as Ortiz Station when he received orders from Interim Governor Pesqueira not to retreat any farther.

Taking a few officers, Obregón cautiously probed ahead until reaching a hill about one kilometer away from the Federal vanguard and devised his plan. He would send about 500 men to attack the Federal center while another 1,200 attacked on his left and another 1,000 men would occupy the sierra on his right. He kept five of his fourteen machine guns in reserve and distributed the remainder appropriately to the three main columns.

After stealthily moving into position during the night, at 3:30 a.m. on May 9, the Constitutionalists hit the soldiers of the 28th hard, forcing them to fall back on Santa Rosa Station. The colonel in charge of the vanguard immediately pulled the 53rd in from the right to support the 28th. Moving parallel to the Constitutionalist lines of attack exposed the 53rd to flanking fire that they could not effectively return, and they made the maneuver with such haste that they abandoned their womenfolk to the Constitutionalists. In the confusion and the dark, lines became mixed with hand-to-hand combat occurring in some places. The Federals suffered significant casualties in the opening moments of the battle.

News of the contact soon reached Medina Barrón who ordered the main body forward, the infantry loading up on trains and about three hundred mounted soldiers riding at a gallop along the rails. The mounted troops arrived first at Santa Rosa Station, where Obregón's machine guns decimated them and forced the remainder to disperse, some not stopping until they reached Guaymas. The trains stopped opposite Aguajito where the infantry detrained and took up position along the railroad embankment until they tied off with the vanguard, which had stabilized its front and was now extended to the east perpendicular to the railroad. The Federal disposition looked like an upside down "L." Medina Barrón marched along the railway encouraging his men who were in the prone position with weapons pointed toward the sierra to the west. By the time he arrived at Santa Anita Station, dead Federals already littered the fields. Surveying the battlefield he asked the commander of the vanguard, "Those hills to the front are ours?" "Yes, General, my battalion has them." "We are good," the general said. But then a captain stepped forward and correcting the colonel said, "Sir, those hills that the general is asking about, the enemy has..."

At that point the general received a wound to the leg, and massaging it said to the colonel "Let's see: reinforce the lines on the berm to the front, the railway, and the bridge," and then turning to the captain the general said, "Captain, communicate to General Gil that it is necessary to advance the artillery." At that point the colonel received two bullet wounds, including one that entered the back of his head and exited to the front destroying his jaw. By 9:30 the station house was already filled with wounded and had taken on the lugubrious appearance of a slaughter house.

On route to the station the Federal mountain battery stopped about two kilometers short of its destination, went into battery, fired in support of the 53rd to clear the Constitutionalists to the north, and then continued to the station. Once at the station the mountain guns went into battery and started to fire at the hills to the west, which were infested with revolutionaries. The Constitutionalists were only about six hundred meters from the Federal batteries, which were near the station and virtually on the firing line executing direct fire missions. Under these conditions the Sonoran sharpshooters shot the artillerymen to pieces; bullets ricocheted off the guns and kicked up dirt all around.

To the rear the rest of the Federal column arrived and then extended to the east so that now the Federal disposition formed somewhat of a square with most of the eastern lateral (that is, parallel to the railroad) missing. Around 11 a.m., Constitutionalist Colonel Salvador Alvarado arrived from Nogales with about 450 men and Obregón sent him to reinforce the right, which covered Aguajito. Aguajito, as the name implies, had a freshwater supply and the Federals had already made several attempt to capture it. The fighting was fierce, especially the front line from the station to the Santa Rosa de Arriba house, which anchored the right of the 53rd.

The battle lines were now set and the shooting continued without letup until around dawn the next morning. Taking advantage of the lull, Medina Barrón ordered the battery of field guns opposite Aguajito to displace to Santa Rosa Station and go into battery between the two sections of mountain guns. As the sun came up, the Constitutionalist sharpshooters again went into action, murdering the artillerymen. Medina Barrón knew he had to clear the Constitutionalists from the closer hills and sent a company of infantry to do the job. The Federal company fought a seesaw effort to capture the hills with Federal guns cannonading the Constitutionalist infantry in support. Eventually, the Constitutionalist infantry prevailed in holding the ground and the action made a profound and lasting impression on Obregón—the Federal artillery had proved useless in inflicting any substantial harm on the Constitutionalist infantry opposing the Federal assault.

As night closed in, the Yaqui war drums started to sound their terrible, rhythmic cadence, signaling impending doom for the Federals. Shooting started to pick up in intensity in close quarter fighting at certain points of the line and several times the Federals had to temporarily abandon their positions before retaking them. Three Constitutionalist assaults and Federal repulses and counterattacks occurred during the night. Meanwhile, some of the Federals began leaving the line and filtering back into Guaymas in the dark, spreading rumors of a disaster. Many of those Federals in the fight had not eaten for thirty-six hours and water was becoming scarce.

The battle continued into the next morning as Obregón's machine guns went to work on the Federal artillery while he considered his situation. In daylight the Federal artillery could not function since Constitutionalists had moved up to within one hundred meters of the Federal batteries and for any gunner to man his piece meant certain death. Nor could the Federal infantry make a push from its current disposition, but the Constitutionalists were running low on ammunition.

Little could Obregón have known that the Federals were already preparing to retreat. As the Constitutionalists' ammunition continued to run low their rate of fire decreased. The Federals misinterpreted the lull as the Constitutionalists' intentions to perhaps quit the fight, but Medina Barrón knew that his men were too tired, hungry, and dispirited to mount an assault and pursuit. He would have to withdraw from the battlefield. It would be difficult, though, because instead of pulling the trains back just out of the zone of combat, on the first day of the battle the cowardly Federal officer in charge of the convoy had ordered them all the way back to Guaymas. Allegedly he had seen a bridge burning and feared that he and the trains might be cut off from the port. Therefore, at 11 p.m. on the night of May 11 the Federals pulled out of the battle and marched back to Guaymas under the cover of darkness. The next morning Obregón received a resupply of ammunition and ordered a pursuit undertaken, but the Federals simply had too great a head start.

The Constitutionalists reported capturing 180 prisoners and recovering 422 dead Federals, along with a vast supply of abandoned materiel and provisions. The Federals admitted to over one thousand casualties but there are no reliable figures for the Constitutionalists since Obregón reported the unbelievably low number of forty-two officers and men lost.

This battle was notable because it demonstrated the uselessness of mounted troops against machine guns and infantry throughout the revolution, as witnessed in the opening hours of the battle. Moreover, it proved how utterly ineffective artillery would be in those situations where it was placed on the firing line, employed in direct fire missions, as opposed to hidden and massed in battery groups (two batteries were simply too few to be more than a nuisance) that could saturate entire zones with indirect fire. Almost as significant, Yaqui Indians who had been so cowed by artillery fire in previous wars were becoming inured to its effects.

A new army model different from those fielded by Mexico in previous centuries also made its debut. The Federals perpetuated the traditions made infamous by Santa Anna: authoritarian officers unskilled in proper tactics who neglected the material needs and comforts of their men, who in turn were untrained vagrants dragooned off the streets and freed from the jails of Mexico City only to desert at the soonest opportunity in the fog of war or the dark of night. Observing the Federal performance, Obregón was rumored to have remarked: "When will we learn the lessons of history?" The Constitutionalists were different. The officers evidenced solicitude for their men, who were eager volunteers skilled in war. Obregón's principal commanders capably handled their missions as he confirmed in his battle report saying, "I feel proud to command a column such as this. I had nothing to order Colonels Cabral, Alvarado and Diéguez, Sosa and Camacho; they operated with true initiative and opportunity." Federals rarely if ever conferred such praise—much less freedom of action—on their subalterns, with whom they competed for position in the Federal Army bureaucracy. After this battle Obregón received a much-deserved promotion to brigadier and from then on he excelled as a commander, giving his subordinate generals outlines of how he expected a battle or campaign to unfold and then giving them the latitude to bring his vision to fruition. Those who didn't rise to the occasion were quietly reassigned. He maintained a loose control, but at the same time he was very unforgiving to those who deviated from his plans or failed to exhibit appropriate élan or competence.

San Alejandro

General Ojeda arrived in Guaymas on May 13, the day after Medina Barrón's column returned to the port. The mere news of his presence had a salutary effect on the morale of the Division of the Yaqui, which he built upon by instilling some organization and order. He resupplied his troops, ordered the rail shops in Empalme to construct armored gondolas out of platform cars and mounted 80-mm. guns on them, and then for the first time he formed the various battalions and corps into brigades. He also concentrated his forces, recalling Colonel Jesús P. Díaz from Torín, where the colonel's position had become untenable after the Federal defeat at Santa Rosa. That move, however, left Colonel Hill in full control of the wealthy Yaqui Valley and allowed him to move up to Cruz de Piedra so that he might now cooperate in military actions with the rest of the Sonora state forces that for the most part had gathered around the Guaymas Valley.

On May 28, Ojeda began his campaign to capture the state capital. Leaving about 1,000 men behind to safeguard Guaymas and Empalme, he set out with the other 2,600 soldiers, eight of his twelve machine guns, and seven of his ten artillery pieces. The sheer size of the column made an impression on Obregón who, once again, estimated the forces opposing him at double their actual number. The Constitutionalists had probably about 4,000 men, including those of Colonel Hill, available for the upcoming battles.

Historians generally refer to the next battle as the Battle of Santa María, but the combatants divided it into two separate engagements with the Constitutionalists referring to the first phase as "the siege of Ortiz" and the Federals calling it the

"Battle of San Alejandro," while both referred to the second action as the "Battle of Hacienda Santa María," which was where the critical fighting took place.

The Federal column departed Guaymas with all due attention to proper security including an extreme vanguard of two hundred riders; a vanguard with an armored scout train containing a section of guns, machine guns, and two hundred infantry; two hundred flankers to each side; the main body; and finally, a small mounted rearguard. As Ojeda's Federals pushed forward, the guns of the scout train engaged in reconnaissance by fire at suspected ambush sites as the small rearguard of Constitutionalists commanded by Colonel Manuel Diéguez retreated, slowly drawing the prey into its web. At night the scout train and cavalry pulled in to join the main convoy as infantry dug in on the hills and lit large bonfires. Due to the conditions of the rails, the advance was slow. Obregón maintained a separation of twenty kilometers between Ojeda's main body in order to keep the Federal general blind and guessing. Meanwhile, the Constitutionalists practiced aerial reconnaissance to keep apprised of the Federal movements.

Upon reaching Ortiz Station, the Federals stopped to repair a fairly sizable bridge they estimated would take more than two weeks to repair; they would advance no farther. In addition to destroying the rails, the Constitutionalists plugged or poisoned water wells, then fell back to Moreno Station (twenty-five kilometers north of Ortiz) and agreed on their plans. General Hill would sever the Federal lines at Anita switch and then advance to Santa María. Another group would cut the lines of communication at Tres Gitos and Batamotal, south of Maytorena Station, and then join Colonel Hill. Colonel Jesús Chávez Camacho would move to Cruz de Piedra to be positioned to harass the Federal rearguard while General Alvarado occupied Aguajito to deny that water source to the Federals. Obregón's reserve remained in Moreno Station under the command of Colonel Diéguez, while Obregón and the rest of his forces occupied the sierra to the west of the rails where they constantly harassed Ojeda's supply lines, especially around Hacienda San Alejandro, which was located just inside the sierra across from Anita Station. All these movements were carried out under the cover of darkness and, surprisingly, came off without a hitch. By the morning of June 19, the Constitutionalists were in place.

For the next six days, the Federals and Constitutionalists fought around San Alejandro, as well as at Ortiz and farther toward the rear in the vicinity of Maytorena Station. The combat does not seem to have been particularly lethal, most probably because of the distances between the lines—the Federals had only consumed a mere 130,000 rifle cartridges but almost 1,300 artillery shells—but the near continuous combat and scorching sun enervated Ojeda's men and water was running low with the source at San Alejandro drying up. Moreover, there seemed to be little hope of repairing the rails. So Ojeda decided to give up on Hermosillo for the time being and return to Guaymas.

Hacienda Santa María

After sending a Colorado major and his men to scout the national road as far as Santa María, where Ojeda hoped to get water for his men, the Federals broke contact at 3:00 a.m. on June 25 and started to retreat. As the sun came up, the increasing heat and the atrocious condition of the sandy road crossed by deep and broken arroyos took its toll on the draft animals, already worn out from a lack of water and poor pasture. The carts loaded down with ammunition could only be moved with great difficulty and much of the baggage was discarded to lighten the load. The light of day also alerted the Constitutionalists to the Federal movements, and Obregón sent an emergency dispatch to Hill to get ready to receive the ragtag Federal column. Diéguez started to move up with the trains and repair the rails while Obregón's men in the sierra quickly displaced to reinforce Hill. As the Federal vanguard came in sight of Santa María, it reported to General Ojeda that Constitutionalists held the hacienda.

Hacienda Santa María was a challenge for any assaulting force since it had numerous berms that presented obstacles for the maneuvering of troops, and its sandy soil absorbed the impact of artillery shells. The main house was built like a fortress and stood only a few meters from the water well. The Federals, desperate for water, futilely assaulted Hill's deployed men as Ojeda's casualty rates skyrocketed, as much from bullets as from the heat. Everywhere Federal soldiers, who only a few months ago had enjoyed the pleasant climes of the central plateau, succumbed to heatstroke. Staff officers who had to move constantly to deliver orders suffered the worst. Ojeda covered the lines kicking and slapping helpless soldiers—nothing more than recruits in uniform who simply could not fight because they had reached the end of their physical endurance—until his hands started to bleed.

At 12:30 p.m. Alvarado arrived at Santa María to reinforce Hill and deployed his men on a firing line divided into a center, left, and right, with a support and reserve. By 2:30 the fighting had become generalized, but there simply was no way for the Federals to dislodge the Constitutionalists. Obregón reached Aguajito and the Federals then blew the bugle call, "enemy to the rear" and attacked in that direction as Federal artillery fired in support at Aguajito. Obregón then sent Colonel José María Ochoa to reinforce Alvarado on Hill's right where he suspected the Federals might try to escape. Still the fighting continued. At 6:30 p.m. Federal buglers blew calls of "enemy to the front!" "enemy to the right!" "enemy to the left!" and "enemy to the rear!" announcing that the Constitutionalists had them completely surrounded. The artillery officers started firing off their remaining shells and disabled their pieces as the rest prepared to break out. At 10:30 p.m. Ojeda ordered the signal to disperse— bugle calls of "enemy to the right" and "enemy to the left"—and the Federals then abandoned their impedimenta and fled as best they could, skirting around Hill's right toward the Sierra de Bacatete. The main body began entering Empalme around 6:30 a.m. the following day. Obregón blamed Colonel Ochoa for allowing the Federals to escape since he did not reinforce Alvarado as ordered; that was the end of his career.

The casualties for the Federals were staggering, officially about 25 percent among killed, wounded, or missing, and possibly as high as 50 percent. Obregón listed his own losses at some thirty killed and about as many wounded, which, again, seems incredibly low. The booty recovered included nine cannons with 2,000 shells, 530 rifles, five machine guns, and 190,000 cartridges, among other items. With help from the workshops of The Cananea Consolidated Copper Company, the cannons were returned to service, although obviously not in pristine condition.

The combined actions at San Alejandro and Hacienda Santa María involved Obregón's use of the natural elements to wear down his opponent and then force him into a classic hammer and anvil maneuver. Alvarado's disposition indicated an ability to employ the ternary system under fire, while Ojeda's abuse of his infelicitous men again provided an example of the Federal Army's martinets and the utter unsuitability of the Federal soldier for modern warfare.

Obregón wanted to press onward and assault Guaymas while the Federals were still in disarray, but his officers convinced him that the men needed a break. Obregón set about placing his men to invest the city and started to reconnoiter, but in all the activity he suffered a heatstroke, and by the time he recovered so had the Federals. Ojeda had his men entrenched on the high ground surrounding the city and could rely on the protective guns of the Federal Navy. Obregón's supply of ammunition was beginning to run out, as he was expending about 15,000 rounds daily, so he considered his options, which he submitted to Governor Pesqueira in a formal proposal. Key to any decision was the ability of the Federal Navy to extract the Federal Army assets and the number of casualties the Constitutionalists would incur in any assault, not to mention the ammunition that would be consumed. Obregón estimated that an attack on the port in a succession of rushes by sector would cost about one million cartridges with few casualties and good chances for victory but would allow the Federals to extricate their men and materiel; an all-out assault would have about a 50 percent chance of success, expend about four million rounds, and the garrison would still probably escape. Keeping the city besieged would free up about half of his forces and require a minimal amount of materiel. When another one thousand soldiers, three batteries of field pieces, and a tremendous amount of ammunition reinforced Ojeda on July 11, it forced the decision to keep the port under siege. Obregón was famous for "his extraordinary quality of knowing how to care for his men so that they were not uselessly sacrificed in a fight," as Miguel Alessio Robles later said.

A CAMPAIGN FOR THE NORTHEAST

The strategic perspective of the Federal Army—concerned about a Norteño advance on Mexico City—mirrored that of the revolutionaries. That is to say, the isolated Northwest concerned them least because of its relative isolation and since there was no direct rail link, the Sonorans would have to travel overland a portion of the way or sail to Manzanillo, which would require a mastery of the seas they did not possess. The Federal Navy controlled the sea lanes and would be able to harass any advance southward along the western coast by the revolutionaries. The invasion of the heartland from Chihuahua was more likely, but still not of an immediate concern because it was a longer stretch than the Northeast route and it had a single railroad with several great engineering works, meaning bridges that could be blown up to impede any advance.

Alternatively, the Northeast not only posed the greatest and most immediate threat, but it also carried the greatest possibilities for a successful offensive due to its numerous advantages: 1.) the shortest routes from Mexico City; 2.) close proximity to the United States railroad centers (at this stage the United States had an embargo against the revolutionaries, but NOT against the Federal Army); 3.) the lack of natural obstacles and rough topography (mountain ranges) perpendicular to the line of advance; 4.) immediate control of Monterrey, arguably the most important industrial city of the republic which could easily become a base of operations by virtue of its railroad links to Nuevo Laredo, Matamoros, Piedras Negras, Saltillo, and Tampico; 5.) the coal mines of Coahuila as a source of fuel; 6.) the port of Tampico for inserting follow-on Federal troops into the region; 7.) the relatively low number of great engineering works that could be damaged by the revolutionaries to impede operations; and 8.) the ability to dominate the theater of operations east of the Sierra Madre Oriental.

In the beginning, the paramilitaries that rallied to Carranza's standard in the Northeast had been relegated to northern Coahuila, just below Monclova to the south and as far east as Candela, which they used as a base of operations to attack the National Railroad running from Monterrey to Nuevo Laredo. Operating out of Candela, Pablo González and Jesús Carranza extended their area of control by capturing Lampazos on March 28. Although the Constitutionalists possessed the border town of Piedras Negras, after the failed attack on Saltillo in early fighting they had only been able to get more ammunition with great difficulty because of the U.S. arms embargo. Their weaponry consisted mostly of .30-30 hunting carbines obtained through the efforts of sympathetic Americans and did not have the range, ease of loading, or ability for sustained rates of fire as compared to the Federal Army's military-grade Mausers. Many revolutionaries went unarmed in the beginning, hoping to get weapons later from Federals defeated in combat. An extremely dangerous tactic for acquiring machine guns that the Norteños (many of them ranchers) employed was to ride up to Federal machine gun emplacements, lasso the weapon, and then ride off with it. For artillery, the Constitutionalists had homemade guns made in railroad workshops using axles. Until a more favorable balance of power could be attained, and to impede the Federal Army's efforts at intelligence gathering, the Constitutionalists kept the railroads leading from Saltillo to the north destroyed, and the rails to the south of the state capital interdicted by brothers Luis and Eulalio Gutiérrez operating out of northern Zacatecas.

In early April, Federal General Fernando Trucy Aubert was ordered to retake Lampazos and continue all the way to Nuevo Laredo, which had been attacked by Constitutionalists on March 18. A couple of weeks later the Federal commander in Saltillo sent Colonel Ricardo Peña with a column of 850 men and one field piece to finish off Colonel Pablo González at his base of operations in Monclova. General Trucy Aubert successfully completed his mission, but Colonel Peña did not, and the Constitutionalists maintained a firm grip on northern Coahuila for the next three months, as one Constitutionalist officer explained, because the Federal cannon fodder was so ill-suited to swift mobilization and offensive operations. The situation would only get worse with Decree 438 that increased the authorized strength of the regular army to 80,000 soldiers and sent the Federal Army press gangs into high gear.

Meanwhile, in the early days of April, First Chief Carranza made a decision that would have far-reaching consequences for control of the Northeast when he sent Colonel Lucio Blanco to march into Tamaulipas and try to take the border town of Matamoros. As Colonel Blanco moved through the northern part of Tamaulipas, the 21st Rural Corps under the command of First Officer Miguel M. Navarrete moved up from the south and on April 22, 1913, attacked the state capital, Ciudad Victoria. Navarrete failed to capture the city. Nevertheless, the dual threat against Matamoros to the north and Ciudad Victoria by these two commands—neither one larger than 250 men—forced the Federals in the state to divide their forces to defend both towns and Colonel Blanco began destroying the rails leading to Matamoros, defeating smaller garrisons, and recruiting.

Matamoros

On June 3, with some nine hundred men Blanco captured Matamoros, defeating a smaller force of about five hundred Federals, Rurales, and volunteers. The Federals never again possessed Matamoros, which quickly became a base of operations for Constitutionalist flying columns sent to harass the Federals and as a resupply and refit base for Constitutionalists as far south as the Huasteca.

The loss of Matamoros caused quite a sense of disillusionment within the Federal Army, an event described by General Guillermo Rubio Navarrete as "inexplicable," since an expert plan for the plaza's defense had been provided to Huerta, but the president failed to act on it. Perhaps he had been too involved in other matters, or he underestimated the Constitutionalist forces operating in the Northeast, or the Federals could not raise enough men to reinforce the plaza. Whatever the larger reasons, the tactical explanations for the Federal defeat were common enough: too few men to defend too long a perimeter, the Federal commander's advanced age and lack of vigor and technical knowledge, and too many inexperienced irregulars and unreliable volunteers. It was a colossal defeat that combined with the loss of Zacatecas three days later forced the resignation of the praetorian Secretary of War, General Manuel Mondragón.

Zacatecas

The revolution in Zacatecas did not begin in earnest until April 10, when Second Officer Pánfilo Natera of the 26th Rural Police Corps took up arms in Nieves, Zacatecas, with the support of sixty men. Soon other Rural Police officers joined Natera, and with them he captured the major town of Fresnillo on the Central Railroad on May 8. Natera's Constitutionalists managed to hold the plaza until mid-May when a column of Colorados organized by Pascual Orozco Jr. on its way from Mexico City to Chihuahua forced them to temporarily escape into northern Jalisco.

After the Orozco column had exited the area, Natera and his fellow Constitutionalists returned to the state and on the morning of June 5, attacked Zacatecas, the state capital, from the south, east, and west. By the next day Natera had captured the city. It was a relatively minor affair, tactically speaking, since the defenders only numbered about four hundred federal and state troops against an estimated twelve hundred Constitutionalists with the usual excuses for defeat: too few troops for too large a perimeter and a shortage of ammunition. In strategic terms, however, the loss of such an important city in the heartland and on the only lines of communication leading directly to Chihuahua was alarming since it opened up Mexico City to attack.

On June 13 Huerta convened a council of war to address the recent military setbacks, namely the loss of Matamoros and Zacatecas. The president and his generals originally planned a very wide-angled, two-pronged offensive in the Northeast. One column would depart San Luis Potosí and drive northward through Saltillo toward Piedras Negras to push the First Chief and his direct subordinates into the deserts of Coahuila, or eastward into Tamaulipas and Nuevo León and into the waiting arms of the second half of the pincer, which was to depart San Luis Potosí and pass through Tampico to retake Matamoros. Huerta's nephew, General

Joaquín Maass, Jr., would lead 2,500 men in the Saltillo operation and General Rubio Navarrete with 1,800 men would retake Matamoros; the two were to be allotted five million cartridges. During the council of war it was agreed to reduce General Maass's column to 1,500 men in order to reassign one thousand men to General José Delgado, who was to march those men northward and retake Zacatecas. Delgado collected the forces of Irregular General Emilio Campa and the remnants of the original Zacatecas garrison in Aguascalientes and then occupied Zacatecas on June 16 after a brief encounter with Natera in Guadalupe, just south of the state capital. After that, Natera retired to the northern part of the state and began taking up track. The direct route by rail from Mexico City to Torreón through Zacatecas was never reopened to the Federal Army. Two days later Huerta's Federals lost Durango City, never again to possess it.

DURANGO STATE

Durango

The Constitutionalist effort in Durango State began with the *pronunciamientos* of leading Maderistas in the Rural Police Corps including Orestes Pereyra (commanding the 22nd) in eastern Durango, and Martín Triana (44th) stationed on the border with Zacatecas. Additionally, Maderistas who had fought in the rebellion of 1910 but had since been placed in reserve or discharged also refused allegiance to Huerta and quickly mobilized their former adherents. Among these jefes were Tomás Urbina in the northern part of the state, brothers Mariano and Domingo Arrieta and Matías Pazuengo in the west, and Calixto Contreras in the east.

The two imperatives for the Federals in the state were to hold Durango, the state capital, and keep the rails open between that city and Torreón-Gómez Palacio. Accordingly, Pereyra and Contreras started to destroy that railroad, fighting a Federal column sent from Torreón at Pedriceña. Thereafter, Pereyra and Contreras swept through the center of state destroying or chasing off all the Federal garrisons until the Federals were reduced to a single axis of control from Durango to Torreón. Federal Brigadier Luis G. Anaya of the 1st Cavalry Regiment (newly promoted for his role in the Citadel uprising) received orders from zone headquarters to take a combined-arms column of five hundred men with two 80-mm. guns to destroy Pereyra and Contreras in Cuencamé. Upon reaching that town, however, he learned that those two commanders were in the western part of the state, so he headed back toward Durango City.

Along the way Anaya detached a column of about two hundred Federal infantry and irregular dragoons at San Gabriel Station. Informed of the presence of about seven hundred revolutionaries in the area, the major in charge of the company-size column marched about twelve kilometers north of the station on March 26, daring to challenge Pereyra and Contreras, who virtually wiped out the Federal command. Anaya rushed to return to the aid of his underling, but arrived too late in the day to be of any assistance. Contreras and Pereyra were still at San Gabriel, however, and engaged Anaya in combat. The fighting was fierce and lasted into the night, with Anaya's artillery firing more than two hundred shells. At 10 a.m. the next morning the Constitutionalists made an enveloping movement and attacked Chorro Station, where Anaya had left his convoy, about twelve kilometers south of San Gabriel. They looted the trains and then burned them to keep Anaya from using them for a retreat. Two hours later, and running short of ammunition, Anaya retreated in echelon to Hacienda del Refugio where he changed into a column of march to reach Durango City on March 29 with the remnants of his command.

With every victory, more and more men joined the Constitutionalists. They continued to tear up track to hinder any Federal reinforcements arriving from Torreón. By April 20, Constitutionalist commanders Pazuengo, Arrieta, Pereyra, and Contreras had the state capital invested and on the twenty-third they attacked. Federal Brigadier Antonio M. Escudero commanded 600 regulars and 120 local volunteers in defense of the city. The Constitutionalist forces of Durango attained a deserved reputation for being the most unruly and undisciplined of forces in the course of the revolution (aside from the Zapatistas), which explained why they did not prevail in the battle. Their attacks came off uncoordinated, which allowed Escudero to exploit his interior lines. Then, a corps of 350 Colorados commanded by General Cheché Campos arrived from Torreón after making the trip on horseback due to the conditions of the railroads and hit the Constitutionalists from behind before entering the city to reinforce Escudero. At that point the revolutionaries retreated and decided to send a commission to interview Tomás Urbina and request the presence of his brigade in the next attack.

After the failed attempt to take Durango City, the ranks of the Constitutionalists continued to swell as "armed campesinos in groups of one hundred, two hundred, twenty, ten, and five showed up daily," increasing their numbers and lifting their spirits. Also, "if upon arrival," as Pazuengo later wrote, Urbina "saw us organized he would be moved to return [with his men], because he had his forces entirely organized and disciplined and he also notified us that if for the taking of Durango we did not have an agreement to give command to one General, the one who inspired the most confidence, then neither would he accept to add his people to a muddle. With this frankness he spoke to us in his communication and as everything that he suggested to us was naturally fair" all the other jefes agreed to his conditions and began concentrating their men in locales around the capital in anticipation of the arrival of Urbina's Morelos Brigade.

From mid-May to June 10 the Constitutionalists prepared for the year's campaign, organizing, fattening up their horses, and overseeing the planting of crops on the haciendas in order to have provisions. Meanwhile, inside the city the Federals recruited locals to serve in the Social Defense corps and assigned them to posts on the perimeter. These volunteers somewhat increased the number of defenders, yet at the same time Campos' Colorados had had a falling out with the regulars and returned to Gómez Palacio, reducing Escudero's available number of men. Once Urbina arrived on the scene, the jefes elected him to serve as commanding general and set the date of the attack for June 17, 1913. They had an estimated 6,000 men against Escudero's 1,300, of whom approximately 1,000 were regulars.

At 11 p.m. a single illumination rocket went off, signaling the beginning of the assault. In mob-fashion, quite atypical for the revolution, the Constitutionalists attacked from virtually every direction. They had the right leg of their pants rolled up to distinguish them from the Huertistas. Almost immediately they overwhelmed the Federals. A fifth column opened fire from inside the city from the very start, shooting the Huertistas in the back and killing the Federal major in charge of a group of Social Defense volunteers, who dispersed immediately. All through the night and into the next morning the fighting continued. Federal forts started to capitulate after rear attacks by the townsfolk and rioting broke out in the commercial district as the residents looted the stores.

Around ten o'clock in the morning it became apparent that the Federals were no longer resisting; they were planning to break out. Some tried to break out to the east and after nearly three hours of combat they succeeded. Escudero and the Huertista governor led that column to the Central Railroad where the area commander sent a train to bring them to Torreón. Another group, those of the Social Defense lucky enough to escape, made a course for Zacatecas and then to Mexico City. A third column composed of Federal cavalry took backroads to reach Torreón, and a fourth command reached Mazatlán, Sinaloa. The Federals suffered something on the order of fifty percent casualties, while those of Constitutionalist were not recorded. Although some of the town's leading citizens blamed Escudero's handling of the battle for the defeat, all sober observers recognized the impossibility of his position. He had had too few men to defend too long a perimeter, especially at night, and with a considerable number of unreliable volunteers against overwhelming numbers and a fifth column.

After the battle, Mariano Arrieta went back to western Durango and later to Sinaloa. His brother Domingo garrisoned the state capital with about five hundred men, while the rest of the forces of Durango headed for Gómez Palacio-Torreón to try to conquer that vital railroad crossroads.

Huerta Loses in Six Months

The general and historian Miguel Sánchez Lamego theorized that by the end of June 1913, Huerta had already lost the war. The deciding factors had been low troop strength in relation to territory covered, in combination with the need to be everywhere at once, protecting cities, towns, and haciendas while keeping the railroads functioning. Further complicating the situation was the need to project a sense of control on the world stage that prevented the Federals from going completely operational, which would have been tantamount to confirming the revolutionaries as equals. The distribution of troops to protect the national patrimony left Federal detachments isolated in smaller groups where the revolutionaries could prosecute asymmetrical warfare—bringing larger concentrations of mobile troops opportunely against stationary and sometimes even stranded smaller garrisons.

In sum, the Federals lacked a national strategy, since the Secretary of War thought the revolutionaries could be easily and quickly brought to heel. The Federals also suffered from a dearth of military maps that, joined to their unfamiliarity with the North, further retarded the formulation and execution of a coherent strategy. Finally, once the decision to go operational had been made, the Federals proved incapable of offensive operations.

The Federal infantry consisted of cannon fodder—vagrants, criminals, political prisoners, and drunks—led by officers advanced in age and/or ignorant in up-to-date tactics with little initiative, which meant that its units could not be trusted to stray far from cities and railroads where strict control could be exercised. As soon as the revolutionaries cut the rails to the rear of any advancing Federal column, something easy for them to do, the forward momentum stalled. The Federals could not hope to match the revolutionary dragoons in mobility because of the preponderance of infantry in its columns and the woeful condition of its cavalry, which left them reliant on irregulars for mounted operations. And frictions often developed between Federal irregulars and regulars, further complicating efforts. Finally, the artillery was the only combat arm where the Federals possessed a clear advantage, yet the only instance where they massed their batteries in significant numbers for large-scale battles occurred in Chihuahua, where by coincidence the guns remained after the campaign against Orozco.

Conversely, the Constitutionalist men and officers were young, virile, and motivated country boys, with superior military intelligence who operated largely on home ground and in close proximity to their base of operations, the United States, with optimal mobility (speed and flexibility) by employing both trains and horses.

By the end of June territorial control for the Federals in the North been reduced to Guaymas in Sonora, Gómez Palacio in Durango, and along three remaining axes of operations: Ciudad Juárez-Chihuahua City in Chihuahua; Saltillo-Torreón in southern Coahuila, with advance forces south of Monclova; and Nuevo Laredo-Monterrey, where they maintained a tenuous hold stretching southward to Ciudad Victoria and Tampico. Finally, at the beginning of July, Torreón and Gómez Palacio were under threat of attack, materially imperiling the Federal pincer offensive getting underway to crush Carranza in northern Coahuila and recapture Matamoros.

First Torreón

After capturing Durango City on June 18, the Constitutionalists recruited, reorganized, replenished their stores, and then at the beginning of July began moving up the railroad toward the twin cities of Gómez Palacio and Torreón. They stopped at Pedriceña and sent forth reconnaissance parties since they had received word that a column had arrived from Mexico City under General Felipe Alvírez with orders to retake Durango City. Also in Torreón were irregulars belonging to Generals Benjamín Argumedo and Emilio P. Campa who had spent April and May organizing in Aguascalientes and were on their way to Monterrey but had been unable to continue their movements because Constitutionalists had cut the rails at Madero Station. By mid-July the Constitutionalists operating out of San Juan de Guadalupe, Durango, under General Cándido Aguilar had also cut the rails to Zacatecas, south of Torreón. Moreover, Pascual Orozco's column had departed Torreón on the evening of July 1, traveling cross country on horseback to relieve Federal General Mercado in Chihuahua, decreasing the number of defenders by about one thousand Colorados. That left Federal General Ignacio Bravo with only about four thousand men at his disposal to defend the cities of Gómez Palacio, Lerdo, and Torreón against the five thousand or so Constitutionalists commanded by Contreras, Urbina, Pazuengo, and Pereyra.

General Bravo did not prepare any defenses ahead of the looming battle, mostly likely because he feared it would indicate a lack of offensive-mindedness to his superiors. Instead, in the early morning hours of July 20 he sent General Eutiquio Munguía with combined arms column of 450 men, two machine guns, and three cannons to seek out and engage the Constitutionalist vanguard. About one kilometer west of Hacienda de Avilés the Constitutionalists attacked Munguía at dawn. In the afternoon, as the battle raged, Munguía reported the forces facing him at two thousand Constitutionalists and requested reinforcements. In the evening Bravo sent him another five hundred men and a battery of field guns, meaning that the commanding general had a quarter of his forces and almost two batteries fighting in the open and subject to capture and annihilation away from the main body of Federals inside the city.

By the morning of the next day the Constitutionalists had overpowered Munguía, and the Federals were in headlong retreat for Torreón. The Constitutionalists gave pursuit but split up with the smaller group of about four hundred Constitutionalists pushing into Torreón where they quickly realized they were outnumbered, became disoriented, and retreated. The majority of the Constitutionalists charged into Lerdo with such momentum that they forced the Huertistas back to Gómez Palacio where the fighting continued into the night, at which point the Huertistas pulled back to Torreón.

The next day, Bravo pulled in the San Pedro de las Colonias garrison, which engaged in a firefight with the Constitutionalists upon passing through Gómez Palacio along the way. For the next week the Constitutionalists continued to attack Torreón from the west and south, employing dynamite as their only means of indirect weapons. These hand bombs and their hunting rifles, however, were no match against Federals who effectively equaled the Constitutionalists in numbers (in this initial phase), had superior weaponry, and were fighting on the defensive with a reduced perimeter, since the clear fields of fire to the north and east virtually precluded any assault from those directions.

At the end of July, while the battle for Torreón was still in progress, Venustiano Carranza arrived in La Laguna after a successful Federal campaign to capture Monclova (see below) had forced the First Chief to flee Coahuila.

When the Constitutionalists first heard that the First Chief was coming, they were very encouraged, but when they saw that he had no arms or ammunition and few people with him they were greatly disappointed. Carranza held a council of war in his railcar the next day, and all agreed to make one final attempt on Torreón. Even though Carranza ordered General Aguilar and his men to participate in the next attempt, they would not have been well-equipped but no one wanted to object to the First Chief about the futility of continuing the battle. The assault began in the early morning hours of July 30 with a máquina loca loaded with dynamite that rocked the entire city as it exploded. The fighting was ferocious and in no time the provisional cemetery that the Federals had established a few short days earlier was already full. But from the relative security of their hasty defensive positions, the Huertistas managed to repulse the attack. That ended of the Battle of First Torreón.

The Federals lost a reported three hundred men between killed, wounded, or missing, while the Constitutionalist casualties remain unknown.

Bravo committed many grave errors, such as placing a large portion of his assets at Avilés where they ran the risk of being annihilated (as would happen in the

Battle of Second Torreón), deploying his artillery on the line of fire inside Torreón, where the pieces were subject to capture, and neglecting to exploit his interior lines such that some Federals fought for days on end while others scarcely fired a shot. The Federals prevailed in this first battle for Torreón almost wholly because of the city's natural advantages for a defense—a river on the north side, clear fields of fire to the north, east and south, and hills to the west and southwest.

Bravo also had the Constitutionalists to thank. They were not duly organized and fought with the lack of discipline characteristic of the revolutionaries from Durango, making uncoordinated attacks at various points on the perimeter. The nominal commanding general, Urbina, fought courageously but not intelligently, making great sacrifices of his men. Pereyra's brigade fought the best, but it was the smallest, while the largest, Contreras' brigade, manifested the least interest in conquering Torreón.

Thereafter, the First Chief appointed General Aguilar as the Military Commander and Chief of Operations for Veracruz and sent him to conquer the petroleum-rich Huasteca. Eulalio Gutiérrez returned to operate in northern San Luis Potosí out of Concepción del Oro, Zacatecas. Lastly, the Constitutionalists of Durango began pulling back just a short distance to keep the Federals in Torreón in check: Urbina to Mapimí and Indé and the other generals to Velardeña (eight kilometers south-southeast of Pedriceña).

The First Chief made a parting comment about the Duranguenses not knowing how to fight and departed to visit revolutionaries in La Laguna, Durango, and southern Chihuahua on his way to Sinaloa and the relative security of Sonora where he would re-establish his headquarters. He alone bore the responsibility for losing Coahuila.

THE FEDERAL COAHUILA CAMPAIGN

At the beginning of July, the Federal Army kicked off three campaigns to turn the tables in the North: Pascual Orozco departed Torreón to reinforce the Division of the North in Chihuahua; General Joaquín Maass drove into northern Coahuila to destroy the Constitutionalists in Monclova and clear the rails all the way to Piedras Negras; and, General Guillermo Rubio Navarrete made his way up from Monterrey to consolidate the rail line all the way to Nuevo Laredo. General Rubio's original mission to retake Matamoros had been canceled, but it did not seem at this point as though his new mission was subordinate, or in service, to General Maass's operation, only that Huerta had decided that keeping traffic open between Monterrey and the border was more important than retaking Matamoros. Still, General Rubio's departure was postponed until Maass could reach Saltillo from San Luis Potosí because Rubio's column would be needed in Saltillo if that state capital were suddenly attacked by the large number of revolutionaries in the area. The delay was having an adverse impact on morale, according to Rubio, and he said that given the opportunity to begin, he could take Monclova in three days. Finally, on June 25, General Rubio received permission from the War Ministry to begin his operation.

Rubio drove northward and after the Constitutionalists attacked Salomé Botello on July 2, Rubio countered the next day attacking Candela, the Constitutionalist base of operations used to raid the rail lines in northern Nuevo León. The Federals swept aside the Constitutionalist security detail from nearby Valladares and drove straight into Candela. The commander of the plaza, Colonel Jesús Carranza, had no hope of holding the town without machine guns, so he fought a delaying action just long enough for Candela's civilian sympathizers to get out. General Rubio requested permission to pursue the fleeing Constitutionalists, but the War department refused, so he left a detachment in Candela and returned to Nuevo León.The Constitutionalists retreated to San Antonio Canyon while Carranza rode the rails to Monclova where he and Colonel Pablo González held a telegraphic conference with Venustiano Carranza. The First Chief amazingly ordered them to retake Candela and came down from Piedras Negras to join them.

González left 150 men to hold Monclova, and with the rest of his command returned with Jesus Carranza to San Antonio Canyon. From there the combined Constitutionalist commands surprised the Federal detachment at Candela and retook the town on July 8. In the meantime, however, Federal General Maass had begun advancing on Monclova, a development that the Constitutionalist garrison reported to González. Therefore, the day after retaking Candela, Colonels González and Carranza, the First Chief, and the rest of the Constitutionalists raced back to Monclova. They arrived just as Maass's Federals were ringing the bells in the town center to announce victory. The First Chief escaped with a few supporters to his hometown, Cuatro Ciénagas, and then to Torreón and ultimately Sonora. González and Carranza retreated to Las Hermanas (forty kilometers north of Monclova) where they trained, organized, and recruited until forced to evacuate the area after the Battle of Las Hermanas on August 16.

A week after that battle, Huerta ordered that "all Federal forces were to advance simultaneously on Hermanas, Sabinas, and [Piedras Negras]...with the object of making the rebels [sic] withdraw to the desert [of western Coahuila] where it will be difficult for them reorganize." Yet for the next six weeks González was able to impede Maass's progress by continually battling his lead elements while destroying the rails to his rear and forcing Maass to waste time trying to keep his supply lines intact. Finally, at the beginning of October, the Constitutionalists abandoned northern Coahuila to attempt a *coup de main* on Monterrey. The Federals marched into Piedras Negras on October 7, having failed to destroy General González and his Constitutionalists. General Rubio would never receive authorization for the operation to retake Matamoros.

Reflecting on the campaign for northern Coahuila, Constitutionalist General Manuel W. González went so far as to say that by tying up the Federals in the Northeast for seven months, General Pablo González had won the war—perhaps an exaggeration. But if General Sánchez Lamego was correct and the Federals had lost the war by the end of June 1913, then Pancho Villa certainly could not claim responsibility for the Constitutionalist victory.

VILLA'S CHIHUAHUA CAMPAIGN

Villa had only been in Ascención about a week when Pascual Orozco's column left Torreón for Chihuahua, continuing a march that had begun in Mexico City. With the rails to Chihuahua City in patchy condition and the Constitutionalists in control of the southern part of the state, Huerta had to send mounted and highly-motivated troops to relieve the beleaguered garrison of Chihuahua City. Since neither of those descriptions fit the Federal regulars, the mission had to be entrusted to Orozco and his Colorados. The going was slow and by July 10 Orozco had only gone as far as Dolores Station, where he set camp. When he learned of a group of Constitutionalists not far from there, Orozco sent two regiments to attack them, which is probably how the Constitutionalist commanders learned of the Colorados' presence. Having made contact with that skirmish, Orozco decided to break camp and continue the march through the night. The Constitutionalist colonel in charge of the Jimenez garrison, meanwhile, abandoned the town to the Colorados—who rushed to occupy it—and fell back to Camargo. Orozco stayed only a short time in Jiménez before pressing onward.

In the early morning hours of July 12, Orozco ambushed and defeated the Constitutionalists of Colonel Maclovio Herrera at Díaz Station (halfway between Jiménez and Camargo) after only six hours of combat. The Constitutionalists escaped to Camargo, which Orozco attacked two days later and again emerged victorious after fighting an estimated 1,500 Constitutionalists commanded by Colonels Rodríguez, Herrera, Chao, and Hernández. The Colorados continued onward and entered the state capital on July 22, accomplishing their mission. Yet Orozco never again directed an operation for the Federal Army, possibly because Huerta reneged on his promise to appoint him military governor of the state.

Shortly thereafter, on July 26, Villa sent a messenger to Guadalupe to tell Toribio Ortega to leave the border and begin moving southward to rendezvous with Villa and his brigade at San Buenaventura. One reason for scrapping the much-anticipated attack on Ciudad Juárez may have been that Mercado had reinforced the border town with another infantry battalion bringing the total defenders to over one thousand men and endangering Ortega's smaller force. But that had happened at the beginning of the month. A more likely motivation was the arrival of Orozco in Chihuahua City.

Ortega reached San Buenaventura on August 12 after attacking a Federal convoy at Ranchería Station that was bringing provisions from the border to the capital city. Mercado had placed Chihuahua City under martial law and instituted an 8 p.m. curfew because the citizenry had grown unruly under the siege-like conditions; no civilians were permitted to enter or leave the city.

As Villa approached San Buenaventura on August 18, Colonel Ortega and his staff rode out to greet him. Upon passing through Namiquipa four days earlier, Villa had been informed of the death of his infant daughter. Additionally, he heard that General Félix Terrazas had occupied San Andrés, where Villa's wife lived, in the expectation that eventually he would have to visit the town to console her. Terrazas' combined arms column consisted almost wholly of Colorados and had been mistreating the townsfolk, whom they suspected of being sympathetic to the revolution in the extreme. Villa's Brigade passed through San Antonio de los Arenales to avoid detection by the Colorados and then, turning easterly toward Bustillos, attacked San Andrés at 5:30 a.m. on August 26.

San Andrés

Alerted by a sentry who noticed Villa's men, the Colorados started pouring in copious fire as Villa instructed his own men to shoot sparingly in order to conserve their limited amounts of ammunition. As the sun came up the Colorados' cannons went into action and Villa could take a better measure of the numbers and disposition of the forces opposing him. Terrazas had some men on a mesa that dominated the town and more on the edge of town along the railroad tracks. Villa spent the day having his men surround the defenders and slowly tighten the circle in preparation for a night assault. As darkness closed in, the Colorados unadvisedly abandoned the mesa for the town as Colonel Ortega's Constitutionalists took possession of the houses on the outskirts of town. Villa's men also got closer to the trains and, by tossing dynamite at the field pieces, they managed to force the crews to break and run. That allowed the Constitutionalists to capture the trains, which in turn denied the Colorados any way of making a rapid retreat except for, of course, General Terrazas, who fled on horseback with his escort. By the next morning the fighting was all over. Villa sent the wounded, including the Colorados, to the hospital. He had all 237 able-bodied Colorado and Federal prisoners shot according the Benito Juárez Law, except the artillerymen whom he permitted to join his brigade since they knew how to handle the captured guns.

The reason for Villa's victory may be as simple as superior leadership or the fact that he had almost twice as many men as Terrazas, if the Huertistas are to be believed. Most telling, however, was the assignment of a section of field artillery to Terrazas' column, which served no tactical purpose and could only be regarded as the first of many such gifts to the revolution.

Second Torreón

A punitive column departed Chihuahua City to avenge the disaster at San Andrés, but Villa sent a detail to blow up the entrance to a train tunnel that bought him enough time to weigh his options. Meanwhile, he retreated to the west, where a group of about three hundred revolutionaries including the First Chief's nephew, Major Carlos Carranza, joined him. He now had another 200,000 cartridges brought from Sonora and initiated a series of random movements to confuse and evade any Federals who might come after him.

Ultimately, Villa left the area west of Chihuahua City and traveled south to join his subordinates at Camargo and then changed his objective from the state capital to Torreón, either because his colonels suddenly made him aware of Torreón as an alternative target, or because the revolutionaries in Durango had sent a commission to request that he come and help conquer Torreón. No matter how Torreón came to his attention, the decision was based on the same sound rationale: Chihuahua City would be more difficult to conquer and it posed no threat to his rear as he moved southward; as a major crossroads and center of industry and commerce, Torreón was much more strategic; he now had a section of guns he could use in the campaign; and he needed the men of Durango to increase the number of his effectives.

From Camargo, the Chihuahuans moved south to Jiménez where Tomás Urbina, one of Villa's former bandit buddies, and his men were waiting and gambling with gold coins pillaged three months earlier during the sacking of Durango City. Villa took a few days to organize his troops and then began mobilizing for Bermejillo where Colonel Herrera was already waiting with his Juárez Brigade. Only recently had Herrera turned back Federals from Torreón at Santa Clara for the second time.

The Federals in Torreón had been cut off for weeks and had been forced to economize their ammunition. A relief column bringing 500,000 rifle cartridges and artillery shells had not been able to reach the city due to the state of the rails but was trying to make its way from the east. The route to the south via Zacatecas was cut and would never again be opened to the Federals. Federal Colonel Antonio Delgadillo had orders to advance northward from Zacatecas with his 1,000-man column if he considered that plaza to be secure, but there was no way such a small column could carry out that mission, unless it consisted of bold irregulars. The new Federal commander of the Division of the Nazas, Brigadier Eutiquio Munguía, was on his own, yet he had been instructed to assume an offensive posture. He had only recaptured neighboring Gómez Palacio most recently on September 6, and then Lerdo the next day.

As Villa entered the area a Federal scout train with three hundred soldiers rolled north to engage the Constitutionalists, encountering them first at Vergel Station (ten kilometers north of Gómez Palacio) on September 15. Those Federals managed to push the Constitutionalists back to around Bermejillo in the first of more than two days of combat, with the most ferocious combat occurring at Hacienda Santa Clara. The growing numbers of follow-on Constitutionalists finally forced the scout train back to Gómez Palacio.

It took many days for Villa's cavalry traveling on horseback to arrive from Chihuahua. On September 22, with his men completely assembled, Villa began moving around to the west through Mapimí and, crossing the Nazas River, turned north to go into camp at Hacienda de La Goma. From there Villa personally went to Velardeña to meet General Calixto Contreras, who agreed to bring his own brigade and another five hundred men belonging to General Domingo Arrieta. At this point, the Constitutionalists had some eight or nine thousand effectives.

To confront the looming threat southwest of Torreón, General Munguía ordered General Alvírez, who had been assigned command of the column designated to recapture Durango City, to advance to Hacienda Avilés. Alvírez was supported on his right by Colorado Generals Emilio P. Campa and Benjamín Argumedo who were marching on the north side of the river through Lerdo toward the Constitutionalists. The two Federal commands combined probably did not amount to more than a thousand men, or about one-third of Munguía's total fighting force. Villa quickly called a council of war to designate a commanding general, and Villa was elected, thus creating the Constitutionalist Division of the North. The date was September 29, 1913.

As the Constitutionalists conducted their meeting, the Federals and Colorados started to cannonade Villa's troops. Villa assigned objectives to the various brigades and counterattacked. On the left the brigades drove the Colorados back into Lerdo and captured that column's single cannon. On the right, they took Hacienda Avilés "by fire and sword" and wiped out General Alvírez's command, killing the general and sending the few lucky survivors racing back to Torreón where they reported the enormity of the defeat. The Constitutionalists shot the prisoners according to the Benito Juárez Law but spared the lives of those artillerymen who agreed to crew the three captured field pieces, again, useless except as a gift to the revolutionaries.

In the meantime, General Munguía ordered General Luis Anaya to take another battery of artillery and some men to reinforce Alvírez, but he had reached that decision too late. When Anaya reported that the remainder of Alvírez's command was straggling in, Munguía told Anaya to keep them outside Huarache Canyon so they wouldn't mix with the rest of the defenders and spread the demoralizing particulars of what became known as the "disaster of Avilés." But it was too late. They had already entered the city.

To receive the Constitutionalist onslaught, General Munguía had followed the example of Bravo in the earlier defense of Torreón and had not prepared any field works, only two lines of fire, one exterior to keep the Constitutionalist artillery at a distance, and an interior line.

In the afternoon of September 30, Villa conducted a reconnaissance in force of the Federal positions with about three hundred men. When Villa's party ran into opposing forces it charged all the way up to the foot of Calabazas Hill under the protective fire of Alvírez's turncoat artillerymen. The Federals fell back through Huarache Canyon as a larger number of Constitutionalists pushed the Colorados out of Lerdo and Gómez Palacio and then applied pressure to the north side of Torreón. On the right, the Constitutionalists of the Zaragoza Brigade dismounted with one soldier holding the horses for six, according to the old custom, and attacked through Las Fábricas Canyon. After nightfall, more dismounted dragoons scaled the hills on the west side of the plaza conquering Federal positions while taking advantage of the darkness for cover, one of Villa's favorite tactics.

By the early morning hours of October 1, the Villistas had taken La Polvorera Hill, and shortly after daybreak took Calabazas Hill. They held the high ground and began pouring in machine-gun fire on De la Cruz Hill, the last position on high ground held by the Federals, immediately west of the plaza. Villa rode up and down the lines brandishing a pistol and encouraging his men, who kept up the pressure and repulsed all attempts at a Federal counterattack. Coming down Calabazas Hill, the Constitutionalists encountered Federals arrayed along the railroad embankment using it as a makeshift battlement. Shortly before noon, the Federals on a section of the line abandoned the fight to eat lunch and Munguía had to rush reserves to the point and personally place men on the line. Munguía sensed imminent defeat and shortly thereafter ordered that two thousand brand new Mausers stored in a warehouse be burned to keep them from falling into the hands of Villa. In the late afternoon, wave after wave of Constitutionalists started pouring through Huarache Canyon and Colonel Herrera's men were attacking along the river on the north side. Famished and weary of combat, Munguía's men "entered into a period of indifference such that the officers could hardly succeed in making themselves obeyed," as the general later complained.

As night started to fall, there was a lull in the fighting and Villa inspected his lines, seeing to the distribution of ammunition and encouraging his men. The Constittionalists rolled up their right sleeves and removed their hats to identify themselves as revolutionaries in the confusion that would necessarily follow the Federal collapse. At 8:00 p.m. the Constitutionalists made one final assault from the north, west, and southwest. Once they had taken De la Cruz Hill, the last dominant point on the Federal line, and the vantage point from which Munguía had been directing the defense, the battle entered its final phase and the pursuit began. Federals scampered down the back side of the hill and abandoned their positions along the rails to flee through the streets of Torreón in an eastward direction. Spaniards inside the city also made good their getaway, sweeping along the Federal soldiers in their flight. In thirty minutes it was all over.

The Federal generals blamed each other for the defeat, but most especially Munguía claimed that Argumedo had fled to Coyote, and since Argumedo commanded the most men (not true, he commanded perhaps 20 percent of the total) his departure precipitated the collapse of the line. Blaming irregulars for defeat became a favorite device of Federal Army generals. Yet based on the testimony of Munguía and Anaya, it would appear that all the Federal generals abandoned their posts, while Argumedo gave the best account of himself in the battle. Indeed, his Colorados were among the most numerous dead according to Villa.

The booty that Villa captured included almost a half million rifle cartridges, eleven cannons, and copious amounts of rolling stock.

In the final analysis, Torreón fell for reasons typical of the Federals: too few men (although plenty of ammunition) against an overwhelming assault force; forces piecemealed into combat so that individual contingents might be defeated in detail; improper use of artillery (distributed to various locations and engaged in direct fire missions on the line of fire where they were ineffective and subject to capture, instead of massed and hidden); failure to construct defensive positions; and despicable leadership compounded by even worse soldiering.

Chihuahua City

After giving his men eight days to rest, Villa left General Calixto Contreras in charge of La Laguna, ordered General José Isabel Robles to block any Federals trying to approach the region from the east in an attempt to retake Torreón and Gómez Palacio, and turned north to destroy the cut-off and surrounded Division of the North. The staffs and what little infantry he commanded made the trip by train, but the mounted contingents had to travel on horseback, and crossing the desert was quite an excruciating journey of privation. Upon reaching Jiménez, the Constitutionalists learned that a Federal combined-arms column composed of about two thousand men commanded by General Francisco Castro was coming to intercept them. Castro had swept aside the forces of General Rosalío Hernández, whom Villa had left behind to protect his rear during the Torreón campaign, and occupied Camargo. His advance forces were at Díaz Station.

As Villa's brigades started pushing forward, Castro's Colorado allies covered the retreat of the Federal column back to Chihuahua City. Villa had already decided on attacking Chihuahua City instead of Ciudad Juárez, even though little had changed since August and the state capital still held little promise of victory for the Constitutionalists: the government's forces occupying it were perhaps twice as large as had been at Torreón with more than half of them irregulars, including such prominent Colorados as Marcelo Caraveo, José Inés Salazar, Antonio Rojas, and Pascual Orozco; there was no single dominant terrain feature that once occupied could compel the city to capitulate, such as Calabazas Hill at Torreón; the Division of the North contained the largest concentration of field and mountain guns deployed in the Federal Army arsenal; and finally, Villa was short on ammunition and did not have enough men to invest the city, much less take it by storm. Still, he thought he had to try.

Obeying informal protocol, Villa stopped in Ortiz Station to send an envoy to request that General Mercado either evacuate Chihuahua or designate the location for an open-field battle that would spare the local populace the ravages of combat. Mercado refused to dignify the missive with a reply.

Mercado had the city defended by a perimeter divided into seven sectors with Federal regulars occupying the important east side that would be the first to receive the Constitutionalists. The Colorados defended the west (leeward to the assault) and southwest sections of the line, including the crucial Chuvíscar Dam sector. If the Constitutionalists cut the city's water supply from the dam, it would compel the Federals to surrender. Mercado also had reserves inside the city and commands stationed on Grande, Coronel, and Santa Rosa hills.

On November 6, Villa's men attacked in three columns across a front running from the northeast to the southwest. In the opening moments of the battle, the Constitutionalists forced those few Federals on Grande Hill (southeast side) to yield possession and pushed back the Federals on Coronel Hill (northeast side), leaving the defenders only in possession of Santa Rosa Hill. Over the next few days Federal communications rockets filled the night sky keeping the forces inside the city apprised of the situation via a mode that complemented the telephone and telegraph, which Mercado used to exploit his interior lines and rush forces back and forth to threatened sectors. Fighting on the south and east sides of the city was fierce, and Villa's men made progress by throwing dynamite only to be stopped and thrown back by a combination of mines, an electrified fence, and the timely arrival of reinforcements supported by the division's many guns. Finally, on November 8, the defenders made a solid push that forced the Constitutionalists back. The next day, they counterattacked and put the remaining Constitutionalists in the area to flight.

The Federals successfully defended the state capital because of all the situational advantages previously mentioned, plus superior communications and the use of interior lines to rush troops to crucial points in the line. Also, far from a fifth column inside the city acting against the Federals, the citizens took a proactive role in aiding them by: providing transportation to move troops inside the city; transporting ammunition, wounded, and the dead; and the local power company furnishing searchlights to illuminate the battlefield at night. Finally, Villa's artillery never came into play because Federal counterbatteries kept them inactive, he lacked shells, and he did not have the crews to man them, especially since a few of the Federal gunners had deserted before the battle and rejoined their comrades inside Chihuahua City. The defectors also told their Federal compatriots that Villa did not have nearly as many men as they had believed, and that they were poorly fixed for ammunition, all of which raised the morale of the defenders.

After the debacle of Chihuahua City, Villa sent all the women, wounded, and infantry south to Camargo. With the rest of his troops he stealthily moved north of the city and captured a Federal train bringing coal to Chihuahua City. Villa had the coal dumped, loaded his men on the empty train, and then employing subterfuge had it return to the north where, in the early morning hours, it entered the station of Ciudad Juárez and disgorged its Constitutionalist soldiers. Villa's men quickly spread throughout the streets of the border town and captured it. Federal General Castro and some of his officers escaped capture by slipping across the border into El Paso. Other Federals were not so fortunate and met their demise.

Tierra Blanca (15 km S. of Juárez)

Villa's turn toward the north had not fooled the Federals inside Chihuahua City who raced after him. Because the lines of communication had been cut, however, the Federals to the south could not warn those inside Ciudad Juárez. Nevertheless, Villa became aware of the Federal pursuit column and almost immediately after capturing Ciudad Juárez began returning troops south of the city as a blocking force. Once he had gathered all his troops remaining in Ciudad Juárez, he marched them south to Tierra Blanca, some fifteen kilometers south of the city, to avoid an international incident that might arise from a battle centered on the city. There was another reason to move away from the border. If he lost the battle because of his shortage of ammunition he would be able to disperse more easily into the countryside from deeper inside national territory.

Villa's forces occupied a line ten kilometers wide that ran on solid ground from Bauche Station on the Northwest Railroad to the east across the Central Railroad and almost to the Texas border. In this instance, the Villistas economized their force by having one man in ten hold the horses. The Constitutionalist disposition forced any Federals attempting to present battle to maneuver over the sandy terrain just south of their lines and kept them away from the water of the Rio Grande.

Just after sundown the lights of trains signaled the arrival of the Federals on the battlefield. Then, at ten o'clock, the Constitutionalists and Federals began exchanging fire up and down the line. Next, at 5:30 a.m. Federal cavalry hit Villa on his right flank attempting to capture the water tanks at Bauche Station. Villa countered that maneuver with a push in the center that relieved some of the pressure on his flank. Federal artillery went into action firing nonstop, and Villa directed his own to return fire initiating a duel of guns. At one point the Federals made a solid push against Villa's left forcing some of his men to retreat as far as the hippodrome just outside Ciudad Juárez. Villa's reserves counterattacked, cutting off some of the Federals who could not regain their own lines fast enough. Action continued throughout the day and into the night before quieting down.

In the morning of the third day the fighting again became generalized. When the Federals decided to unload their artillery so that it could support a push by the infantry, Villa decided that the time had arrived and ordered his dragoons to charge. They quickly overwhelmed the artillerymen, who abandoned their guns in the sand, and forced the infantry to break and run. The dragoons ran down the hapless Federals, who tossed their weapons and played dead or hid in the dunes, or tried to surrender only to be shot with pistols. The Federal trains began backing up in retreat, abandoning the infantry and artillery, all except one train that the Constitutionalists managed to stop, while the mounted contingent fled to the south. Only nightfall ended the day's massacre and saved the Federals and their Colorado allies from utter annihilation. Irregular General José Inés Salazar, who held overall command of the column, had lost probably fifteen hundred men and all his artillery. He ordered General Caraveo to cover the retreat and burn all the railroad bridges along the way.

The severity of the defeat and Salazar's failure to recapture Ciudad Juárez meant that Mercado was entirely cut off from ammunition, provisions, and money—which he desperately needed—that might be sent to him from Mexico City via the United States. In a council of war, therefore, Mercado proposed abandoning the capital city for Ojinaga. His Colorado generals and colonels argued against it. It would be better for them to fight it out in Chihuahua City and if defeated retreat into the sierra, but Mercado would not listen. (Huerta claimed that Mercado had been ordered to move south to Jiménez, but that order, if indeed given, made absolutely no sense since it would not have improved Mercado's position in the least.) He evacuated the city dragging the Huertista civilian officials and sympathizers in train as far as the railroad extended, to San Sóstenes Station, and then across the hills and desert to Ojinaga in what became known as the "caravan of death." The Colorados went too, reluctantly.

Although Villa always considered Tierra Blanca his greatest battlefield victory, General Maclovio Herrera claimed that he alone had made the decision for the decisive charge that carried the day, and many contemporaries backed him up. Additionally, Juan N. Medina claimed that the idea for the Ciudad Juárez *coup de main* was his idea. If both are true, it would seriously diminish Villa's reputation as a skilled general.

After occupying Chihuahua City and tending to duties of administration, Villa left to take command of his troops battling Mercado at Ojinaga. On January 10, 1914, his Constitutionalist Division of the North destroyed the Federal Division of the North and forced the survivors to cross over the Rio Grande to safety and into the awaiting arms of the U.S. Army, which imprisoned them at various forts. Villa was the first Constitutionalist theater commander to destroy an entire Federal division in the field, and his Constitutionalists controlled the entire state of Chihuahua.

NORTHERN SINALOA

During the month following the Battle of Santa María, Obregón tended to political matters and addressed logistics for the upcoming drive southward. Federal General Pedro Ojeda likewise was busy, preparing for an offensive to retake Empalme in order to push back the siege lines and cut communications between the Constitutionalists in the northern and southern parts of the state and deny them the valuable garbanzo bean harvest from the Yaqui and Mayo river valleys. In an operation carried out over the two days beginning on August 24, the Federals managed to capture Empalme. The Constitutionalists kept up pressure on the area throughout September and October, but never retook Empalme. Yet neither could Ojeda break out from the Guaymas-Empalme area, so that soon complacency, demoralization, and desertion became his chief preoccupations.

Around the same time Ojeda carried off his Empalme operation, the 2nd Brigade of the Federal Division of the Yaqui mounted an amphibious operation designed to quash the Constitutionalists in Sinaloa by attacking their headquarters in

San Blas. Three hundred Federal regulars and volunteers commanded by Lieutenant Colonel Teodoro Valdivieso left Mazatlán aboard the gunboat *Tampico*, intending to land at Topolobampo and drive inland toward San Blas, while a slightly larger second column under Colonel Miguel Rodríguez left Sinaloa de Leyva traveling by rail and road for San Blas. The two forces hoped to catch General Ramón F. Iturbe's Constitutionalists at San Blas in a pincer and disorient and disperse them. Valdivieso landed unopposed at Topolobampo on August 27, but the next day he encountered stiff resistance from General Iturbe who personally directed the effort to drive Valdivieso and his Federals back into the sea. After two days of combat, however, Iturbe could not prevail and had to retire to Los Mochis. Still, Valdivieso was mortally wounded, had to be evacuated, and died a few weeks later. The other Federal column fared much worse.

Colonel Rodríguez divided his battalion-size column into roughly equal contingents of about 250 men, one group traveling by rail, the other taking the highway, and departed Sinaloa de Leyva the same day that Valdivieso disembarked at Topolobampo. Rodríguez reached the rail station at San Blas and pushed the Constitutionalist advance forces back with little effort. He then sent a messenger to locate the contingent traveling by rail and received the unsettling news that its commander had been killed after which his command had scattered and began filtering back to Sinaloa de Leyva. Rodríguez continued on and reached Topolobampo on September 2 without encountering any resistance along the way and assumed command of what remained of Valdivieso's command. The Federal offensive stalled.

General Ojeda was not one to give up so easily, so he dispatched Colonel Heriberto Rivera with another column of 350 men and a section of 80-mm. guns to land in Topolobampo and fulfill Valdivieso's mission to destroy the Constitutionalists in San Blas. At the same time, northern Sinaloa state was demanding the attentions of General Obregón. Venustiano Carranza was due to arrive in El Fuerte on the Kansas City-Mexico & Orient Railroad that ran through San Blas to Topolobampo. Obregón wanted to secure the latter port in order to receive supplies from California, and he was properly concerned for the safety of the First Chief. With that in mind, Obregón immediately ordered Colonel Benjamín G. Hill to take six hundred men and reinforce San Blas. Obregón accompanied Hill's command, met Carranza in El Fuerte, and then escorted him back to the relative safety of Sonora. At the time, most of the Sonoran troops were in the area of Maytorena Station, and they provided security for Carranza as he passed by Guaymas to reach Hermosillo safe and sound.

Down south, Hill attacked Rivera at Los Mochis and sent him reeling back to Topolobampo where Iturbe's men pressed the attack. Shortly thereafter Rivera loaded all the remaining Federals on the *Morelos* and sailed for Mazatlán, possibly because the jefe de las armas of that port had grown concerned about the numerous

Constitutionalists operating in his area. Abandoning the port of Topolobampo, however, left Sinaloa de Leyva as the Federal's northern-most outpost in the state and on October 5, recently-promoted General Hill and Iturbe's battalions easily overpowered the four-hundred man garrison. Obregón then returned to Sinaloa from Sonora to take charge of the operation to capture Culiacán.

Culiacán

The Constitutionalists continued moving southward from Sinaloa de Leyva, but the going was slow because of the conditions of the rails and burned bridges, which had to be negotiated and/or repaired, and the need to send out security details. Joining the Sonorans and Sinaloans would be General Mariano Arrieta, who crossed the sierra with his Duranguenses to participate in the taking of Culiacán, the state capital. At the same time, Constitutionalist General José de la Luz Blanco took a regiment toward the coast to the port of Altata to provide security against any Federal seaborne operations and at the same time to be in a position to receive supplies sent on the water.

Obregón performed the requisite reconnaissance, gave his principal commanders their objectives and prepared to assault the city from the east, west, and south. He probably had about three thousand men and a battery of guns against a Federal combined-arms column with less than half that number. Federal Brigadier Miguel Rodríguez had divided his seven-kilometer-long perimeter according to the cardinal directions and distributed about three hundred men to each of the four sectors. He maintained a small reserve and his four guns were far enough inside the lines to avoid easy capture.

The Constitutionalists had taken up position for the assault when two events forced Obregón to postpone his attack. First, a Federal patrol got lost in the bush and coming across a Constitutionalist outpost got into a close quarters firefight. Upon returning to their lines, the Federals happened upon Obregón's artillery without infantry support and removed the breechblocks from one section of guns and were attempting to do the same to the other section when forced to flee upon the arrival of a group of Constitutionalists. Second, a Federal gunboat was reported trying to unload reinforcements at Altata.

The gunship docked and a shore party made a brief reconnaissance and collected some Federal stragglers, but finding no Constitutionalists loaded back up. The ship set sail for Boca del Río at the mouth of the Culiacán River and with that threat gone, at 5 a.m. on November 12, the Constitutionalists finally attacked Culiacán. Blanco's men arrived from the security mission on the coast and rode the rails to the section in the city where they were to detrain and join the fight. This was one of the few instances of troops being delivered railborne into action during the revolution. The train, meanwhile, had lost its steam and could not egress after delivering the troops. The railroaders took fire and incurred casualties. With constant pressure and overwhelming numbers the Constitutionalists eventually subdued the defenders, who began to abandon the plaza at around 2 a.m. on November 14 and fled to the southeast. Obregón ordered Blanco's cavalry to give chase toward the coast, sent Diéguez's infantry by train and Arrieta's cavalry along the highway for the same purpose, and told Constitutionalist General Juan Carrasco besieging Mazatlán to be on guard for any retreating Federals who might enter his area of operations.

Diéguez caught up to the Federals in retreat at Quilá on the San Lorenzo River and entered into combat, but when the Federals broke contact he could not follow because his men were too exhausted. Arrieta, too, had to quit because his horses were too worn out. Blanco, on the other hand, reached San Dimas on the coast and was in a prime position to destroy the Federals, although he was probably outnumbered two-to-one. After a brief two-hour firefight he simply watched as the Federals loaded their ships over the course of two days and sailed to safety. Obregón, who had been trying to communicate with Blanco for three days, was not impressed and fired Blanco, who never again commanded troops for the rest of the revolution.

The entire state of Sinaloa with the exception of Mazatlán was in Constitutionalist hands, and so Obregón returned to Hermosillo to begin planning his final campaign to conquer the rest of the west coast and Mexico City. He needed to accumulate arms and ammunition that had become difficult to obtain due to the U.S. arms embargo, and he also had to begin moving rolling stock south of Empalme, which was difficult because that rail center was in Federal hands. His chief of military trains, J. Lorenzo Gutiérrez, however, came up with an ingenuous plan. By laying down sections of track from Maytorena Station toward Cruz de Piedra, moving rolling stock forward, taking up the track from the rear and laying it down again ahead of the train, in essence hopscotching in sections, the Constitutionalists bypassed Empalme and moved rail assets south for the upcoming 1914 campaign.

TAMAULIPAS

After being forced out of northern Coahuila at the beginning of October 1913, General Pablo González moved down to Alamo Nuevo and used the pueblo as a staging area for his planned attack on Monterrey. In the interim General Jesús Carranza marched down along the Rio Grande to Hidalgo and then turning inland destroyed the rails in order to isolate the Federal garrison in Nuevo Laredo.

Afterward Carranza rejoined the main body of Constitutionalists at Alamo Nuevo. From there, González divided his command into three columns; the first, under the command of Antonio I. Villarreal, moved southward just to the west of the National Railroad, crossed it at Los Morales Station to destroy the rails, wound up in a three-hour firefight with the Federals there, and then continued onward to reach the town of General Zuazua, northeast of Monterrey. González personally led the second—and largest—column, which moved down much farther to the west of Villarreal in order to cut the rails and attenuate the possibility of any Federals arriving from Coahuila to reinforce Monterrey. Villa had just taken Torreón two weeks earlier and forced what remained of the Federal Division of the Nazas to regroup at Letona Station, a mere 167 kilometers west of Monterrey. González's men fought the Federal garrison at La Mina and then began destroying the rails, eventually winding up at San Nicolás, just north of the objective. General Carranza, commanding the third column, remained in Alamo Nuevo to collect the last of the stragglers fleeing Coahuila and then departed for Monterrey heading straight down the National Railroad. Carranza's men fought the Federal crew repairing the damage done earlier by Villarreal's men at Los Morales and then pressed onward, taking the same route as Villarreal and joining him in General Zuazua.

At this point, General Lucio Blanco, who had been in Matamoros since the beginning of June, received orders to begin marching westward over the rails to participate in the attack on Monterrey. Blanco, however, refused to comply, mainly because he did not recognize González as his superior officer. It would also appear that he had grown comfortable in his environs. The First Chief intervened and ordered Blanco to join the operation, but the general remained obdurate, so Carranza ultimately told Blanco to turn over command of his brigade and report to Constitutionalist Army headquarters in Sonora.

González could not wait for Blanco to get into line, and so he proceeded with his plans to attack Monterrey encountering the first small group of Federals, whom he virtually annihilated, at Villa del Carmen on October 20. Continuing southward, González's men burned the bridges around Villa Escobedo before turning back northward to deal with the rather large garrison at Salinas Victoria. The commanding general sent a messenger with orders for Colonels Carranza and Villarreal to join the attack on Salinas Victoria, but he did not wait for them to arrive before beginning the battle. Fighting into the morning of the next day, González finally forced the Federals to board their trains and retreat to Monterrey just as Villarreal and Carranza were arriving on the scene. The Constitutionalists raced after the Federals, but they had expended a good quantity of their ammunition such that most units only had an average of forty-five rounds per soldier, and other units had about half that number. The pursuit, therefore, did not accomplish much.

Monterrey

In the morning of October 22, the Constitutionalists reached the town of San Nicolás de los Garza and noticed a group of Federals to the west on Topo Chico and neighboring hills. In a move that sounds eerily reminiscent of the "disaster of Avilés," the Federal commanding general inside Monterrey had detached a group of about 350 regulars and irregulars with two 80-mm. guns to reconnoiter the Constitutionalist positions and defeat them, if possible, but strictly enjoining them not to get too far away from the main plaza. Nevertheless, there the Federal irregular general was, ensconced on Topo Chico with his guns on the crest, where it would be difficult to extract them. In short order the Constitutionalists assaulted the hill, captured the two guns, and put the Federals to flight. These were the first proper field pieces ever possessed by what would become known as the Constitutionalist Army Corps of the Northeast.

González now prepared to assault the city proper. Riding into San Nicolás he telephoned Federal General Adolfo Iberri and asked him to exit the city to spare it from any collateral damage and to fight on open ground, but the Federal refused. González again divided his forces into three groups, Carranza attacking the northwest side, Villarreal due north in the center, and Francisco Murguía commanding the third that would move around and attack from the east and, crossing the river, also assault the south side. Blanco's men still had not arrived but there was no time to lose. Their unenviable dearth of ammunition meant that the Constitutionalists would have to take what they needed from their Federal Army foes. Inside the city about 1,700 Federals prepared to battle the Constitutionalists who numbered an estimated 1,000 to 2,500 men, although the actual number was probably closer to the lower end.

Fighting commenced at around six o'clock in the morning of October 23. The Constitutionalists on the north side pushed forward quite rapidly, and by ten o'clock they had captured the most important landmarks, including the train stations and the Federal Army barracks. In that latter location they recovered a significant amount of ammunition, which they immediately began carting off to Apodaca, site of their field hospital. After this initial success, however, the offensive stalled as the Federals fell back on key strongholds and turned the streets of the state capital into rivers of lead. The battle continued through the night and into the next day. Around 3:00 p.m. González ordered his staff officers to inspect the various sectors and report back. The returns from that exercise revealed that a large number of his men were highly intoxicated. The Constitutionalists had turned the battle into a party, raiding liquor stores and cantinas and even pressing an errant musician into service to play songs then in vogue. With his men impaired, the lead elements of a Federal relief column entering the city from the southwest, and no sight of Blanco on the horizon, González had to quit the battle. The Constitutionalists blew up all the ammunition they could not carry or that did not match the calibers of their own weapons—along with the Federal barracks that housed it—and withdrew in an orderly manner shortly after nightfall to San Nicolás and Apodaca. Unfortunately, some of the more intoxicated Constitutionalists remained inside the city, oblivious to the retreat, and were captured and later executed. The Constitutionalists sent their wounded to hospitals in Matamoros.

The main reasons given for the Federal victory at Monterrey were a perimeter size that corresponded well to the number of defenders, compounded by the Constitutionalist mistake of assaulting along a broad front instead of choosing a particular point of attack, and the number of highly inebriated Constitutionalists. The defense in depth also received some praise from General Sánchez Lamego, although it is hard to understand why: Iberri had uselessly sacrificed a section of 80-mm. guns and the command of Brigadier Miguel Quiroga (who was killed in the battle) was decimated.

The day after the battle, Colonel Cesáreo Castro sent a dispatch to inform González that he was on his way from Matamoros with three hundred men and would be joining his column. The messenger also told González that effective September 27, 1913, he had been promoted to general of brigade and appointed commander of the Army Corps of the Northeast. Finally, the Northeast had the same unity of command as had existed in Chihuahua since the end of May and in Sonora since March. With the copious booty collected from Monterrey, González next set his sights on conquering Ciudad Victoria and the entirety of Tamaulipas State.

Ever since the Constitutionalists had taken Matamoros, Huerta had seemed content to let them keep it and much of the state of Tamaulipas, which had drawn the ire of some officers in the Federal Army. But there may have been sound reasoning behind Huerta's position. The Wilson Administration had been consistently hostile to Huerta, and after he disbanded the national legislature in October the Americans placed an arms embargo on his regime. Tensions had reached such heights that he feared an American land invasion, and so the Constitutionalists in Matamoros, in a strange and ironic twist of geopolitics, actually blocked that possible point of entry and allowed the Federals to economize their forces in the Northeast, with the two brigades of the Division of the Bravo remaining tied to the Saltillo-Piedras Negras and Monterrey-Nuevo Laredo axes of advance. The primary mission of the Division of the Bravo seems to have been to prepare to repulse an American invasion, with a secondary mission to subdue the Constitutionalists. And there may have been a tacit agreement between Blanco and the Federals to maintain the post-June status quo, which González's attack on Monterrey and invasion of Tamaulipas undermined.

Ciudad Victoria

González's Constitutionalists slowly began making their way down the Gulf Railroad toward Ciudad Victoria conquering minor towns and pueblos. He left the men of the old Blanco brigade to hold the northern part of the state and keep the Federals in Monterrey distracted while he simultaneously ordered smaller groups of nominal Constitutionalists to join his command for the campaign. It had been reported that the Federals in Monterrey were making a push and repairing the rails toward Nuevo Laredo and Matamoros in what appeared to be a coordinated effort, but González believed he had sufficient forces in the north to be able to withstand any operations that the Federals might attempt, so he continued to drive southward.

Rather belatedly, the Federals in Monterrey sent a column of one thousand mounted soldiers under General Ricardo Peña to give chase after González, whereupon the jefe de las armas of Ciudad Victoria sent another group of about 450 soldiers headed north in hopes that the two commands might form a pincer against González and destroy him. Instead, the Constitutionalists managed to defeat the two commands in detail. Peña caught up to, and did battle with, the Constitutionalist rearguard at Hualahuises on November 3. The Constitutionalists held the high ground and caught some of the Federals in a lane between two fences where they poured in murderous fire. Peña tried several flanking movements but each was turned back and then General González ordered his soldiers to break contact and continue the march in order to conserve ammunition for the main objective. Peña did not pursue and the next day, González's rearguard defeated the column from Ciudad Victoria at Garza Valdez.

A few days later the two defeated Federal columns linked up and made their way to Monterrey. Peña's cavalry belonged to the Division of the Nazas and Tamaulipas was in the Division of the Brazos' area of operations. More importantly, his cavalry was needed for the operation to retake Torreón.

Meanwhile, González pressed on toward Ciudad Victoria. After sternly admonishing his troops not to get drunk and repeat the debacle of Monterrey, on November 16 he ordered his principal commanders to attack the capital city on a line running from the north to the east. The fighting was fierce because the Federals held the high ground in several locations and had dug in with a series of defenses and trench works around key landmark edifices that afforded clear fields of fire. Moreover, the Constitutionalists had no artillery worth mentioning, which meant they had to get close enough to lob dynamite at the major defensive positions. Nevertheless, by the early morning hours of November 18 the Federals had been reduced to the interior of the city. The Federal commander ordered a great quantity of arms and any other materiel that could not be carried to be burned, had the breechblocks of his four field guns removed, and abandoned the plaza for the sierra to the southwest. González sent several columns in pursuit and these managed to catch up to the fleeing Federals at several points in what developed into minor skirmishes.

The Federal defeat at Ciudad Victoria was attributed to a shortage of ammunition, a hodgepodge of Federal Army units that broke down unit cohesion and *esprit de corps*, an overweight reliance on fickle civilian volunteers who fled or switched sides almost at the outset of combat, an aged commander in ill health, and too few men for too large a perimeter, all against a sizeable opposing force close to its own supply lines to the United States.

On the same day that his fellow Federals abandoned Ciudad Victoria, General Rubio Navarrete was arriving at the train station in Monterrey with a trainload of some fifteen hundred raw recruits for the Division of the Brazos. Before he left Mexico City, the Secretary of War had sternly warned Rubio against using these troops for offensive operations. They were to be distributed among the various detachments along the rails and trained in the school of the garrison. The commander of the Federal Division of the Brazos, General Joaquín Téllez, overrode those orders. The situation for the Federals in Tamaulipas was desperate because they had never received orders to go operational nor was there time for them to do so now; they were scattered throughout the state where they could easily be overwhelmed in detail. Additionally, the troops that Villa had left under General José I. Robles to watch over the rails east of Torreón were strong enough to threaten Saltillo and Monterrey from the west and, supported by the Constitutionalists of General Eulalio Gutiérrez to the south, had even sent messengers to González to propose a combined operation to attack Saltillo. Téllez's hand was forced.

Téllez ordered Rubio to move southward and relieve the besieged garrison of Ciudad Victoria, if it had not fallen. If the Constitutionalists had conquered the city, then Rubio was to link up with what remained of the Federal forces and return to Monterrey, or if the Federals had taken a course for the south, Rubio was to draw the attention of the González's Constitutionalists away from them and to the north. General Téllez gave Rubio a section of field artillery, a battery of machine guns, and 150 sappers. Together with these and the recruits from Mexico City, he was to move south along the rails as far as Garza Valdez, beyond which the rails had been destroyed, detrain the recruits, gather together the various Federal commands and stragglers in the region, send the empty trains back to Monterrey, and continue marching southward. Téllez promised to load the rest of Rubio's brigade on the trains and send it to join him, but he never did. And Rubio only had about twenty-five mounted soldiers to carry out various duties of security and communications (the telegraph only functioned as far south as Tinaja) after he abandoned the rails at Garza Valdez, meaning he would be functioning essentially blind.

Once González became aware of the presence of General Rubio's column, he sent several brigades to attack him, which they did on November 23 at Santa Engracia. Shortly after the fight had begun, Rubio received a message from Téllez. The commanding general of the Division of the Brazos was upset because the Secretary of War had taken away five hundred more men, and so he told Rubio not to count on receiving any reinforcements. Also, the Ciudad Victoria garrison had reached the San Luis Potosí-Tampico rail line and the safety of Tampico, so his mission was over and he should retreat as soon as possible. To pull out of a battle and retreat without being annihilated in the process, however, would have been a supreme accomplishment for the most seasoned of troops, and Rubio only had a handful of veteran soldiers among his two thousand men. Nevertheless, Rubio managed to stave off the Constitutionalist attack and then make a forced march back to his trains at Garza Valdez, collect another three hundred or so Federals, entrain, and return to Monterrey on November 29. The Constitutionalists gave chase, but botched the pursuit.

Eleven days later, on December 9, Federal General José Refugio Velasco's Division of the Nazas managed to push aside General Robles' Constitutionalists and retake Torreón.

Although Rubio's Santa Engracia mission turned out to be relatively uneventful, it revealed that many Federals (all those in the state of Tamaulipas) still had not gone operational as late as the fall of 1913. It also demonstrated the growing interdependence of operations between at least two theaters of operations at this stage in the revolution. If Villa had been correct during the summer in asserting that each theater commander functioned of his own accord, that view certainly was not applicable by the end of the fall, as Mariano Arrieta proved by crossing the sierra from Durango to participate in the attack on Culiacán. This growing realization was formalized and memorialized in the first strategy prepared by the Constitutionalists at the end of November. In that document, the First Chief, ecstatic over Villa's capture of Ciudad Juárez, maintained that within ten days the rest of Chihuahua would fall, and so with González and the Constitutionalists in San Luis Potosí blocking any Federal reinforcements from Mexico City, on November 27 he proposed that the forces from Chihuahua and Durango should use Torreón as a base of operations to conquer Saltillo and Monterrey. At the same time, he requested that the revolutionaries of the South, specifically those of Puebla and Michoacán states, put pressure on Mexico City. Although he gave González latitude in selecting his own objective, he considered the San Luis Potosí-Tampico line most appropriate, probably because Luis Gutiérrez was well-positioned to impede the Saltillo-San Luis Potosí route.

Carranza's grandiose plans were not to be; it simply was too early. The Federals retook Torreón before Villa could complete his conquest of Chihuahua State.

Therefore, Pablo González set about conquering the last two Federal strongholds in Tamaulipas—Nuevo Laredo and Tampico—at opposite ends of the state. After reorganizing and refitting his Army Corps of the Northeast and redesignating most of his brigades as divisions (although until the summer of 1914 they also continued to be called brigades, which better reflected their true troop strength), he assigned three divisions to attack Tampico. Taking the rest of his troops north, he planned to personally take Nuevo Laredo.

Nuevo Laredo

González sent the Blanco Brigade under its new commander, Colonel Andrés "Death" Saucedo, to make a feint against Monterrey, destroy the rails and telegraphic and telephonic communications for large stretches to prevent any Federals from approaching Nuevo Laredo from Monterrey, and then wheel north. At that point, according to his written instructions, "If you think that you can successfully attack the Laredo plaza you will execute that operation, otherwise you will establish your Headquarters in any of the points mentioned that you may have taken, giving account to this Headquarters in order to draft orders for a combined movement of forces through there for an attack on that latter stated plaza." However, Saucedo failed in his mission and reinforcements successfully reached Nuevo Laredo to double the number of Federals inside the plaza just as González arrived from Matamoros with another brigade and assorted minor units. The next day, Saucedo persisted in his dereliction of duty by entering into combat five hours after the designated time of attack. The battle started on January 1, 1914, and concluded the next day in defeat for the Constitutionalists. González attributed Saucedo's recalcitrance to the reassignment of his friend and mentor, General Blanco, and referred Saucedo to the First Chief for a court-martial. The Army Corps of the Northeast continued to suffer the adverse effects of Blanco's disobedience.

Aside from Saucedo's ineptitude, the reasons for the defeat could be attributed to the Federals' superior firepower, inexpungable trenches, and magnificent fields of fire.

Tampico

The attack on Tampico likewise failed precisely because just before the battle began the Federal garrison that had been defeated at Ciudad Victoria arrived to double the number of defenders inside the city. With a perimeter reduced by swamps, and supported by the port's coastal artillery and the guns of Federal Navy, which also delivered another 650 or so soldiers, the Federal position was virtually impregnable against the assault force. The reported 2,000 Constitutionalists (this number was most likely exaggerated) barely matched the Federals in numbers and did not possess even the appearance of unity of command.

4. THE CONSTITUTIONALIST REVOLUTION, JANUARY – AUGUST 1914

In December 1913, Federal General Eduardo Paz retired. As a gift to the institution he so loved, on his way out the door he delivered a strategy proposal to the Secretary of War for defeating the Norteños. Up until this point, the Federal Army had not formulated a formal strategy, mostly because back in February the leading generals figured they would be able to snuff out any resistance rather quickly. But now it had become painfully obvious that the revolution was growing and advancing.

Paz began his analysis of the failed Federal campaigns in the North by first confessing to the Federal Army's base deficiencies, namely, the woeful quality and reduced quantity of its soldiers and officer corps to cover such a large expanse of territory, which in turn required them to be detached in groups that could easily be overwhelmed by comparatively larger and more mobile bands of Constitutionalists. In essence, instead of mobilizing for war, the Federal Army had assumed a constabulary role that properly should have been delegated to the states. To these defects Paz added the Federals' ignorance about the geography in the North and institutional predispositions that had officers more engaged in bureaucratic processes than taking the fight to the enemy.

Map 4 Federal Army Strategy, 1914

In light of these factors, Paz suggested that the role of policing should be turned over to the states, and the Federal Army should mass for maneuver warfare, designating key cities oriented along an east-west line as staging areas that could quickly render aid to one another. Almost as important, these cities should be located close to the 21° parallel (10,000 men in Guadalajara; 5,000 in Lagos; 10,000 in Guanajuato/Querétaro; 5,000 in Zimapán/Tamazunchale; 5,000 in Tantoyuca; and 5,000 in Tuxpan) in order to draw the Constitutionalists away from their base of operations along the border and into unfamiliar territory, which carried the added benefit of allowing the Federals to economize their forces. Conversely, abandoning the Tepic, Aguascalientes, San Luis Potosí, and Tampico line to the Constitutionalists would force them to squander an estimated 200,000 troops in garrison duties across the vast northern territory and maintain 30,000 men to hold the front. In other places in his report, Paz suggested that the Federals might want to hold the Tepic, Aguascalientes, San Luis Potosí, and Tampico line, which would have reduced both the opportunities for the Federals to economize their forces and for their concentrations of soldiers to render mutual aid since they would have been separated by greater distances.

In Paz's opinion, if the Constitutionalists tried to carry out their stated goal of conquering Mexico City, they would either accomplish it with a general advance across the entire front or by using a turning movement at a particular point along the Federal axis of defense. In the first case, Paz considered that the Federal "defense-offense" would prevail, but it would require excellent scouting and intelligence services. If the Constitutionalists attempted a turning movement at any major point along the front, it would have to negotiate difficult territory and be exposed to a flanking movement by the next nearest concentration of Federals.

Secretary of War Aurelio Blanquet submitted Paz's strategy for study to General Eduardo Camargo, who then returned with his own proposal. Camargo did not believe the Constitutionalists capable of dislodging the Federals from their strongholds at Monterrey, Saltillo, Torreón, Nuevo Laredo, and Piedras Negras, much less from the "second line of resistance" consisting of Tepic, Aguascalientes, San Luis Potosí, and Tampico. He did seem, however, to like Paz's idea of various axes of defense, so much so that he proposed a total of three. The main base of operations would run through Durango, Torreón, Saltillo, Monterrey, Linares, and Ciudad Victoria, although as of January, 1914, the Federals only held Torreón, Saltillo, and Monterrey, and the first of these had only recently been recaptured at the beginning of December. The second axis would be formed by Cuernavaca, Iguala, Chilpancingo, and Acapulco in the South to face the Zapatistas.

Camargo also identified a sort of peripheral front running through Nuevo Laredo, Piedras Negras, Lampazos, Guaymas, and Mazatlán that "fortified and reinforced as much as possible, are garrisons that will demand a rebel main force to face them that will take away from their attempts on the interior of the country, and form new lines of [Federal] operations for subsequent outcomes in those regions and so Guaymas, Topolobampo [penciled in], Mazatlán, and Culiacán, form a flanking base to aid Durango." Yet, of these four cities mentioned as the "flanking base," the Federals only possessed Guaymas and Mazatlán, and the forces in those two cities (both of which the Constitutionalists ultimately invested and bypassed) were in a virtual state of siege at the beginning of 1914. Furthermore, to propose that cities in Sonora and Sinaloa provide "a flanking base to aid Durango," which was on the other side of the Sierra Madre, suggested a gross ignorance of geography. Nevertheless, Camargo advocated a line running from Guadalajara through Tepic, Mazatlán, Culiacán, and Guaymas that could be provisioned by sea, and he essentially argued for a fourth front to be developed along the border by sending flying columns recruited from "among thirteen and a half million inhabitants, interested in the defense of lives, honor, and interests," that would cut off the Constitutionalists "from their base of supplies and resources, which are the towns on the border," and drive them southward into "the interior of the country" where "their morale is terminated, as much by the distancing from their lands, as by the exhaustion of the ammunition that will be spent." This proposal essentially reduced strategy to a numbers game: the Norteño states only had a population of at most three million, while the rest of the country had more than five times as many people; surely the Center would prevail in fielding a larger army. And to put his plan into action would require the full authorized (not actual) strength of the Federal Army: 150,000 men.

Both plans suffered from major weaknesses. Camargo never addressed the core issues raised by Paz, specifically, the lack of knowledge about the terrain in the North and the Federal Army's perennial inability to recruit troops, much less qualified officers, and certainly not on the scale required by his plan. Paz, meanwhile, assumed that the Constitutionalists would make cross-country turning movements that could in turn be outflanked. Yet the Constitutionalists always advanced along the rails, if they were available, and the Federals had never shown any inclination or ability for mobility (defined as speed and flexibility of movement). They were even more tethered to the rails than the Constitutionalists. Additionally, his defense-offense required excellent situational awareness, which the Federals had never possessed, and a maneuverability of large units that the Federals had never demonstrated. Alternatively, Paz discounted the Constitutionalists' abilities for command and control, believing they would only be able to field one force of 10,000 men, when in fact they were able to put together several mobile army corps that met or exceeded that number.

In the final analysis, the Federal Army's poor command and control capabilities growing out of its outdated eighteenth-century culture explain why Paz's recommendation was never put into play. As he cogently remarked, "Those [Federals] who cannot conceive how 45,000 men can be concentrated, commanded, and directed, reject the cases expounded up to this point, and believe that with light columns of 1,000 to 3,000 men, operating without coordination and broken up into small detachments far away from the main body they can obtain victory." That is essentially what the Federals did; it was a self-inflicted policy of "divide up and be conquered" in which the Federals virtually assured they would always and everywhere be outnumbered and outclassed. If the Federals brought a brigade, the Constitutionalists responded with a division. If the Federals had a division, the Constitutionalists sent an army corps, and so on. Metaphorically speaking, the Federals consistently brought knives to gunfights.

At the beginning of 1914 the Constitutionalists also started to contemplate a national strategy, now that Villa possessed all of Chihuahua. General Álvaro Obregón proposed leaving enough men to keep Guaymas, Mazatlán, and Tepic invested, and then with around four or five thousand men—depending on the number of arms that could be smuggled in light of the U.S. arms embargo—he would cross over into Chihuahua and participate in the taking of Torreón. If that city proved impregnable, he proposed to bypass it and leave blocking forces at Matamoros Station, Coahuila, (eighteen kilometers east of Torreón) and in northern Zacatecas. Cut off from their supply lines, the Federals would have to make for the border across the desert where they would be decimated by Constitutionalist flying columns and the natural elements. If the Federals tried to move southward to reestablish their lines, they would run into Constitutionalists who would assume a defensive posture on ground of their own choosing. The first instance was a perfect application of the indirect method, and the second, southerly scenario, combined a strategic offensive with a tactical defense—both very advanced concepts. Yet, the plan seemed impractical on several accounts.

First, Torreón sat right along Villa's lines of communication. It was the communications hub of all north-central Mexico and could not be easily bypassed like the port cities of Guaymas and Mazatlán. Additionally, the Federal Division of the Nazas was of much better quality owing to its large number of highly-motivated and capable irregulars and in a better position to cooperate with neighboring Federal divisions than the Division of the Yaqui that Obregón had been fighting. Finally, it put all the Constitutionalist eggs in one basket, or theater, and needlessly so. Military considerations, however, were most likely not at the forefront of Obregón's thinking. Villa had just invited the First Chief to come to Chihuahua and Carranza had accepted. Obregón probably did not want to allow those two to get chummy and leave him on the outside.

Carranza rejected Obregón's strategy and instead gave him orders to "exterminate the ex-Federal [Constitutionalists considered the Federals to be outlaws, not legitimate troops of the federation] troops in Sonora, Sinaloa, Jalisco, Aguascalientes, and Colima and the Territory of Tepic [Nayarit], attacking the ports of Guaymas and Mazatlán straightaway or whenever you judge it opportune in view of military, political, and commercial interests." Carranza also gave him authority to print vouchers and requisition materiel, but he instructed him to leave enough men to keep the Federals in Guaymas immobile and the Yaquis in southern Sonora under control.

Pablo González also made his own request. He still wanted Monterrey, which was the key to the Northeast, and he was worried about the Federals in Nuevo Laredo who threatened Matamoros, the Army Corps of the Northeast's base of operations. González offered to send about two thousand men to disrupt the rails around Saltillo to keep that garrison immobilized. Doing that would allow Villa to send one thousand men to destroy the rails north and south of Torreón to keep the Division of the Nazas in check, and with the rest of his Division of the North take the old Colorado invasion route through Sierra Mojada, Cuatro Ciénagas, and Monclova.

Once Villa had captured Monclova, González with 4,000 men, twenty machine guns, and two field pieces would conquer Lampazos. The two Constitutionalist forces would then drop down to attack Monterrey in cooperation with one another. The main point of contention with González's plan is that he did not address what, if anything, he intended to do about the Federals in Piedras Negras or Nuevo Laredo.

Villa responded that the conditions of his horses after the Battle of Ojinaga did not permit him to undertake such an operation as envisioned by González, but he would soon attack Torreón, and after that he would assist González in taking Monterrey. The First Chief essentially accepted Villa's offer saying that once Villa was ready to march against Torreón he should inform González so that the Army Corps of the Northeast could begin its operation to take Monterrey, thus drawing the attentions of the Federal Division of the Bravo away from assisting the Division of the Nazas at Torreón. In the meantime, the First Chief wanted General Francisco Coss's brigade to begin destroying the rails south and then northwest of Saltillo, at Hipólito Station. Afterward, Coss would take up position between Saltillo and Monterrey to keep the two cities incommunicado. In concluding, Carranza wrote to González: "Taking Torreón, General Villa will advance on Saltillo or Monterrey, according to whichever is most convenient to attack first, but in any case you will be allied or in communication with him so that both your forces contribute to the success of this operation." This exchange was the first instance of planned inter-theater cooperation in the revolution.

Over the next two months, Villa continued to recruit and organize in Chihuahua. He and all Constitutionalists were greatly aided in this endeavor by the lifting of the U.S. arms embargo against the Constitutionalists on February 3, 1914, while the embargo against Huerta and his Federal Army continued in force. With a sizable war chest, the Constitutionalists ratcheted up their purchases in the United States and brought over materiel that had been stored in warehouses in the United States in expectation of just such an event. Until this point, the hands of the Constitutionalists had been tied, but now the gloves would come off, so to speak. The Federal Army would have only six months left before its dissolution.

SOUTHERN COAHUILA

At the beginning of February, 1914, the headquarters of González's Army Corps of the Northeast issued orders for the positioning of troops to carry out the operation to clear the Federal Army from the Torreón-Monterrey axis of operations. General Cesáreo Castro's 4th Brigade and General Francisco Murguía's 2nd Brigade would rendezvous in northern Nuevo León and attack the Federal garrison at Lampazos. Thereafter, Castro would move southward destroying the rails as far as Salinas Victoria where he would establish his headquarters. Murguía would cross over to Candela, in far eastern Coahuila, and ascertain the military condition of Monclova. If he deemed Monclova too well-defended to attack, he was to pass through San Gerónimo Canyon and travel along the eastern side of the National Railroad to Villa del Carmen. From there he would detach a party to destroy the rails between Monterrey and Saltillo at Villa García. If, on the other hand, Murguía judged Monclova vulnerable, he was to attack it and then begin moving southward toward Saltillo, completely destroying the rails as far as Reata. Turning eastward, he would continue taking up track between Anhelo and Paredón until reaching Villa del Carmen where he would remain in anticipation of the attack on Monterrey. His advance forces in Villa Escobedo and Salinas Victoria would allow him to communicate with Matamoros. General Teodoro Elizondo's 3rd Brigade remained to the northeast of Monterrey as a blocking force against any Federal attempt on Matamoros, occupying area towns such as Pesquería, Marín, General Zuazua, and Doctor González, where the brigade had its headquarters. Two smaller columns of Constitutionalists occupied the towns around to the southeast of Monterrey. They were to be joined later by General Jesús Agustín Castro and his 8th Brigade, although this never happened because of the brigade commander's lack of martial spirit. The 1st Brigade of General Antonio I. Villarreal, who also held the post of Governor and Military Commander of Nuevo León, covered the northern part of the state from Los Aldamas and Los Ramones, the latter being the Constitutionalist temporary state capital. Villarreal had the bulk of the army corps' artillery, which was not saying much. To the south, the 6th Brigade under General Alberto Carrera Torres interdicted rail traffic between San Luis Potosí and Tampico, while General Eulalio Gutiérrez, officially part of the Constitutionalist Army Corps of the Center, did the same in southern Coahuila, tearing up the track between the towns of General Cepeda and Parras, to the west of Saltillo, and in northern San Luis Potosí state. Finally, General Coss in command of the 7th Brigade had already received his orders directly from the First Chief in January, as previously stated.

As the Constitutionalists prepared to put their strategy into action, the Federal Division of the Bravo decided to launch an attack from Nuevo Laredo and Monterrey against González's base of operations at Matamoros. They hoped to recapture the port of entry and then reestablish communications farther south between Monterrey and Tampico. Consequently, at the end of January a column consisting of one thousand men commanded by Federal Brigadier Ignacio Muñoz began pushing eastward into Cadereyta. A few days later, another Federal force pushed Villarreal's men out of Los Ramones, although General Elizondo's 3rd Brigade subsequently retook the town. Yet in mid-February, the Federals forced

General Elizondo out of Doctor González and then on February 22 pushed his men out of Papagallos, which left Villarreal's eastern flank exposed. Two days later the Federals again captured Los Ramones. The loss of these towns prompted a general pull-back farther north so that the Constitutionalists could solidify the line.

At the same time, General Jesús Carranza had become worried about a Federal column under Brigadier Gustavo Guardiola y Aguirre moving down from Nuevo Laredo to attack Matamoros. Carranza's men had taken over control of the customs houses along the border and the part of the Federal pincer from Nuevo Laredo threatened the very finances of the Army Corps of the Northeast. Therefore, Carranza urged González to give precedence to the operational situation over the strategic plan (to capture the Torreón-Monterrey axis) and defeat the two prongs of the Federal pincer in detail, before they could join forces.

On February 25, Villarreal and Elizondo counterattacked and retook Los Ramones encountering only light resistance, and General González calmed Carranza's nerves by pointing out that Guardiola's Federal column would sense General Murguía and Castro's men in northern Nuevo León and pull back. Indeed, only the day before, on February 24, Castro had attacked Lampazos while Murguía destroyed the Federal detachment at Rodríguez Station, which may explain why Guardiola delayed his mission for almost three more weeks. Keeping rail traffic open between Nuevo Laredo and Monterrey remained a higher priority for the Federals than conquering Matamoros.

In the first week of March, Generals Castro and Murguía began moving southward to fulfill their missions preparatory to the attack on Monterrey. That allowed Federal General Guardiola to begin traveling down the Rio Grande to attack Matamoros. To counter this threat, General Villarreal sent his mounted Constitutionalists on horseback to Ciudad Mier while his infantry rode the rails to Ciudad Camargo, then marched to join the cavalry and men from Jesús Carranza's brigade. Once united, they moved northward and defeated Guardiola's Federals at Ciudad Guerrero (twenty-five kilometers north of Mier on the Rio Grande). This battle was relatively small, involving maybe one thousand men on each side, but it was notable because Major Carlos Prieto, an engineer by trade, in command of the army corps' artillery successfully handled his pieces, conducting all three types of fire mission: *con precisión* at a Federal cannon, *segadora* at the deployed infantry, and *progresivo* against the column in retreat. At one point, however, his task had been made difficult because some of Carranza's men resorted to the old ranchero tactic of *echar las rastras*, dragging branches behind horses to kick up the dust and make an enemy think that a column of reinforcements was arriving. That trick had merely obscured Prieto's view of the battlefield, and so he ordered the men to go somewhere else if they wanted to keep up the ruse. At any rate, the Constitutionalists had scant need of trying to demoralize the Federal column, which already evidenced an atrocious *esprit de corps* and had suffered mass desertions from the moment it moved into open country—which observers had always predicted would happen to any Federal column that attempted to leave its garrison or depart from the rails. The Federals returned to Nuevo Laredo and never again attempted offensive operations from that border town, and the commanding general, Guardiola, faced a court-martial for his inability to accomplish a fool's errand.

Unfortunately for the Constitutionalists, Generals Castro and Murguía, instead of executing their assignments, exceeded their authority with disastrous results. Remember that Murguía was supposed to attack Monclova with his own command if he believed he could take the city; a subsequent refinement of his orders stated that he should detach no more than four hundred men to make the attempt. Instead, once he got into eastern Coahuila he invited Castro to join him in attacking Monclova; they did so and on March 11 were defeated. Four days later a pursuing Federal column again defeated the two Constitutionalists brigades at San Buenaventura (twenty kilometers northwest of Monclova). González sent final orders on March 20 to the two generals to hurry into place for the upcoming battle, but Murguía had to flee to the safety of far western Coahuila after having exhausted all his ammunition. For the next month his brigade was out of the fight. Castro, meanwhile, did manage to make it back to the Monterrey area at the end of March, operating out of Sombreretillo, but his men were worn out. The imprudence and disobedience of the two generals placed the entire operation for southern Coahuila into jeopardy.

Fourth Torreón

Villa spent two months accumulating war materiel, recruiting and organizing, and tending to matters of public policy ahead of the Torreón campaign. Of particular interest to Villa was the artillery, crucial to his favored cavalry-artillery tactical combination. Colonel Benjamín Bouchez with workers from Cananea, Sonora, oversaw the manufacture of artillery shells in the Industrial Mexicana foundry in Chihuahua, where craftsmen also made replacement breechblocks for those removed by their previous owners in the throes of defeat. The First Chief also sent his Undersecretary of War, the celebrated artillerist General Felipe Ángeles, to take charge of the tremendous number of guns that Villa's men had taken from the Federal Division of the North.

In the second week of March, the lead elements of Villa's Division of the North began moving southward from Chihuahua City and surrounding towns for Yermo Station, the staging area for the Laguna campaign. Villa and Ángeles followed on March 16 to assume command of the men in the field. Villa's division could properly be called an "army corps" since it had eight "mixed" brigades of cavalry and infantry, one artillery brigade with forty field and mountain guns, twenty-five machine guns, a Sanitary (medical) service, general staff, and headquarters escort, for a total of about 11,000 men.

As Villa moved down into La Laguna, the First Chief ordered other major units in the area to cooperate in the campaign, specifically assigning General José Isabel Robles the job of cutting the rail lines from Saltillo and Paredón to Torreón. Carranza also put all the forces of the state of Durango at Villa's service. At the same time, Villa reiterated his request to have General Pablo González put pressure on Monterrey to distract the attentions of the Federals, and promised to help subdue Monterrey after taking Torreón.

Villa's Division of the North began moving toward Gómez Palacio from the north while the forces of Durango began closing in on the area from the northwest and southwest. With those forces, Villa had close to 16,000 effectives at his disposal. Opposing them would be the Federal Division of the Nazas under General José Refugio Velasco with about 7,500 men and eighteen field and mountain guns, a whopping thirty-five machine guns, a rumored three million cartridges, and a vast number of artillery shells. Approximately one-third of the government soldiers were irregulars, whom some observers considered superior combatants even to Villa's men. Additionally, the newly re-created Federal Division of the North was making its way up from Mexico City to Chihuahua with about 1,100 men.

To prepare for battle, General Velasco had established a ring of trenches completely surrounding Gómez Palacio that incorporated La Pila Hill to the north and La Cruz Hill on the south side, both of which contained artillery. The much smaller plaza of Lerdo to the southwest of Gómez Palacio also had a long trench to the northwest that made use of the San Juan Isidro cut, which tied off on the Nazas River. The Nazas separated Gómez Palacio, Durango, to the north, from Torreón, Coahuila, on the south side. For Torreón, the Federals had constructed an exterior arc of defenses that made use of the hills to the west/southwest and an interior line of defense at the foot of the reverse slope of those hills. The north and east sides of Torreón possessed excellent fields of fire and therefore were less heavily defended. Velasco had also taken the precaution of issuing a public notice that in the event of a battle no citizens inside the city would be allowed on the streets in groups larger than three people, and under no circumstances were individuals allowed on the rooftops. If any fire was reported coming from a house, it would be demolished along with all the inhabitants inside—a prophylactic against any fifth column that might materialize. All three plazas were connected by streetcars that Velasco intended to use in order to exploit his interior lines, but since the Federal general had the bulk of his effectives in Gómez Palacio, it seems clear that he intended the battle to play out in that city.

Strategy and Tactics of the Mexican Revolution

As Villa's Constitutionalists closed in on Gómez Palacio, they overwhelmed and utterly destroyed Velasco's outpost of 300 Rurales at Bermejillo on March 20. The Federals at Tlahualilo put up light resistance and then fell back to Hacienda Sacramento. The evacuation of those two towns, Bermejillo and Tlahualilo, forced Benjamín Argumedo to abandon Mapimí and leave it to General Tomás Urbina's approaching troops. It took another two days for the Constitutionalists to conquer the defenders at Hacienda Sacramento, which was reinforced mid-battle by Federal irregulars pulled in from neighboring San Pedro de las Colonias. Additionally, the Constitutionalists' mountain guns roughed up during the trip and using faulty ordnance failed to come into play. The fact that Federal reinforcements for Sacramento had arrived from the east concerned Villa and he once again sent an appeal to General Pablo González to put pressure on Saltillo and Monterrey to detract Federal attention from his own area of operations. Yet, as previously pointed out, while González probably had as many men as Villa, they were spread out over a much larger territory covering multiple axes of advance, rushing to intercept Guardiola marching against Matamoros, preventing those in Monterrey from joining in that operation, or licking their wounds after the attack on Monclova. The only brigade that González had in the area, Francisco Coss's 7th, already had an assignment and at any rate numbered too few to change the operational situation appreciably.

After observing the protocol *de riguer* of requesting that Velasco surrender the plaza—which the Federal general refused—Villa's Constitutionalists pressed onward. His cavalry deployed along a front ten kilometers wide perpendicular to the railroad and prepared to attack. Observing the deployment, the Federals inside their defenses scrambled to their positions. The Constitutionalist cavalry was supposed to advance to within four kilometers of the Federal trenches, dismount and, maintaining proper spacing, advance toward Gómez Palacio under cover fire from their artillery. After an hour of waiting, however, the men grew impatient, so the dragoons took it upon themselves to begin slowly advancing at a walk, then a trot, and finally a gallop. Crashing into the Federal lines, some made it into the houses of the city while others fell victim to the murderous artillery, machine-gun, and rifle fire coming from La Pila. The Constitutionalists were unaccustomed to delivering such an assault borne of unbridled enthusiasm, but soon awakened to the unvarnished realities of modern firepower they became disconcerted and bunched up, especially in the center where they increased the efficacy of the Federals' guns. At that point, the Constitutionalist artillery could not enter into action without endangering the lives of friendly forces close up. The Federals put up a good fight and took apart the assault. Fighting continued into the night, but by the next morning, March 23, the Villistas had retreated.

In the next attempt, Ángeles emplaced his artillery on each side of the rails, and then the entire division again moved toward Gómez Palacio, this time dismounted and with the Constitutionalist artillery firing in support. As soon as the Constitutionalists came into view, the Federal guns atop La Pila opened fire. Villa's men moved forward in perfectly formed lines. In no time the number of his casualties approached five hundred, so he instructed General Maclovio Herrera to move around to the west opposite Lerdo and see if he could take that smaller town. Leaving his horses tied up at San Ignacio Hill, the young general rushed off with his men to execute the maneuver at around seven o'clock in the morning. Inside Gómez Palacio, General Velasco observed this development and sent Colonel Federico Reyna's irregular cavalry to charge Herrera's dismounted men in an attempt to roll up Villa's right flank and capture the Constitutionalist guns on San Ignacio Hill. In response, Villa rallied his escort and led a counter charge that completely destroyed the Federal irregular column and allowed Herrera to move into position for the assault on Lerdo.

In the afternoon, Villa learned that his men had finally taken Sacramento, so he pulled back to wait for the arrival of those victorious forces. After nightfall, the Villistas resumed the attack along three fronts with Lerdo on the right and Gómez Palacio in the center of the line. By 9 p.m. Herrera had achieved his objective, taking Lerdo. At this point, facing the Federal artillery that continually harassed his rail lines and, with superior ordnance, kept Ángeles' guns out of action, and given the clear fields of fire that he confronted around to his left, Villa decided to suspend combat and wait for the troops of Durango and La Laguna to arrive. Most of the Constitutionalists spent the following day resting and pasturing their horses under Federal harassing artillery fire until the order came for the brigades to fall back to Vergel. Herrera remained in Lerdo and was resupplied. The Federals used the next day to clean up the battlefield and continued to receive stragglers from Sacramento and other outposts that filtered into Torreón, while Villa evacuated his wounded to hospitals in Jiménez, Camargo, and Chihuahua City.

By March 25 Villa was ready to resume the attack. His men took up position to assault Gómez Palacio from every direction except from the southeast. After nightfall the Constitutionalists began attacking. Villa preferred to fight at night because it disoriented and discouraged the Federals and gave their draftees the opportunity to slip away and desert. Shortly after the assault began, General Tomás Urbina's men from Durango started arriving and Villa threw them into the line. By 9 p.m. the assault on La Pila began. The fighting was ferocious and the light show fantastic as rifles and guns lit up the hill and Constitutionalists exploited the intermittent moments of darkness to get right up to the five forts and fire into the loopholes. In some cases they even disarmed the Federals by yanking their rifles out of the forts by the barrels and lobbed dynamite into the trenches. The brigades on the Constitutionalist left, however, had not yet entered into action and they never would, apparently because those commanding generals lacked a proper sense of urgency. On La Pila the Constitutionalists had taken two forts and Villa threw General Calixto Contreras' Duranguenses, just arriving on the scene, into the fray in hopes of taking the other three forts. But then Federal General Ricardo Peña led a bayonet charge that forced the Constitutionalists from the hill at 4 a.m. At that point Colorado General Benjamín Argumedo carried out a counterattack to capture the Constitutionalist guns. The move disconcerted the artillery's infantry support, but in short order calm was restored and Argumedo's irregulars repulsed. On the Constitutionalist left General Toribio Ortega pulled back at around 6 a.m.; by now his brigade had lost about half its strength between killed and wounded.

The fight for La Pila was termed "the greatest of the feats of war ever recorded in Mexican military history," by General Sánchez Lamego. General Peña was killed and Federal General Eduardo Ocaranza was seriously wounded.

Inside Gómez Palacio the Federals began removing everything of value to Torreón on March 26 as Federal infantry, their faces blackened by powder after days of combat, followed in trains. Constitutionalist General José I. Robles finally arrived with his brigade and then Federal irregular cavalry suddenly appeared about eight hundred meters to the front, deployed, and prepared to attack. Federal artillery thundered away and then, almost as quickly as they had appeared, the irregular cavalry withdrew. It had been a screen to cover the retreat of the remainder of the troops to Torreón. Villa's Constitutionalists now possessed Gómez Palacio and Lerdo.

The Constitutionalists then surrounded the city of Torreón, and beginning in the evening of March 28 started to battle the Federal defenders, who fired their guns from the hills and exploded dynamite they had previously buried in the approaches. The Constitutionalists scaled the hills, engaged the gunners and their infantry support, and conquering Santa Rosa to the north and Calabazas and La Polvorera hills to the west of the plaza, disabled or turned the guns on their former owners. The Federals counterattacked and retook the hills to the west but not Santa Rosa. On the east side the Constitutionalists penetrated the first streets of the city capturing two Federal barracks and sixty-five draft mules belonging to the artillery.

Then on March 29, a small holding force of Constitutionalists in San Pedro de las Colonias reported the arrival of three train loads of Federals from the east, some 1,600 men strong. This column composed of Federal regulars and Colorados had been intended to participate in the operation to take Matamoros, but since Villarreal had defeated Guardiola at Ciudad Guerrero those troops were now available to reinforce Torreón. General Pablo González's good news became Villa's own bad luck and he fumed, once again, over González's failure to block those reinforcements. He could not rely on the Army Corps of the Northeast, so he sent

Generals Ortega and Rosalío Hernández with their brigades to hold back the Federal reinforcements until he could finish the task at hand.

Attempting to force the capitulation of the Federals inside Torreón with a new sense of urgency, the Constitutionalists continued to tighten the noose around Torreón. Villa brought in reinforcements from Chihuahua City to assault across the river on the north side as the Federals cannonaded the Constitutionalists on Santa Rosa Hill and fired randomly at Gómez Palacio. The Federals continued to resist with a tenacity that surprised the Constitutionalists, but they had also given several indications that they intended to try to break out of the city. Accordingly, on the evening of April 1, Villa instructed General Robles on the left to pull back and leave the route to Viesca open.

In the early morning hours of the next day, the Constitutionalists to the west of the city using dynamite as grenades finally conquered the forts on Calabazas and La Polvorera Hills, but the Federals once again retook Calabazas because there were too few Constitutionalists to hold it. The Federal guns resumed cannonading Gómez Palacio and Santa Rosa Hill, shooting wildly to cover the retreat from the plaza that had begun. At 5:00 p.m. fires and explosions rocked Torreón and covered it in smoke as Argumedo's mounted irregulars took up position in front of Huarache Canyon, similar to what had occurred prior to the evacuation of Gómez Palacio, and allowed Velasco and the rest of his division to flee. Villa's plan for "creating a silver bridge for the enemy" to escape had worked.

Velasco claimed he had to quit the fight because his division was exhausted, hungry, and down to sixty rounds of ammunition per man, which pro-Huertistas blamed on the U.S. arms embargo. According to Velasco:

> The Revolution had great moral and material aid inside and outside the country, and on the other hand, the Federals only had the hate of almost everyone, the almost complete lack of resources, great difficulty in communications and in most all cases, the lack of cooperation of our very own men.

Velasco was considered by many to be one of the better Federal generals. He visited his men in the trenches, directed counterattacks, oversaw the construction of fortifications that improved his men's morale and chances of success, declined to engage in persecution of a local populace overwhelmingly sympathetic to the Constitutionalists, and saw to the orderly evacuation of the city instead of simply saving his own skin, as many Federal generals did upon losing a plaza. But running out of ammunition was frequently an excuse exploited by Federals who lost battles (it should also be pointed out that the U.S. Army Attaché to Mexico observed that fire discipline in the Federal Army was atrocious), and many suspected that the explosions before abandoning the cities were caused by Velasco blowing up his own magazines of ammunition in excess of what he thought he could distribute and carry.

Villa later faced criticism for not picking a single point on the Federal line to focus his assaults—although one wonders just how valid that criticism is against an opponent with good interior lines—and then carrying off the assaults in an uncoordinated fashion, in sequence rather than simultaneously. What could be called "surround and pound" was fast becoming Villa's signature tactic. He invested plazas with superior men and materiel and then hammered the defenders into submission, which resulted in high casualty counts for his enemies and his own men. In this case, however, he allowed Velasco to escape when he could have completely annihilated the Division of the Nazas. He would have to fight those same men again at San Pedro de las Colonias, but his plan to allow the Federals to evacuate echoed Sun Tzu: "When you surround an army leave an outlet free. Do not press a desperate foe too hard. Such is the art of warfare."

Yet, if Villa had done what General Pablo González had recommended back in January and attacked Monclova and then Saltillo-Monterrey, those Federals in Torreón would have had to choose from three equally undesirable options. Either they would have died on the vine, cut off from their supply lines, or they would have moved toward Saltillo to reestablish their lines and had to fight on the offensive at a point of the Constitutionalists' choosing. Or, the Division of the Nazas could have retreated toward Zacatecas across open country where it would have been decimated by desertions and hit and run tactics carried out against it by area Constitutionalists. The successful execution of González's operation would have been similar in its effects to what Obregón had proposed, would have achieved the epitome of what became known as the "indirect method," and it would have resulted in fewer casualties all around.

In fact, "far from suspecting the presence of Federal troops in" San Pedro de las Colonias, as he later confessed, Velasco had initially headed for Zacatecas. At the time Federal General Guillermo Rubio Navarrete was directing the effort to repair the railroad between Torreón and Zacatecas. But when Velasco learned from a civilian in Viesca that a powerful Federal column was gathering in San Pedro, he decided he would have a better chance to reach that town.

San Pedro de las Colonias

After conquering Torreón, Villa had given his men only one day to rest before he started forwarding troops to reinforce the two brigades under Ortega and Hernández already at San Pedro de las Colonias. He hoped to defeat the Federal column there before it could link up with Velasco's fleeing division. By April 8, he had joined the fight with the rest of his Division of the North and put his surround and pound tactic into practice. Even with his entire division engaged, however, Villa was unable to prevent Velasco from using Argumedo's irregular cavalry to make contact and, on April 10, safely enter the Federal perimeter while under attack by Villa's Division of the North.

The arrival of Velasco's division brought the total number of Federals at San Pedro to around nine thousand men. Villa apparently judged this development as "a good military act"—even though it defied the rule of "divide and conquer"—since it would afford him the opportunity to destroy the single concentration of Federals in one fell swoop. And that is precisely what happened.

San Pedro was one of the rare open field battles of the war against the Federals given that their perimeter extended beyond the city limits and into the surrounding countryside, which lacked any meaningful terrain features. Yet there had been little time for the Federals to entrench and they were so demoralized, not to mention in arrears in pay, that even the jefes lacked any *esprit de corps*, and the customary application of draconian disciplinary procedures failed to instill any semblance of obedience among the enlisted. Moreover, this time Villa could count on a fifth column inside the plaza and the Army Corps of the Northeast for cooperation. The First Chief ordered Murguía to abandon northern Coahuila, which he mistakenly believed had been cleared of Federals, and attack Saltillo from Monclova while General Francisco Coss's 7th Brigade put pressure on Saltillo, and Eulalio Gutierrez's Constitutionalists in northern San Luis Potosí cut the rails to the south of the capital of Coahuila. At the same time, González's 1st, 3rd, and 4th Brigades were closing in for the attack on Monterrey, the capital of Nuevo León.

Realizing the hopelessness of their situation, the Federals fired the town of San Pedro, most likely as retaliation against the fifth column. And once again Velasco relied on the mounted irregulars of Generals Argumedo and Juan Andrew Almazán to try to cover the flight of the regulars by ordering them to execute a north and south pincer attack against the Constitutionalists. This time, however, the magnitude of the defeat and impossibility of completing or even surviving the mission caused the irregulars, who reportedly were in a state of extreme inebriation, to carry out the mission with atypical ineffectiveness. The Constitutionalists recognized these two signs—the appearance of the Federal mounted irregulars and the burning of the town—as preparatory to a retreat and pressed even harder. The Federal lines collapsed, lost all order, and quickly devolved into a "save yourself if you can" headlong flight. One Federal jefe committed suicide rather than be taken alive. Others pistol whipped their own men to try to get them to reform their lines. Terrified, the Federal artillery fired canister into the commingled lines of Villistas and forces in the hope of buying time to extricate their pieces. Ultimately, the Federals abandoned ten guns and limbers and left their numerous wounded to the mercy of the Constitutionalists.

Villa's men did not give chase, most likely because they were too busy trying to fight the fires in the city. They also had to provide succor to the inhabitants, who had been prohibited from leaving their homes to get food for ten days, and tend to the human tragedy that was the Federal Army. Most of those hapless soldiers had been dragooned into service and sent north to fight for a cause they did not believe in and to do a job without the training or means to accomplish it. Villa's sixteen thousand soldiers had annihilated the nine thousand or so mostly raw (untrained), green (untested in combat), and/or disorganized (those who had just fled Torreón) Federals from three divisions (Divisions of the Bravo, North, and Nazas).

Two days after the battle, the First Chief ordered Villa to give pursuit and told González to send his nearest general to attack the vanguard of the retreating Federal column. General Coss reported that his men registered several small-scale clashes with the Federals and destroyed the rails between Hipólito and Paredón and also between Paredón and Saltillo. Conversely, Villa's pursuit force quickly returned to San Pedro after deciding there was no way to catch up to the Federals after their two-day head start. Yet the Federals remained at Hipólito Station (160 kilometers east of San Pedro) concentrating and reorganizing until April 20—a week after the battle—when they finally left for Saltillo, meaning Villa's men could easily have caught up to them. More likely, Villa was exhausted after a month of combat and wanted to rest, refit, and recruit in preparation for his upcoming campaign to conquer Zacatecas.

TYING UP THE NORTHEAST

The Federals who had fought at San Pedro started arriving in Saltillo on April 23 to the news that two days earlier the U.S. Navy had invaded Mexico, capturing the port of Veracruz. The Huertistas tried to make political hay of President Woodrow Wilson's imbecilic decision to attack Veracruz in order to prevent the landing of arms and ammunition purchased by Huerta's government. Federal Army generals in every corner of the country sent envoys to invite the closest enemy commanders to join forces and expel the gringos from the fatherland. The First Chief so tired of receiving requests for instructions from his jefes as regards these entreaties that he ultimately issued a blanket communique saying that if any more Federal officers approached his generals for any reason other than to discuss unconditional surrender, they should be shot immediately—white flag of parley be damned. From the standpoint of uniting the various factions in common cause under the Huertistas, the American invasion had little effect, on that hand.

On the other, the fear that the Federals and Constitutionalists *might* join forces against the Americans compelled the U.S. government to once again institute an arms embargo against all belligerents in Mexico and tighten its border surveillance against smugglers. The intervention also drove numerous Mexicans from the South to join the Federal Army to fight the Americans. The Federals, however, cynically marched those new recruits off to fight the Constitutionalists and Zapatistas, all of which adversely impacted the Constitutionalist war effort, quite the opposite of what Wilson had intended. In the short-term, however, the invasion materially altered the strategic situation in the Northeast, as will be seen.

Monterrey and Allende

By April 21, Constitutionalist General Francisco Murguía had received his shipment of arms and ammunition and began moving toward Monclova, per instructions received from the First Chief. But Monclova still had too many Federals so he headed northward, in the direction of the border, to attack more vulnerable targets.

Also, leaving Jesús Carranza's small brigade to secure the border against any Federal offensive from Nuevo Laredo, on April 2, General Pablo González began positioning his 1st, 3rd, and 4th Brigades, and other smaller columns around Monterrey for a planned April 20 attack. General Jesús Agustín Castro's 8th Brigade was also supposed to participate in the battle, but refused to comply. Preparatory engagements at Cadereyta over two days beginning on April 14 and at Salinas Victoria on April 16 took place as the Constitutionalists drove in area commands, compelling those two and other outlying Federal garrisons to fall back on Monterrey. At Cadereyta the Federal resistance had been light, and the Huertistas attributed their defeat to a lack of expertise among the commanders, the absence of unity of command between the jefe de la plaza and the commander of the locally-based flying column, and the overwhelming number of opposing forces. The Huertistas put up a much better fight at Salinas Victoria, but likewise succumbed to superior forces after the Constitutionalists turned back trains bearing reinforcements from Monterrey and from the north. As the Constitutionalists closed in, the Federals sent a 1500-man force from Saltillo to relieve Monterrey. General Francisco Coss's 7th Brigade intercepted and defeated the relief column just south of Ramos Arizpe (ten kilometers north-northeast of Saltillo) on April 20. That very day González finally began the formal attack on the state capital of Nuevo León with about four thousand men against an estimated three thousand Federal Army defenders. The U.S. invasion occurred in the middle of the battle.

Strategy and Tactics of the Mexican Revolution

The Federal commander in Monterrey, Brigadier of Engineers Wilfrido Massieu, had overseen the construction of formidable fortifications to be able to resist the Constitutionalists and had rather handily repulsed all attacks. After three days of fighting, it was said that the Federals had deployed only half their effectives while the remainder rested under the protection of their artillery. On the third day, however, the Constitutionalists began making some progress, penetrating the first line of defenses and forcing the Federals into the interior of the city, similar to what had happened during the failed October attack on Monterrey. Then, in the early morning hours of April 24, the Federal guns started firing wildly at the Constitutionalists' positions. At the same time the Federal commander sent runners to tell the sector commanders he was pulling out, and then he abandoned the plaza through the Canyon of Santa Catarina to the southwest in hopes of reaching Saltillo. In the opinion of some leading Huertistas, the Federals had evacuated the city without cause because they appeared to be winning the battle. But there may have been another reason at play that forced the Federals to capitulate.

The Federals had long known that an American invasion of Mexico would include a landing at Veracruz and probably Tampico in combination with a sweep down from Texas through the Northeast. The Veracruz operation, however, was not executed according to the "General Mexican War Plan" but rather as Special Plan "A," dated April 16, 1914—and a truncated version of that limited operation because the U.S. military did not receive the required two weeks advance notice. And Constitutionalist General Luis Caballero's message to the U.S. Navy offshore at Tampico not to interfere with his planned attack on that city probably spared the Americans the bother of an armed invasion of that port, as called for under Special Plan "A."

The Federals could not know the targeted nature of the American operation, and accordingly on April 21 put full countermeasures into effect: the commanding general of the Federal Division of the Bravo, Joaquín Maass, ordered the garrison commanders in Piedras Negras and Nuevo Laredo to set fire to those towns and begin moving southward to kick off a scorched earth campaign. General Teodoro Quintana followed those orders and torched Nuevo Laredo and repaired to Villaldama, and eventually Paredón. That left the Federals in Monterrey exposed and on the frontier of the government's forces facing the Constitutionalists—and possibly the Americans—and so the Federal commander had ordered the plaza evacuated. Or, if one believes the Constitutionalists, on April 23—the same day General Massieu learned of Maass's orders to Piedras Negras and Nuevo Laredo—they penetrated the outer perimeter and it was only a matter of time before they conquered the rest of the plaza. Either way, Nuevo Laredo was abandoned, and now so was Monterrey either because Massieu lost his nerve or the Constitutionalists compelled him to quit the fight.

The Constitutionalist Revolution, January – August 1914

The Federal commander of Piedras Negras, General Luis Alberto Guajardo, disobeyed the order to set the plaza ablaze, although he did evacuate that border town and move down to Allende. When General Murguía learned of Guajardo's change of venue, he decided to attack him. It is tactically significant to note that by backing up to the border, the Federals were able to economize their forces since they did not have to guard the side of the plaza facing the United States. Additionally, any Constitutionalists contemplating an attack against a border town would be dissuaded from attacking the side opposite the border for fear of bullets landing in U.S. territory. Now Guajardo and Murguía's forces were in a similar geopolitical position and of the same relative strength in numbers, although unequal in quality. Accordingly, Murguía's 2nd Brigade attacked and destroyed Guajardo's command on April 27 at Allende and remained along the border for the next two weeks recruiting and organizing.

Upon reflection, one could make a good argument that the U.S. invasion handed the Northeast over to the Constitutionalists, albeit it in an indirect manner, by changing the strategic calculus for the Federal Army. On the other hand, the Constitutionalists were rolling and undoubtedly would have conquered the Northeast anyway, and soon.

The Federal defenses at Monterrey although formidable had been far from perfect. An inspection by Federal Colonel Carlos García y García and General Jacinto Guerra in March had revealed numerous egregious deficiencies including: the lack of cleared fields of fire and enfilading fire or zones of concentrated fire, particularly as regards high-speed avenues of approach; subsequent lines of defense such that the exterior lines dominated the interior ones in elevation (which may explain why the Federals had mined the exterior blockhouses, as the Constitutionalists alleged but General Sánchez Lamego dismissed); defensive works that resulted in a misallocation (undesirable grouping) of troops; concertina wire not far enough away from the forts; and artillery pieces placed too far forward where they were subject to destruction or capture. To these, Guerra also made qualitative points typical of the Federal mindset that bemoaned the immobility of those troops assigned to the forts and the overworking of the garrison forces in patrolling, outpost, and fatigue duty, among others, caused by the extended perimeter. The Federal perimeter then was more than fifteen kilometers in length and, given the assumed garrison of 3,150 men, too weak to sustain a "serious" attack.

Guerra's suggested remedies sounded like a general corrective that could have been applied to every defense of a plaza by Federals: the fixed fortifications should be designed to accommodate only the bare minimum garrison troop strength; only garrison troops should be made to construct said fixed fortifications while any mobile columns presently inside a plaza under attack should prepare field (that is, "hasty") fortifications and/or reinforce the garrison troops in their prepared defenses.

Massieu probably made some corrections, but a complete redo of the defenses ahead of the battle would have been impracticable, and it seems clear that Monterrey could not have held out for an extended period of time.

General Massieu finally reached Saltillo on April 30. His column of Federals and Huertista sympathizers from Monterrey had been hounded along the way, decimated by General Coss's Constitutionalists who trailed the column like a pack of hungry wolves, picking off the stragglers and the weak. Others fell out due to dehydration, hunger, and exposure. Those lucky enough to reach their desired destination found the rest of the Division of the Bravo preparing to make another arduous retreat over open country through northern San Luis Potosí.

Tampico and Paredón

After taking Monterrey, General Pablo González wanted to capture Tampico before the U.S. Navy was obliged to attack the port to protect foreign citizens. The city had a sizeable community of American expatriates due to the regional petroleum industry, and the locals were none too pleased about the invasion of Veracruz and felt no compunction about publicly venting their ire with threats and intimidation. Constitutionalist General Luis Caballero's 5th Brigade had kept Tampico under a loose siege ever since the failed attempt to capture the city back in December, but General González had told him to desist in any attacks during the battle for Monterrey. He had been concerned that if Caballero captured the port any fleeing Federals might head north and reinforce the Monterrey garrison. Accordingly, on April 24, the day of the Federal evacuation of Monterrey, Caballero resumed the attacks for four straight days. Also on April 24, General Alberto Carrera Torres's 6th Brigade, which had been assigned the mission of constantly harassing the San Luis Potosí-Tampico rail line, destroyed a sizeable Federal column of 1,800 men at La Herradura that was on its way to reinforce Tampico. During the course of the following week Carrera Torres assailed and virtually annihilated the remaining Federals trying to hold the western part of the San Luis Potosí-Tampico line in engagements at La Herradura, Montaña Station, Villar Station, and points between. That left the Federal garrison at Tampico cut off to all points north and west by the Constitutionalists and to the east by the U.S. Navy.

González began moving forces down to Tampico for the final push to take the city. Cesáreo Castro's 4th and other minor brigades and corps moved to join Caballero's 5th Brigade. Jesús Agustín Castro's 8th Brigade, which had wintered in eastern San Luis Potosí, also received orders to join the fight, but the bulk of those forces did not arrive until the final moments on the last day of the battle. Villarreal's

1st in Monterrey, Elizondo's 3rd Brigade to the east of the state capital, and Coss's 7th Brigade to the west constituted González's rearguard against the Federals in southern Coahuila, while General Murguía's 2nd remained in northern Coahuila. At the same time, the First Chief directed General Villa to attack the Federals at Saltillo, but Saltillo belonged to the Army Corps of the Northeast's area of operations and Villa, in his mind, had already done enough to aid González. Besides, Generals Villa and Ángeles deemed Zacatecas to be of much greater military importance than Saltillo, which had the added inconvenience of being far from Villa's lines of communication that stretched back through Chihuahua City to Ciudad Juárez. Nevertheless, the First Chief wanted the entire Northeast cleared of Federals, but González's brigades were otherwise disposed and Villa had repeatedly promised to help take Saltillo. Ergo, he accepted the mission, albeit reluctantly.

On May 9, González's estimated four thousand Constitutionalists began the assault against Tampico, a port city with many area lagoons and swamps that limited the avenues of approach, which in turn were blocked by zigzagging trenches reinforced with forts, supported by Federal Navy gunboats on the Pánuco River, and defended by an estimated three thousand Federal soldiers. Over four days, the Constitutionalists crawled under the barrage of 101-mm. naval shells, coastal and field artillery, and machine guns to approach the Federal forts and trenches and fight, holding the ground gained at night against the maddening onslaught of insects. Then, in the early morning of May 13 a torrential downpour began flooding the trenches, forcing the Federals out of their defenses as a northern wind drove rain into their faces, limiting their visibility. The Federals also claimed to have begun running low on ammunition and therefore had to abandon the city, first to the west. Elements of the 6th and 5th Constitutionalists Brigades, however, blocked the path of the fleeing Federals and forced them to move southward through Veracruz and into the area of operations of Cándido Aguilar's Constitutionalist 1st Division of the East.

After campaigning in the Huasteca at the end of 1914 (subsequent to his reassignment after the Battle of First Torreón), Aguilar had spent the winter in Tamaulipas organizing and equipping elements of his 1st Division of the East before starting to advance southward on April 5. Moving through San Luis Potosí and into Veracruz, Aguilar gathered in the minor commands he had left in the area, conquering the plazas of Ciudad Valles (April 12), Huejutla (April 22), and continuing toward the coast. Unbeknownst to him, his activity had the unintended effect of driving some Federals from the region into Tampico, where they strengthened the garrison and made General Pablo González's job all the more difficult. Alternatively, on May 11, in the midst of the Battle of Tampico, Aguilar began his attack on Tamiahua. Tamiahua was a chokepoint for reinforcements and supplies headed up the coast that may have been a factor in the Federal defenders of Tampico running out of ammunition. Tamiahua finally fell on May 17, which in turn prompted the Federal evacuation of Tuxpan. With the loss of those two important plazas, the Federals fleeing Tampico after the battle had to turn inland and head for Mexico City, never to fight again.

The conquest of Tampico left the last remaining Federals in the Northeast operating along a single axis running from Paredón through Saltillo, with a large concentration of Federals in Paredón, including many of those who had evacuated Nuevo Laredo and Monterrey. On May 11, Villa began moving toward Hipólito Station, which he used as a staging area. At General Ángeles' urging, he sent a column under General Toribio Ortega through Zertuche Canyon in the direction of Ramos Arizpe—where Pascual Orozco was rumored to have two thousand Colorados—to cut off any Federal retreat from Paredón. Meanwhile, the rest of the Division of the North headed toward Paredón in two columns, one mounted led by General Villa and the other consisting of artillery under the command of General Ángeles. The concentration of Federals at Paredón totaled an estimated five thousand men which made no sense as a blocking force for the 8,000 Federals in Saltillo against the Division of the North, because it was too small, or as an advance force, because it was too large. Hence, General Sánchez Lamego surmised that the Federals there were intended for a future campaign to try to retake northern Coahuila or Monterrey; they also could have been anticipating a U.S. invasion force. Regardless, an estimated total force of 13,000 Federals occupying the Paredón-Ramos Arizpe-Saltillo line seems wildly inflated, and the true number was probably was about half that size.

At any rate, once the objective came into sight on May 17 the Villistas deployed in a line of foragers approximately four kilometers wide and charged, scarcely giving Ángeles time to place his guns into battery, which only got off two shots. The assault also surprised the Federals, who may have trusted the destruction of the rails to the west too much, or maybe they had their eyes fixed to the east and reconquering Monterrey, or north to the Americans and the border. It was a massacre. The Villista cavalry quickly got under the Federal guns, which were elevated on a rise about two kilometers behind the infantry, and smashed their lines. The Federal cavalry fled the field abandoning both the infantry and artillery in the most cowardly fashion. Virtually all the Federal officers perished in the fighting or were executed afterward. Many of the Federal enlisted men, who had been told they would be fighting American invaders, decided to join Villa's Division of the North as infantry, which Villa desperately needed.

Those few Federals who managed to get past General Ortega's blocking force arrived in Saltillo just in time for the evacuation. General Joaquín Maass abandoned Saltillo for San Luis Potosí and the Division of the North began moving into the capital after briefly engaging Argumedo's Colorados on May 20. Villa telegraphed González to come take control of the city, and after briefly meeting with his counterpart, his division began returning to Torreón to prepare for the Zacatecas campaign. He had heard an entirely unfounded rumor that the First Chief intended for the Division of the North to continue toward San Luis Potosí, further distancing Villa from his lines of communication and denying him the glory of taking Zacatecas, the gateway to Mexico City. General Cesáreo Castro's 4th Brigade, which had already begun returning to the north from Tampico to participate in the attack on Saltillo before the battle at Paredón had taken place, garrisoned the capital of Coahuila for the Army Corps of the Northeast.

THE HEARTLAND

The only areas that remained to be conquered for the Constitutionalists on the way to Mexico City were the heartland and the West. General Álvaro Obregón's Army Corps of the Northwest would handle the West, but for the first task, the First Chief intended to fully develop the Constitutionalist Army Corps of the Center, composed of General Pánfilo Natera's 1st Division of the Center and the newly formed 2nd Division of the Center to be commanded by his brother Jesús Carranza and formed from units raised in the states of San Luis Potosí and Hidalgo and others carved out of González's Army Corps of the Northeast. The First Chief intended for Natera to conquer the plaza of Zacatecas and for his brother's 2nd Division to take San Luis Potosí, once it received enough arms and ammunition.

Zacatecas

To carry off the attack on Zacatecas, the First Chief put the troops from Durango State belonging to the Arrieta brothers at the disposal of Natera. When Villa learned that Natera had been given the mission that he so craved, Villa felt slighted. Moreover, using Torreón as a gauge, he did not think that the combined forces of Natera and Arrieta would be sufficient for the task. Nor would he be in a position to render aid in the event of their defeat, because he needed to reprovision, his men and horses needed to rest after the recent two months of campaigning, and the rails from Torreón to Zacatecas had not been repaired.

Nevertheless, on June 11, Natera and the Arrietas opened the Battle of Zacatecas, but they could not crack the Zacatecas nut, just as Villa had predicted. After two days of combat Natera reported that he needed another three thousand men. Accordingly, the First Chief requested that Villa send artillery and three thousand men under General Robles (significantly, his brigade had been recruited in Zacatecas) to Natera, but the next day he increased the request to five thousand men and some .30-30 cartridges. Villa gave Carranza a number of excuses why he could not comply, and when Colorado General Argumedo arrived with reinforcements for the Federals inside Zacatecas on June 14, Natera had to discontinue the battle. This act of insubordination on Villa's part—especially contemptible since some of the brigades in his "Division of the North" technically belonged to the Army Corps of the Center—would be the genesis of the later schism in the Constitutionalist Army.

The day after Natera broke contact with the Federals in Zacatecas, Villa's Division of the North began mobilizing. First, Contreras' brigade, then Urbina's and the artillery of Ángeles, and finally the rest of the division departed Torreón, arriving at Calera (twenty-five kilometers north of Zacatecas), which became the staging area for the upcoming battle. So long were the convoys of Villa's division that the rails from Calera were clogged with rolling stock almost as far back as Fresnillo. Upon arriving, the Constitutionalists began scouting the area. Soon Arrieta and Natera realized that the entire division was coming, not just a brigade or two. They knew about the troubles between Villa and the First Chief, but those issues did not concern them; destroying the Federals did.

Slowly and methodically the Constitutionalists began moving into position. The Federals occupied the many hills around the city, yet their position was anything but secure. First, they only had about two batteries of field guns—woefully few for such an epic battle—emplaced high atop El Grillo and La Bufa Hills, such that the Constitutionalists were under the guns almost from the outset of the battle. Second, the Federal commanders were so technically deficient that their positions were on the topographical instead of the military crest, which created an undesirable condition known as sky lining, and they used stones to provide cover, which was discouraged in field manuals precisely because they could chip under fire and create projectiles that could harm the defenders. Third, their artillery pieces were exposed, in the open and not hidden. Last, and most importantly, they did not possess enough troops for such an extended perimeter. General Sánchez Lamego estimated their total troops to start the battle at some 1,800 men, which seems unbelievably low, while Constitutionalist assessments placed the number of Federal defenders from ten or twelve thousand to a whopping 22,000. The true number based on empirical evidence such as Federal casualties and booty recovered, as well as statements from Federals present for the battle, suggests peak strength of some 10,000 men. The Constitutionalists had roughly twice as many men as the Federals and dozens of field and mountain guns, but they had to use the faulty ordnance manufactured in Chihuahua City.

Initially, Generals Urbina and Ángeles commanded the Constitutionalists, performing reconnaissance, placing troops, and contesting key terrain features beginning on June 19. In the evening of the next day the Federals received another relief column of about two thousand men and a battery of artillery from San Luis Potosí, although instead of making a flanking movement, as called for in General Eduardo Paz's strategy proposal, General Antonio Olea's column had headed first south and then returned back to the north by rail, probably to avoid Constitutionalist General Eulalio Gutiérrez's area of operations. Upon arriving, Olea suggested to the general in command, Luis Medina Barrón, that he should abandon the plaza for the more defensible Palmira Canyon, about twelve or fifteen kilometers southeast of the city, but Medina Barrón had orders to hold the city at all costs.

On June 22, two more columns of Federals approached from the south to reinforce the embattled Federals of Zacatecas, one of them commanded by Pascual Orozco. Neither of the two commanders advanced past Palmira (twenty-five kilometers to the south) because they deemed it too dangerous to attempt a break through. Also, on the twenty-second, General Villa finally arrived on the scene and with General Ángeles agreed upon the final details for the next day. Essentially, all the units of the Division of the North would attack the hills to the north, west, and south of the city simultaneously (the watchword being no "partial" combat that would permit the defenders to exploit their interior lines), converging on the plaza and forcing the Federals to retreat toward the neighboring town of Guadalupe to the southeast and into the awaiting arms of the Constitutionalists of the 1st Division of the Center. The Federals had already suffered tremendous casualties over the four days of combat—the last three with insufficient food and water—and many of them considered the battle already lost. It seemed to them that the Constitutionalists had only wait for Villa to arrive before delivering the death knell so that he could receive the glory for the victory.

The next morning, almost all the Constitutionalists began advancing on foot toward the city. At 10:00 a.m., Ángeles' hidden batteries opened fire and walked in shells ahead of the infantry, at last revealing their locations to the Federals who fired back in counterbattery. The Constitutionalists easily took position after position from the Federals who fired their weapons to the point of overheating in that target-rich environment. The Constitutionalists to the north overtook the most vulnerable point on the Federal line—the gently sloping heights of Santa Clara occupied by Argumedo's irregulars—as Natera's men captured the train station on the south side. The Federals began retreating into the heart of the city where they rushed from one point of danger to another, aimlessly, like so many sheep. Ultimately, the Federals tried to break out in the one direction left open to them, toward the southeast. Here, however, they only found death as General Arrieta and Natera's men from elevated positions on both sides of the road to Guadalupe opened fire on the fleeing Federals. It was like shooting fish in a barrel, and soon the road had been turned into a charnel house seven kilometers long, so full of dead men and horses that it was impossible to pass with a carriage, and even on foot one could not avoid stepping on cadavers to pass. Virtually the entire Federal command had been killed, wounded, dispersed, or captured, all except three hundred who reached the safety of Aguascalientes. One of Villa's staff put the Federal casualties at 6,000 killed, 1,500 wounded, and 4,000 captured, which seems quite possible since as of June 29 General Natera reported that 4,837 corpses, almost all of them ex-Federals, had been recovered and yet many more dead still covered the landscape. The Constitutionalists probably suffered about one thousand casualties, with the brigades of Generals Maclovio Herrera and Manuel Chao suffering the most due to the rashness of General Herrera in the first days of combat.

In the final analysis, the Battle of Zacatecas was just one more example of Villa's tactic of surround and pound against a foe that was ignorant in the art of war, demoralized, and outnumbered.

On June 26, Villa's division began making its way southward, with many finding quarters in Guadalupe. The very next day, however, Villa changed his mind and his men began returning to the north. Carranza had put a freeze on coal for the trains of the Division of the North and blocked delivery of his arms and ammunition through Tampico. Villa retaliated by impounding the offices of the First Chief in Ciudad Juárez and imprisoning its personnel, as the rift between the two continued to widen. Others have maintained that Villa also turned back because he had become worried about an attack on his lines of communications by González's Army Corps of the Northeast from Saltillo, although these concerns were grossly overstated. He needed more ammunition and other materiel, and under the circumstances further campaigning would have been unthinkable. The Arrietas, too, left Zacatacas to return to their base of operations in Durango. The Army Corps of the Northeast, Center, and Northwest would have to carry out the final destruction of the Federal Army.

Obregón in the West

After seeing the First Chief off to Chihuahua in March, Obregón returned from Nogales to Navojoa, where the Army Corps of the Northwest had been recruiting men, procuring horses, training, and waiting for ammunition to arrive from the United States. Toward the end of the month he sent Carranza his campaign plans. He would send General Manuel Diéguez with six thousand men to defeat the Federals around Tepic and seal off the Federals in Jalisco, freeing up General Ramon F. Iturbe's three thousand Sinaloans to attack Mazatlán backed up by a similar number of Sonorans and all the army corps' artillery. General Salvador Alvarado remained in Sonora keeping the Federal Division of the Yaqui bottled up in Guaymas and pacifying the rebel Yaquis while General Plutarco Elías Calles, a paper tiger of the first order, remained behind as the jefe de las armas of Hermosillo.

On April 14, 1914, General Obregón ordered the rest of his army corps, ten cannons, and ten machine-gun crews to entrain at Navojoa and join the rest of the operations units concentrating at Culiacán. En route Obregón stopped at Topolobampo where he ordered an aerial bombing of the Federal transport vessel *General Guerrero*. None of the bombs hit the mark, but it may have been the first aerial bombing in naval history.

At the very moment Obregón was extending his radius of action to the south, General Lucio Blanco and his cavalry had orders to leapfrog Diéguez's infantry on the way to Tepic and form the extreme vanguard of the army corps. Obregón, meanwhile, stopped in Culiacán for almost two weeks to attend to matters of administration. During this time he received word of the American invasion of Veracruz. Acting in a rather impulsive manner, Obregón recommended that the First Chief declare war on the United States and formally pronounce the Federal Navy outside the law, "pirates." The Federal Navy concerned Obregón for two reasons, one logistical the other operational. First, for a stretch of one hundred kilometers north of Mazatlán, the railroad ran very close to the coast where the Constitutionalists trying to move men and materiel to the south would be subject to its guns. Second, the Federal Navy could also land shore parties to destroy the rails and bridges, plant bombs and, conversely, evacuate Federal Army men and materiel.

On April 29, Obregón departed Culiacán for Mazatlán, where the Constitutionalists had been holding the Federals under a state of siege since October 1913. With the added forces that Obregón brought with him, the Constitutionalists tightened the circle and stepped up the attack. On May 5, Obregón ordered his artillery to begin targeting the ships of the Federal Navy that supported the defenders of the port. Just like at Guaymas, the Federal Navy increased the risk and complexity of the attack and attenuated the possibility of a great prize, since the men and materiel inside the plaza might be evacuated at the last moment. Also on May 5, Obregón learned that the garrison of Acaponeta, on the border with Tepic, had surrendered to Diéguez and Blanco either because those Federals thought the Constitutionalists had accepted their offer to join forces against the Americans, or out of an intense fear of the Constitutionalist Yaqui soldiers. Obregón had the prisoners sent to the penitentiary at Hermosillo.

The next objective in the general offensive was Tepic. Diéguez had specific instructions to attack the plaza from the north with infantry and artillery while Blanco's cavalry moved south to secure the bridge over the Santiago River and block any Federal reinforcements from Guadalajara and prevent the Federals inside Tepic from escaping. The commander inside the capital city, General Domingo Servín, was one of the most competent in the Federal Army. He had a mobile cavalry force for pursuit and counterattacks, successive lines of defense broken up into sectors, an infantry reserve to exploit interior lines, and a mobile scouting service that held open his intended route of egress or, alternatively, that could guide in reinforcements. The battle began on May 14, but after three days of combat and running low on ammunition, and overwhelmingly outnumbered seven thousand to his own two thousand, the Federals began to retreat. Blanco did not do as ordered, and the Federals escaped doing tremendous damage to the Santiago Bridge along the way.

At that point the door to Guadalajara was open and Obregón had no time to tarry. He could not justify the loss of men and expenditure of assets balanced against the possibility that the navy might evacuate the Federals and any potential booty from Mazatlán, so he "proposed to leave this port in the same conditions as that of Guaymas—besieged—and undertake the advance with the main body of the army corps to cut off their lines of communication running through Manzanillo and thus force the Federals defending it to abandon it." It was a perfect example of the indirect method and predated "island hopping" by the Americans in World War II by decades. Obregón had other reasons to press onward: the First Chief was beginning to suspect Villa and Ángeles of disloyalty at this early stage and urged Obregón to get to Mexico City as soon as possible. Therefore, leaving General Iturbe with three thousand men, five cannons, and three machine guns to contain Mazatlán, he departed with the rest of his troops on May 18.

Obregón arrived in Tepic the next day. While repairs to the Santiago Bridge continued, he attended to administrative matters, namely procuring the more than two hundred carts and two thousand mules he would need to carry his impedimenta from Ixtlán del Río, Tepic, the terminus of the Southern Pacific Railroad, to San Marcos, Jalisco, on the National. In the meantime, he sent Colonel Jesús Trujillo with three hundred dragoons to destroy the railroad between Quemado Station and the Colima state line, and also at Zacoalco inside Jalisco State, in order disrupt the Federals' supply lines from Mexico City to Manzanillo and ports on the West Coast. After completing his mission Colonel Trujillo was to rejoin the army corps, taking whatever route he judged most opportune. The raid also had the added benefit of drawing the attentions of the Federal Division of the West to the southwest and away from direction of approach of Obregón's Army Corps of the Northwest, which started departing Ixtlán del Río, Tepic, on June 10 with the battalions of General Diéguez in the vanguard. Obregón brought up the rear with the artillery, impedimenta, and reserve infantry, departing Tepic four days later.

Obregón did not reach Etzatlán (ten kilometers east of San Marcos) until June 24 because of frequent rainfall, which also hampered communications. General Diéguez was waiting for him with news that the Federals had dispatched a strong column from Guadalajara to seek out Obregón and defeat him on their own terms. Based on that information, Obregón ordered all his men to concentrate at Ahualulco (thirteen kilometers southeast of Etzatlán), where he scouted the terrain and then deployed his infantry in a giant parabola using the railroad as the vertex with a brigade of cavalry held in reserve. He dangled Colonel Trujillo—who had returned after successfully completing his raid—as bait in the extreme vanguard, hoping the Federals would attack his small regiment and then, giving pursuit, run headlong into the Constitutionalist infantry. General Blanco, with the main body of Constitutionalist cavalry, was located farther to the south around Ameca with orders to be prepared to cut off the retreat of any defeated Federals. Such a battle plan reflected his favored strategic offensive combined with a tactical defensive, a difficult maneuver that required an aggressive adversary.

Obregón and his troops waited in that disposition for several days until he learned that his plans had been foiled unwittingly by a fellow revolutionary. General Julián Medina, a nominal Constitutionalist, had refused Obregón's orders to join the main body of the army corps as it entered Jalisco and took it upon himself to destroy a bridge to the Federal rear. That caused the Federals to back up to Orendáin to reestablish their lines of communication, thus eliminating any possibility of an engagement around Ahualulco any time soon. Moreover, by now, July 1, Obregón had learned that Villa had pulled back to Torreón, which freed up Federals previously allocated to block the Division of the North to reinforce Guadalajara. Obregón did not have time to spare. Consequently, he created a new campaign plan.

General Diéguez with seven battalions of infantry would swing around to the north/left and wipe out the Federals at La Venta, east of Orendáin. When the Federals turned to reestablish their lines, they would be forced to engage Diéguez in a battle of reverse fronts. At that point, Obregón with Hills' infantry and regional cavalry from Tepic would attack Orendáin from the west. Caught between the two Constitutionalist forces, the Federals would be crushed and flee. Meanwhile, Blanco's cavalry division would take a wide berth around La Primavera Forest and appear to the southeast of Guadalajara between Hacienda del Castillo and La Capilla on July 6, to confuse the Federals remaining in Guadalajara and cut off any escape they might attempt. That plan, with some minor hitches in timing, was exactly what happened. The bulk of the Federal Division of the West was destroyed at Orendáin on July 7, and the next day Obregón entered Guadalajara after what remained of the Federal division had evacuated it and headed for El Castillo (twenty kilometers south-southwest of Guadalajara) on a collision course with General Blanco's cavalry.

El Castillo

The Federals pushed Blanco's Constitutionalists out of the *casco* of Hacienda del Castillo where they established their headquarters and tended to their wounded. Around 10 a.m., the Constitutionalists returned and the 3rd Brigade, composed mainly of new recruits from Jalisco, dismounted and attacked the main compound in three waves. The Federals waited until the Constitutionalists were about one hundred meters away and then employing a familiar stratagem started to wave a white flag. When the Constitutionalists let down their guard, the Federals loosed a volley of deadly machine-gun and rifle fire accompanied by grapeshot. "The brigade reeled like a drunken man under the impact, turned, and ran in disorder for any place of shelter," Colonel Thord-Gray remembered. The Federals then counterattacked but did not get far because the 1st Brigade, composed of battle-hardened Yaquis, covered the retreat of the 3rd and soon the Federals blew recall. At noon, the 1st Brigade delivered a new assault with the 1st Regiment on the right, the 2nd on the left, and the 3rd in reserve. The force of the push and the freakish death of the Federal commander, General José María Mier—killed almost at the outset of the attack by a random bullet to the heart while he was surrounded by others and engaged in conversation in a hallway of the hacienda—precipitated a rout as the Federals broke and ran in disorder.

General Obregón's campaign of maneuver to capture Guadalajara surely seemed bold and imaginative, but one must remember that he had an entire army corps at his disposal against General Mier's division of despondent, untrained, and poorly-led Federals.

A week after El Castillo, Interim President Victoriano Huerta resigned from office, turned over the government to the lawyer Francisco S. Carbajal so that he might negotiate terms of surrender, and went into exile with his former Secretary of War and Navy, General of Army Aurelio Blanquet. The new Secretary of Defense, General of Army Corps José Refugio Velasco, ordered the remaining Federals in Aguascalientes, San Luis Potosí, and Querétaro to evacuate those states and begin concentrating around Mexico City for one last stand. Accordingly, the Federals began taking up railroad track between Lagos and Encarnación de Díaz and on July 17 began evacuating the capital San Luis Potosí. That same day, the Division of the Yaqui started boarding ships and setting sail from Guaymas for the port of Manzanillo with the intention of linking up with Federal forces in Colima State, forcing its way past Obregón's army corps, and then entering Mexico City to participate in the final stand.

The sudden appearance of the Division of the Yaqui in Manzanillo caught the attention of Obregón who sent a blocking force of some two thousand cavalry and infantry that destroyed the remaining Federals in Colima and then invested Manzanillo. Manzanillo could not sustain such a large force of Federals and eventually the Division of the Yaqui again boarded its ships and set sail for Salina Cruz where its soldiers were mustered out of service after the final defeat of the Federal Army.

Obregón returned to the business of repairing the rails into the heartland and sent a small cavalry force of about 1,400 men to scout ahead. The reconnaissance force destroyed a small Federal division at Hacienda Temascatío on July 31 in what was the last significant engagement against what remained of Huerta's once-proud Federal Army. The Constitutionalists continued to filter south with the Army Corps of the Northeast passing through the Bajío (the breadbasket region just to the north of Mexico City) to points around the nation's capital accompanied by the 2nd Division of the Center, which also had elements passing through the Huasteca. Farther to the east, Aguilar's 1st Division of the East headed for Jalapa, the capital of Veracruz.

The Army Corps of the Northwest and the Army Corps of the Northeast finally linked up at Querétaro on August 1, 1914, according to the First Chief's long-standing strategy. A week later the Federal Army and Navy signed the Second Act of the Treaties of Teoloyucan and ceased to exist—the factors against continuing to resist far outweighed any chances of victory, or even a negotiated surrender.

5. CIVIL WAR, SEPTEMBER 1914 – MAY 1915

The 1914-1915, civil war, phase of the Mexican Revolution was the most complex strategically and operationally. Gone was the Federal Army, which had evidenced no ability to maneuver and had largely limited its participation in general actions to the defense of plazas in the North. Only remaining now were the revolutionary armies, fully mobilized at the outset of hostilities and prepared to contest control of the entirety of the national territory, from Baja California to Yucatán. The Constitutionalist and the soon-to-be-formed Conventionist armies were also large and mobile, with the result that ensuing engagements resembled more of what was being seen across the ocean on the battlefields of Europe: epic in scale, long in duration, and often fought over strategic rural terrain. Moreover, between September 8, 1914, and October 19, 1915, the United States did not impose an arms embargo on any of the factions. The U.S. government was content to let Mexico descend into fratricidal war until one victor with a clear ability and mandate to control the vast majority of countryside had emerged. The Wilson administration based this policy decision on the North's experience in the U.S. Civil War when it had argued against European recognition of the Confederacy based on the percentage of territorial control.

In the case of Mexico, the civil war would begin and end in the Northwest.

Back in June 1914, Governor José María Maytorena of Sonora had competed with the jefe de las armas of Hermosillo, Colonel Plutarco Elías Calles, for power and control of the military. In order to avoid an escalation of events, Calles had taken his battalion out of the state capital to Naco on the northern border.

Two months later, in August, Maytorena granted amnesty to all the ex-Federals in the Hermosillo penitentiary and promised to recognize their previous ranks if they agreed to serve in his militia. At the same time, his supporters arrested General Salvador Alvarado, the Constitutionalist general who had been in charge of the siege of Guaymas, and pronounced themselves in rebellion against the First Chief. Villa and Obregón tried, unsuccessfully, to negotiate a truce that would bring Maytorena back into the Constitutionalist fold.

Then on September 23, Villa definitively split with the First Chief and Maytorena ordered his men in Nogales to attack Calles. A new war was on the horizon as Carranza commanded Eulalio Gutiérrez and Pánfilo Natera to cut the Central Railroad between Torreón and Zacatecas. Gutiérrez in return suggested that Natera should take up position in towns in northern Zacatecas and the Army Corps of the Northeast should place blocking forces in Cuatro Ciénagas, Reata, and Parras, Coahuila, to contain Villa. But it was Villa's men and his allies who drew first blood.

The Maytorenistas destroyed Calles' advance forces at Martínez Station on September 25 and pressed forward to Naco, Calles' headquarters, and surrounded the town. The following day, General Calixto Contreras held a meeting with the Arrieta brothers in Durango while his men started to encircle the city, forcing the brothers and those who remained loyal to Constitutionalism to flee the state capital for the western part of the state. Next, Villa's men pushed the Herrera brothers out of southern Chihuahua and into northwestern Durango where the Herreras proceeded to get in contact with the Arrietas in order to lend mutual support.

OCTOBER

As Villa and Maytorena tried to remove any opposition in their own backyard, on October 1, the Constitutionalist generals met in Mexico City to form a new government. With the Villistas in rebellion, however, everyone knew that there was no way to consolidate power and so the delegates agreed to invite all wayward Constitutionalists to participate in the convention and move it to Aguascalientes, neutral ground.

The invitation was accepted and the date of October 10 selected for the opening of the Convention of Aguascalientes. Governor Maytorena, however, did not wait to see what resolutions might come from it and ordered his 3,000 Yaquis to attack General Benjamín Hill and Colonel Calles in Naco beginning on October 4.

Strategy and Tactics of the Mexican Revolution

An observer described the Yaqui tactics as something akin to what might be seen in an old western movie: "the Yaquis twisted like snakes, dragging themselves through the undulating plains, behind the scrubland..." Suddenly the air around plaza exploded with fire as the cannons sounded and the Yaquis opened fire and the Constitutionalists responded with bugle calls, small arms fire, battle cries, and the occasional swear word.

Ex-Federals recruited by Felipe Ángeles and brought in from Baja California operated the artillery in support of the Yaquis, but the Maytorenista guns did not perform well using shells manufactured in the shops of Cananea: the powder charge was deficient resulting in muzzle velocity at half of what was required, and the increased weight of the projectiles caused irreparable fatigue in the cannons. Moreover, the gunners proved untalented.

Calles had no artillery and his men could only return small arms fire sparingly because armed mainly with .30-30s, they feared their carbines might overheat. But at least in those early days of October the moonlight exposed the Yaqui attackers and after twenty hours of combat the Constitutionalists succeeded in repulsing the first attempt on Naco.

The Constitutionalist defenses consisted of crenulated and (mostly) covered trenches that were constructed with sidewalks and gutters inside, interlocking fields of fire, loopholes, communication trenches to the interior of the plaza (headquarters, hospital, ammunition stores), machine guns intermittently placed around the perimeter, forts near the international border line to anchor the left and right, wire to keep attacking infantry at a distance, electrical and contact mines, and spotlights. Later, the Constitutionalists were able to import a few cannons. Telephone lines linked the units on the firing line to Calles' headquarters, which in turn could communicate with Hill's command post, and beyond with the Douglas and Bisbee switchboards. Pipes carried water from the railway tank to the firing line.

After the initial attack failed, Governor Maytorena sent his wounded back to the hospitals in Cananea, Nogales, and Hermosillo, smuggled ammunition from the United States through Osborn Station, Arizona (seven kilometers east of Naco where there was no official port of entry), brought in more reinforcements, including 425 men who sailed from Santa Rosalía, Baja California, and reorganized.

Around this same time the Villistas began moving into southern Coahuila, between San Pedro de las Colonias and Hipólito Station. In response, the First Chief could only continue pleading with his generals to cut the rails in southern Coahuila and northern Zacatecas in order to isolate the Villistas in Chihuahua and Durango states and order that Calles be supplied with arms and ammunition.

The Constitutionalists also finished assimilating the assets acquired from the Federal Army arsenals, most especially those in the nation's capital but also those of the former Division of the Yaqui, which had surrendered in Salina Cruz, Oaxaca.

On the eve of the opening of the Convention of Aguascalientes, the Arrietas asked the First Chief to send ten to fifteen thousand men to invade Durango if the Villistas did not pull out of the state. They also demanded the return of private property and asked that the Convention respect Mariano and Domingo's appointments to the posts of jefe de las armas of Durango and commander of the Division of Durango, respectively. Carranza could not mobilize so many men at this point and instead urged the Arrietas to go to Sinaloa and then invade southern Sonora and catch Maytorena's troops in a pincer with General Hill on the northern border at Naco, since it would be easier to supply them on the coast. The 6th Battalion of Sinaloa had already departed Mazatlán on October 6 for El Fuerte to join the "Guerrilla de Alamos" stationed there. The 3rd Battalion had left a day earlier for San Blas, and the 2nd was billeted in Morcorito. The Expeditionary Column of Sinaloa designated to invade southern Sonora would consist of soldiers from these units. The loyalty of the Constitutionalist battalions in northern Sinaloa, however, was not assured and the Arrietas decided to remain in western Durango and cooperate with the Herrera brothers to contain Villa.

The First Chief also wanted General Pablo González to concentrate his divisions in Puebla to be ready to move, but his men were busy battling the Zapatistas, especially to the northwest and southwest of the state capital, and González could not leave it to the regional Constitutionalists to do the job because they were too poorly organized.

The next assault on Naco came on October 10, the date the Convention of Aguascalientes opened. In that fight, Maytorena's Yaquis entered the United States after nightfall and using railroad cars for cover fired on the Constitutionalists. Only with great difficulty did General Hill rally his men and turn back the sneak attack to his rear. Since border security as practiced by the American soldiers had been woefully deficient, the incident was a supreme embarrassment for the colonel in charge. After that, Hill had to assign men to the northern side of the plaza, string wire, build redoubts, and use searchlights to illuminate the borderline.

As Maytorena continued his offensive, the Convention tried to broker a truce among all factions and Pablo González, for one, appeared to heed it, ordering his soldiers to stop battling the Zapatistas, who had been invited to send delegates to the assembly. Maytorena did not respect the cease-fire. On October 16, his Yaquis slowly crawled into position using the cover of darkness and cleared the wires, detected all the while by the Constitutionalists, who suddenly lit up the field with searchlights and opened fire at point-blank range. Constitutionalist officers detonated mines to the front of their positions, blowing Yaquis to pieces. It was a disaster for the Maytorenistas. Those unfortunate Yaquis still inside the wire at daybreak could not decide whether to move forward or backward until a Constitutionalist counterattack sent them fleeing.

Over the next five or so days the Maytorenistas continued to pour in harassing fire and then abruptly pulled back to Villa Verde and San Pedro. Calles was incorrect in thinking the battle was over, but that withdrawal did end the first phase of combat.

The Convention continued trying to pause the fighting in the Northwest and also warned the various factions against maneuvering troops to gain a strategic advantage, but to no avail. In that third week of October, General Villa ordered the stretch of the Central Railroad between Aguascalientes and Zacatecas handed over to his Division of the North, and the Maytorenistas began sealifting ex-Federal men and materiel from Baja California via Guaymas to Naco. To hinder the arrival of Maytorena's ex-Federals, Hill and Calles planned an expedition to destroy the rails between Naco and Nogales and north of Hermosillo.

The First Chief then admonished the leadership of the Convention, whose sovereignty he had never acknowledged, not to give orders regarding military matters to his subordinates and reminded his supporters to ignore any such directives coming from Aguascalientes. The First Chief also ordered Maclovio Herrera to take his brigade to Mazatlán, apparently still wanting to carry off the invasion of Sonora from Sinaloa or perhaps to quell the rebellion brewing in the latter state. However, it took at least a week for communications from Mazatlán to reach Herrera, who was deep in Durango on the other side of the sierra, and in the meantime the young general attacked Parral. The result was devastating for Herrera's brigade, which lost all the guns of his field battery and about two-thirds of his effectives—mostly deserters who rushed to rejoin the Division of the North.

NOVEMBER

The Bajío

On November 1 the Convention elected General Eulalio Gutiérrez to the post of interim president and gave the First Chief an ultimatum: appear before the assembly by November 10 or be declared in rebellion. Carranza had the good sense to ignore the order to appear, and as soon as the deadline passed Villa's men began moving from Aguascalientes, which they had occupied in violation of the city's declared neutrality, toward Guanajuato. General Alberto Carrera Torres, whose division garrisoned Guanajuato, had recently declared that he would side with the Convention against the First Chief and Constitutionalism. As Villa's brigades moved down through the Bajío on their way to Mexico City, General Teodoro Elizondo's 3rd

Division of the Northeast was overwhelmed and forced to capitulate or join the Conventionist Army; Elizondo chose the latter. None of this loss of manpower particularly bothered General Jacinto B. Treviño, the First Chief's former chief of staff, who later claimed that he positioned Carrera Torres and Elizondo's divisions in the first line of Constitutionalists facing Villa's Division of the North precisely because those were the worst-performing in the Constitutionalist Army.

As the Villistas continued to advance, General González ordered his sappers to destroy the rails between San Juan del Río and the city of Querétaro, blow up bridges, and with the remainder of his troops he pulled back toward Tula, Hidalgo. González also asked Constitutionalist Secretary of War Ignacio L. Pesqueira to have the Arrietas and Herreras in Mazatlán begin harassing Villa's supply lines in Durango to slow the Division of the North's march through the Bajío. Pesqueira, in turn, requested that the First Chief order General Iturbe to invade Sonora from Sinaloa and open a second front in that state and relieve the pressure on General Hill.

Northwest

The First Chief had been trying to get Herrera to invade Sonora for weeks. General Herrera, however, flatly refused to accept the Durango or the Sonora mission, saying that Iturbe had enough men to do the job without his brigade, and he preferred to attack Chihuahua, entering from the east and linking up with his brother, Luis, who would invade the state from the northwest. Shortly thereafter, Maclovio Herrera left for Tampico—on the other side of the country—to begin recruiting a force east of Chihuahua while his brother brought their brigade over the sierra to Culiacán, Sinaloa. Meanwhile, Iturbe's subordinate, General Ángel Flores, made several attempts to kick off the invasion of Sonora but his "Expeditionary Column of Sinaloa" always seemed to encounter some setback. Conversely, Juan Cabral, Ramón V. Sosa, and Jesús Trujillo had declared their support for the Convention and had already returned to the North to invade their native Sonora from Chihuahua.

In Naco, Hill and Calles continued to receive arms and ammunition such that by November the number of their machine guns had grown from two to fourteen. The Maytorenistas returned to the environs of Naco on November 8 with almost two batteries of artillery but did not attack in force until November 18. This time Maytorena's ex-Federals took over the brunt of the fighting from the Yaquis, which became apparent not just because of the predominance of artillery in the assaults but also because they attacked in the unmistakable echeloned formations dictated by the ternary system and in the daylight: the Federals could not fight in the dark like the Yaquis because the enlisted men might desert. When the frontal assault failed, the ex-Federals tried flanking movements and then forced their reserves into the fray, pushing them forward *a caballazos*. But the ex-Federal cannon fodder in the support and reserve refused to reinforce the chain of shooters and at length, and after many casualties, the assault failed.

The next day, the Yaquis returned to the fight, eschewing flanking attempts so often employed previously in favor of alcohol-fueled direct assaults "carried out with indescribable tenacity and boldness, furious, well-sustained...and superb," in the words of one observer. Indeed, so vicious was the fighting that the Constitutionalist morale started to flag, and a desperate General Hill requested money from the First Chief to begin recruiting Colorados, who were active along the border and promised to raise ten thousand men. Colorado General Marcelo Caraveo, in particular, promised to lead a column of two thousand men to capture Ciudad Juárez if the Constitutionalists could give him twenty-five to thirty thousand dollars and supply him with arms and ammunition. Nothing came of these efforts, but within the year several prominent Colorado officers had been commissioned into Calles' 4th Division of the Northwest.

Strategy

At the time, the true danger to Constitutionalism was not in the Northwest but rather the Villista juggernaut pushing toward Mexico City.

The first of Venustiano Carranza's generals to recommend a strategy for winning the war was General Jacinto B. Treviño. Under the terms of his proposal dated November 3, 1914, the Constitutionalists would position: a 40,000-strong army of operations along the Lagos-León-Irapuato-Querétaro axis of operations that would seek a definitive pitched battle north of Lagos while avoiding the defense of plazas; 5,000 to concentrate around San Luis Potosí; 16,000 along the Toluca-Mexico-Puebla line to face the Zapatistas; and 10,000 for garrison duties. Finally, the Division of the Bravo would attack Torreón from the Saltillo-Monterrey area. The main flaw in Treviño's plan was that the Constitutionalists only had an estimated 60,000 troops, which left no room for the troops allocated to the Division of the Bravo or the garrison troops. Nevertheless, this strategy, which had much in common with what the Federal Army's general staff had recommended for defeating Villa and the Constitutionalists earlier in the year, was essentially the one that would play out over the course of 1915.

Map 5 Constitutionalist Strategy and Movements, November – December, 1914

General Álvaro Obregón also floated his strategy idea. He proposed destroying short sections of track in the Bajío that would impede the progress of Villa for about forty days but still allow the Constitutionalists to stop retreating and, with quick repairs to the rails, go on the offensive. González would remain in Mexico City and, after being reinforced with four to five thousand men, defend that city until forced to abandon it at the last moment. Ángeles, Villa, and Zapata would move in to occupy the capital, and Obregón predicted a falling out between those personalities. Meanwhile, Obregón would move 8,000 infantry, 80 cannons, and 900 mules around to Salina Cruz, Oaxaca, set sail from that port for Manzanillo, and then move inland to surprise Villa's rearguard as General Antonio I. Villarreal simultaneously attacked Torreón. Iturbe would march from Sinaloa into Sonora and, pressured by General Hill in the north, Governor Maytorena would be forced to go into exile in Tucson, Arizona, just as he had done in 1913. General Manuel M. Diéguez was already concentrating his division at La Barca on the Jalisco-Michoacán state line and would march on Irapuato as soon as the loyalty of General Joaquín Amaro in Michoacán could be determined. That determination became impossible because Amaro deferred to his commanding general, Gertrudis Sánchez, feigned ignorance about the current crisis, and professed his resolve to defend the integrity of Michoacán State.

No overarching strategy would be reached by the Constitutionalists until weeks later because Villa was barreling toward Mexico City, the loyalty of many Constitutionalist generals had not been ascertained, and Wilson's Americans were still in the port of Veracruz. Only on November 13 did Washington reach a definitive decision to evacuate the city on November 23, but they had announced departure dates previously only to recant later. A certain degree of stasis was needed before key decisions could be made.

As it happened, the Americans did evacuate the port of Veracruz on November 23, and Constitutionalist General Cándido Aguilar's soldiers moved in to occupy it. Those Constitutionalists in and around Mexico City who decided to side with the First Chief loaded as much materiel from the city's warehouses as possible on trains and fled for Puebla, Tlaxcala, and Veracruz states, all except for General Murguía. Murguía was in Toluca, on the west side of the capital, and Obregón feared that General Lucio Blanco, loyal to the Convention and in charge of a significant cavalry force in the Federal District, would try to cut off his retreat to the east so he recommended that Murguía take his division west—which also conveniently played into Obregón's recommended strategy. Eventually, Murguía linked up with General Diéguez in Jalisco state after battling Amaro in Michoacán and losing half his men and all his artillery along the way.

To draw attention away from the Constitutionalists in Puebla and Tlaxcala, General Jesús Carranza ordered Pascual Morales y Molina to move against Morelos State from his base in Chilpancingo, Guerrero, as soon as he had enough men, an unlikely event since Morales y Molina was no match for the Zapatistas opposing him.

On November 24, the Zapatistas moved into Mexico City on the heels of the last of Obregón's departing soldiers. General Lucio Blanco decided to remain in the nation's capital and serve the Convention. He had had a major disagreement with Obregón going back to the Western campaign over the proper use of cavalry. Blanco came from the Northeast where mounted units prevailed as the dominant arm, but Obregón believed that "once its principal function, which was to impose itself via the clash of arms with the white arm, or rather saber, had been abolished [the cavalry was] reduced to the function of reconnaissance and covering retreats, which not at all justified the maintenance of an arm so costly and difficult to sustain." General Treviño, who uttered those words, believed that Obregón had been in the right, but the horse tradition in the Army Corps of the Northeast and Villa's Army Corps of the North held sway in those organizations.

The defection of so many former Constitutionalist friends and allies to the Conventionist Army had a disorienting effect on key Constitutionalists, who struggled to discern their next moves. At the end of November, General Obregón visited the old Castle of Perote, west of Jalapa on the narrow gauge Interoceanic Railroad. The fort had been built to oppose any foreign invaders marching inland from Veracruz to Mexico City, but it could just as easily perform the same function against Conventionists in Mexico City trying to attack the port of Veracruz. Returning to Veracruz, Obregón received permission from the First Chief to scout the Isthmus "in the remote case that we might have to withdraw to that region," although more likely he still intended to carry out his harebrained scheme to attack Villa's rearguard by coming up through Manzanillo.

In Pachuca, General Pablo González, too, was trying to determine what course of action he should take. He narrowed his options down to three: move south to join the rest of the Constitutionalists in Puebla and Veracruz, and possibly sail to Tampico; try to reach Murguía and head west; or march north through the Huasteca to Tampico. He chose the latter option because his jefes from the Huasteca told him that the roads through the Huasteca were excellent, and he could make the journey with all his carts, automobiles, and impedimenta in ten days—none of which proved true—and he would be able to reinforce Tampico and rebuild his Army Corps of the Northeast around Villarreal's Division of the Bravo and Caballero's 5th Division. Tampico and the Huasteca held tremendous strategic significance as: "a base of future operations against the other ports of the Gulf littoral, especially Veracruz, [and] the source of inexhaustible resources of money that the dominion of Tampico and the petroleum zone would produce for" whomever held it, according to Carranza's Chief of Staff, Juan Barragán. The First Chief approved González's plans, but it must have been at the last minute because his forces evacuated Pachuca on November 29, while under attack by the Villista vanguard. The First Chief also returned César López de Lara's brigade to General Luis Caballero for the defense of Tampico. López de Lara had been in Puerto México attached to Jesús Carranza's 2nd Division of the Center.

DECEMBER

In the first days of December, General Villa finally entered Mexico City. General Ángeles had tried to impress upon him beforehand that Emiliano Zapata needed to place his Surianos under the orders of a unified Conventionist Army command, but Zapata refused. He thought that his Liberating Army of the South should act independently from Villa's Army Corps of the North; he only needed arms and ammunition, which Villa promised to provide. Without the Zapatistas to fill out his army, Villa started recruiting ex-Federals. And he would have to craft a strategy that did not count on Zapata's men.

Map 6 Conventionist Strategy and Movements, September – December, 1914

Villa consulted with Ángeles, who urged the immediate annihilation of Obregón since he was the "hat rack" holding up all the other Constitutionalist "hats," and Villa needed to crush him before he could get organized. But Villa considered Puebla and Veracruz to be Zapata's area of operations and did not want to offend his Suriano partner. Besides, he had Diéguez in Jalisco and Villarreal in Coahuila threatening his long lines of communication and blocking access to coal for his trains. He also needed to try to conquer Tampico before the Constitutionalists solidified their hold on that port. Therefore, he decided that General Ángeles should return to Torreón and attack Villarreal's Division of the Bravo. Supporting Ángeles in the Northeast would be Urbina and Chao departing Aguascalientes and moving east toward Tampico, although the conditions of the railroad to San Luis Potosí significantly delayed their progress. In the meantime, the Convention assigned the objectives of Ciudad Victoria and Tampico to Generals Alberto Carrera Torres and Magdaleno Cedillo, respectively. Villa would personally attend to Diéguez in Jalisco and he would leave Obregón to Zapata and the Constitutionalists in the Northwest to Maytorena and the youthful Rafael Buelna. Villa's Conventionists intended to push on all fronts from the center of the country outward, driving the Constitutionalists into the sea.

Essentially, Villa's strategy for the upcoming war would be a rehash of the Constitutionalist strategy of 1913, with commanders assigned various theaters and tasked with subduing the enemy within their area of operations. It was a sum-of-the-parts rather than a holistic strategy. But such a plan could never work in the milieu of 1915, now that military operations engulfed the entirety of the nation's territory, from Baja California to Yucatán, with competing priorities, limited resources, and an interconnectedness unprecedented for Mexico, in which results in one theater would profoundly impact others. In truth, after November 10, the Convention functioned politically and militarily more like a confederacy of regional interests than as a sovereign body.

West

Villa wasted no time in loading his men on trains, on December 12, and driving on Guadalajara. Without putting up a fight, General Diéguez abandoned the state capital and moved the Constitutionalist seat of government to Ciudad Guzmán. In so doing, Diéguez reasoned that the numerous armed and marauding bands of "the reaction" would concentrate in Guadalajara where they would begin squabbling among themselves (which sounded similar to what Obregón believed would happen among the Conventionists inside Mexico City) and be susceptible to one decisive blow. Whether or not that made sense, it provided political cover for evacuating the second city of the republic.

East-Northeast

To isolate and capture Ciudad Victoria, Conventionist General Carrera Torres sent a column through Galeana to destroy the railway between Carrizos and Tinaja to the north of Ciudad Victoria. At the same time, he instructed General Pedro Ruiz Molina to send a small force along the Miquihuana-Jaumave axis to clear the Constitutionalist rearguard from Santiaguillo (halfway between Miquihuana and Jaumave), and to detach a second contingent from Ocampo and Xicoténcatl to destroy the railway between Tampico and Ciudad Victoria to halt any Constitutionalists pushing to the north and west from Tampico. The latter column would also have the added benefit of protecting Conventionist General Magdaleno Cedillo's flank as he marched on Hacienda el Ébano. Constitutionalist General Manuel Lárraga had been defending the road to the west of Tampico against Cedillo for months, being forced by Cedillo's superior numbers to fall back to El Ébano, while Constitutionalist General Luis Caballero protected the Tampico and Ciudad Victoria line from any attack by Carrera Torres.

On December 19, Caballero's Constitutionalists foiled Carrera Torres's plans when they forced his Conventionists to give up Jaumave and retreat back toward Tula. That setback made Cedillo's left vulnerable to a Constitutionalist push through Xicoténcatl. The next day, Pablo González reached Tampico bringing with him 2,750 men, some of whom reinforced El Ébano. When Cedillo attacked El Ébano on December 23, he was repulsed thanks in large part to the men who had accompanied General González through the Huasteca. It would appear that Generals César López de Lara and Manuel Lárraga were planning to abandon that critical chokepoint when General Andrés Saucedo, separating from González's main column, arrived with reinforcements.

Cedillo blamed his defeat on a lack of arms and ammunition, artillery, and men. The Conventionist Secretary of War assured him that reinforcements were on the way and advised him that in the interim he should assume a defensive posture. On Christmas Day, Eduardo Carrera passed through San Luis Potosí on his way to reinforce Tula and breathe new life into Carrera Torres's campaign to take Ciudad Victoria.

Carrera Torres's far left was supported by General Ángeles, who departed Mexico City a few days after Villa, on December 17, and headed for Torreón. From there his troops started advancing on Saltillo along the Coahuila and Pacific Railroad, which was such a ramshackle railway that even in good times people referred to it as the "Coahuila and Patience." Upon reaching Balboa Station, Ángeles determined that his trains could advance no farther because the track had been so thoroughly destroyed, so he sent his cavalry division commanded by General Emilio Madero ahead on horseback while he returned to Torreón with his infantry division. Madero defeated a small brigade-size force at Parras on December 26, and Ángeles advanced along the northern rail line as far as Marte Station where conditions of the rails again forced him to stop. He rang in the New Year there at Marte. Only thirty-five kilometers, including fifteen kilometers of destroyed track, separated him from the lead forces of Villarreal's Division of the Bravo, now under the operational command of Maclovio Herrera, who had given up his dream of invading Chihuahua on his own. Herrera's colleague, Constitutionalist General Domingo Arrieta, was only waiting for the Division of the Bravo to attack Ángeles to begin moving from western Durango to assist in that operation.

Northwest

In Naco false rumors that the Sinaloans had captured Guaymas reached the ears of General Calles, who speculated that his position would fall in ten days under the daily and promiscuous shelling by the Maytorenista cannons. He inquired about the estimated time of arrival of the Expeditionary Column of Sinaloa, which, unbeknownst to him, was still stuck in northern Sinaloa at the time. Almost as bad, Calles had finally received a section of Bethlehem guns from the United States, but an ignorant German filibuster ruined one with the first shot and damaged the aiming device of the other.

Conditions around Naco were also insufferable for Maytorena's Yaquis, who had started dying from exposure in the cold December weather. Dead Yaquis swept away by the water from torrential rains floated into the Constitutionalist lines. The ex-Federals ran out of French-made artillery shells and Cananea needed petrol to sustain the mining industry, a key source of government funding. In dire straits, the Maytorenistas started to resort to tricks and stratagems. Yaquis made pathetic cries to entice sympathetic Constitutionalists to come to their aid so that they might kill them. The Maytorenistas also launched a máquina loca at the plaza, without success, and tried to stampede cattle toward the Constitutionalist lines, but these were turned back by glaring searchlights and American troopers. All the while the Maytorenistas were tunneling underneath the defenses, but according to one estimate, at the rate they were progressing the tunnel would have taken over two years to complete. The Maytorenistas then attempted to suborn the Constitutionalist commanders and offered amnesty to Constitutionalist soldiers. Finally, they tried to fire on Naco, Arizona, in an attempt to get the Americans to close the port of entry or possibly to provoke an armed intervention, all to no avail. In spite of Calles' pessimism, a stalemate had been reached.

Center

In Puebla and Tlaxcala the Zapatistas had been quick to mobilize against the Constitutionalists of General Salvador Alvarado, whom the First Chief had appointed commander of the Veracruz-Puebla line after his release from jail in Sonora (a conciliatory gesture by Governor Maytorena while the Convention had been in session). Zapatista Generals Marcelino Rodríguez, Agustín Cortés, and Aurelio Bonilla, coming up from the southwest, and Zapatista Generals José Sabino Díaz, F. Muñoz Santarriaga, and Colonel José Hernández, among others, attacking from the northwest formed a pincer against the plaza of Puebla. Obregón did not return to Veracruz from his scouting trip to the Isthmus of Tehuantepec until December 11, even though for days the Constitutionalists had been fighting in the vicinity of San Martín Texmelucan to the northwest, to the south at Atlixco, which they had been compelled to abandon on December 8, and even around the plaza of Puebla.

On December 12, the Zapatistas pushed those Constitutionalists around San Martin Texmelucan all the way back to Puebla after a day of hard fighting that began at 4 a.m. at Hacienda San Antonio de Chiautla. In those early morning hours, General F. Muñoz Santarriaga's Zapatistas had deployed along several lines and advanced against the Constitutionalists positioned on the rooftops of the hacienda and in the trees. At about seven hundred meters range the defenders opened fire on the Zapatistas, who outflanked the Constitutionalists and forced them to break and run. Zapatista General José Sabino Díaz and Colonel José Hernández did not arrive soon enough with their cavalry to cut off the retreating Constitutionalists, who made it to San Martín Texmelucan. After a couple of hours spent cleaning up the battlefield, the Zapatistas attacked that latter plaza, dislodging the Constitutionalists and forcing them to concentrate in Puebla City.

The loss of San Martín Texmelucan rattled General Alvarado's nerves. He stated, quite hysterically, that Zapata had five thousand men advancing along the San Martín line and another six thousand coming up from the southwest around Atlixco, both gross exaggerations. More cogently, Alvarado suspected that news of Villa's campaign in Jalisco had been a ruse and that his twenty thousand Norteños and sixty cannons were probably headed his way. He pleaded desperately to be reinforced by Obregón's Yaquis currently in Tlaxcala (Santa Ana Chiautempan, Apizaco, and Huamantla) since the two divisions under his immediate command belonged to Generals Cesáreo Castro and Francisco Coss, Northeasterner dragoons. He needed infantry since mounted troops were not appropriate for withstanding a siege.

None of what Alvarado had to say made any sense. Just the day before, he had stated that General Pablo González's march on Tampico had drawn Villa back north and now was the time to go on the offensive; today he was crying that he was about to be overrun. The First Chief called Obregón, who had just been appointed "Chief of the Army of Operations to march on the capital of the Republic," to the telegraph station to try to talk some sense into Alvarado, but to no avail. Obregón would leave Veracruz the next day for Puebla but he would not arrive in time. On December 15, a frantic Alvarado ordered Puebla evacuated, essentially handing it over to the Zapatistas. Obregón's first task as chief of the Army of Operations would be to take it back.

JANUARY

Puebla, December 29-January 5

The new year got off to a quick start in Puebla. The commander of the Constitutionalist Army of Operations, General Álvaro Obregón, developed a plan to retake the state capital with a three-pronged attack. Taking the infantry and artillery, he rode the rails northward to Apizaco, Tlaxcala, which was garrisoned by Constitutionalist General Gabriel Gavira, and then turned southward to attack Puebla from the north. Cavalry commanded by General Cesáreo Castro marched directly against Puebla from the east through Tepeaca and Amozoc. Finally, General Salvador Alvarado took two thousand dragoons and moved around the south of Puebla, cut the Zapatista lines of communication just to the north of Atlixco, and then appeared on the west side of Puebla. These movements had the effect of keeping the Zapatistas, who suffered from a weak command and control structure, on the horns of indecision.

While Obregón was still in Apizaco, the Zapatista commander of Puebla State, General Francisco Mendoza, proposed a pincer against Obregón's forces with an attack against Apizaco from Mexico City while he attacked through Amozoc, Puebla, a prospect that Obregón had feared. But nothing came of it. On January 5 the Constitutionalists approached the city of Puebla from three sides and with overwhelming numbers, attacked and retook it, sending the Zapatistas fleeing in the only direction left to them, to the south.

After recovering, the Zapatistas again formed their previous fronts to the northwest, around Huejotzingo and San Martín Texmelucan, and to the southwest at Atlixco in order to block the Constitutionalist routes of advance against Mexico City and Morelos and threaten the city of Puebla, which they never again possessed.

Not long afterward, the Constitutionalists received news of a falling out among the Conventionists in Mexico City, just as Obregón had predicted. The President of the Convention, General Eulalio Gutiérrez, had fired Generals Villa and Zapata and accompanied by his supporters fled the capital to Pachuca. Taking these developments into consideration, the First Chief ordered Obregón to make haste for Mexico City.

Avoiding San Martín Texmelucan, Obregón's Army of Operations marched northward through Tlaxcala in the direction of Pachuca, defeated a group of Zapatistas sent to intercept it at Irolo Station, and then turned southward to occupy the nation's capital, virtually unopposed, on January 29.

Gutiérrez had an undetermined number of soldiers, possibly 4,000, with him in Pachuca and was on his way to join his supporters in San Luis Potosí. These "Gutierristas" formed a new, third, faction in the civil war between Conventionists and Constitutionalists.

The defection of Gutiérrez was unwelcome news for Villa, who received it in the train station of Silao. Villa was returning from Ciudad Juárez after brokering an end to the Battle of Naco in cooperation with U.S. Army General Hugh Scott. Continuing toward Mexico City, Villa stopped in Querétaro where he encountered General Teodoro Elizondo, who had surrendered with his 3rd Division back in November.

Elizondo was trying to return to San Luis Potosí, either to join the Gutierristas or González's Constitutionalists. Assisted only by his escort of Dorados, Villa stopped the division in its tracks, relieved its disloyal leaders, and placed it under the command of José I. Prieto, a feat that the Dorado Juan Vargas called a true "*Panchovillada.*" Elizondo managed to escape with just his staff and escort.

With the defection of Gutiérrez and now Elizondo, Villa did not know whom he could trust and suddenly became aware of the vulnerability of his lines of communication. He pulled back with the troops accompanying him to Irapuato, a crossroads where the rail from the north meets those leading to Guadalajara and Mexico City, and ordered his troops commanded by General Agustín Estrada to evacuate Mexico City and join him there. This turned out to be propitious, because after a few days in Irapuato he learned of the loss of Guadalajara to the combined forces of Generals Diéguez and Murguía.

Battle of El Cuatro

After linking up in early January, Diéguez and Murguía began moving north to attack Guadalajara, and Villista Generals Rodolfo Fierro, Calixto Contreras, and Julián Medina deployed just south of the capital city to meet them. On January 18, the Constitutionalists sent two thousand dragoons around the right to interrupt the Villista lines of communication to Mexico City while two thousand more dragoons on the left hit the Villista right flank anchored on Gachupín Hill. At the same time, Diéguez's infantry battalions assaulted through the center. The Constitutionalist cavalry on the right accomplished its mission first, capturing La Capilla and El Castillo and pushing the Villistas back to San Martín. Then, with the 21st Battalion in "hammer" and maintaining an interval of one hundred meters between battalions, the Constitutionalist infantry advanced fast enough to get under the guns of the Conventionists before suffering any significant casualties, pushed the Villistas off El Cuatro Hill and Alamo Heights, and captured twelve of the Villistas' fourteen guns in the process. The Constitutionalists on the left achieved their objective last, clearing El Gachupín late in the day.

Villa was furious with General Fierro, whom all blamed for the defeat. The Constitutionalists, without any artillery, had cut the Villistas' lines of communication on their left, hit them on the right, and then made a push in the center, keeping the Villistas off balance and constantly reacting to their initiative. Moreover, Fierro failed to maintain reserves and provide sufficient infantry support for his artillery. Fierro, in turn, complained that Contreras' brigade and the regional forces of Jalisco did not fight with the élan characteristic of the Division of the North. Indeed, according to some Villa partisans the greatest casualty at El Cuatro had been the Division of the North's reputation of invincibility.

Thus, five events in the third week of January combined to substantially change the strategic complexion in the Center: 1. the occupation of Guadalajara by Diéguez and Murguía on January 18 put pressure on Villa's rearguard, 2. facilitating the advance northward of Obregón, who had departed Puebla as a result of 3. Gutiérrez's defection, and then 4. fought a battle against the Zapatistas at Irolo Station on January 23, which in turn pressured Gutiérrez to abandon Pachuca and continue moving north through the Bajío to link up with his supporters in San Luis Potosí, who, 5. forced out of San Luis Potosí by the approach of Generals Chao and Urbina, simultaneously pushed southward to link up with Gutiérrez.

Villa remained in Irapuato to collect the stragglers dispersed from the Battle of El Cuatro and sent General Estrada—in Querétaro with four or five thousand men—and General Francisco Carrera Torres from Guanajuato to begin moving toward San Luis Potosí to attend to his most immediate problem: the Gutierrista linkup. Estrada and Carrera Torres located their quarry in northern Guanajuato on January 28, and in fighting over three days quashed the movement in what became known as the Battles of San Felipe Torres Mochas. Gutiérrez escaped through eastern San Luis Potosí to arrive in southern Nuevo León, where he established his headquarters with a few hundred remaining soldiers. Most of the Gutierristas reverted back to the Convention while others went south to rejoin Obregón and the Army of Operations, or east to rejoin Pablo González and the Army Corps of the Northeast.

Northeast Strategy

The year 1915 began full of promise for Pablo González. The Division of the Bravo in southern Coahuila was poised to march on Torreón and, after having stopped the Conventionist drive on Tampico at El Ébano just before Christmas, the soldiers under González's immediate command has assumed the offensive and were pushing westward toward San Luis Potosí. General Caballero also reported that he had defeated Conventionist General Alberto Carrera Torres's men trying to take Ciudad Victoria in five separate engagements and had pushed them back to Tula. Caballero now prepared to focus his attentions on defeating Carrera Torres's second column approaching Ciudad Victoria from Galeana. At the same time he had General César López de Lara at Xicoténcatl ready to drive a wedge through the seam between Carrera Torres and Cedillo with a push toward Villa Quintero. In that favorable environment, González informed the First Chief that he intended to take 2,500 dragoons from Tampico through Ciudad Victoria to Saltillo, and then travel down the National Railroad and attack San Luis Potosí from the north, since he considered terrain along the eastern route between Tampico and the city of San Luis Potosí unfavorable for operations. Within the week, two reverses forced González to scotch those plans.

Ciudad Victoria, January 5-9

First, after receiving reinforcements at the end of December, Carrera Torres again renewed his offensive, taking Jaumave on January 3. Two days later Carrera Torres attacked Ciudad Victoria with two thousand men. The Conventionists initially experienced a fair amount of success, reducing the Constitutionalist toehold to one section of the city. Yet Carrera Torres did not have enough ammunition or Mausers to finish the job and appealed to the Conventionist Secretary of War for more of both. He also urged high command to get the offensive against Saltillo and Tampico underway since he was sure the Constitutionalist defenders inside Ciudad Victoria were receiving reinforcements from those locations. In the morning of January 9, after battling for seventy-nine hours straight, the Conventionists made one final push but a Constitutionalist battery emplaced near the Chapel of Guadalupe silenced the Carrerista artillery and then helped clear the Carrerista trenches, ending the battle. If the received battle narrative is true, it would be typical of battles in 1915 in the sense that the Conventionists had plenty of men but not enough war materiel to accomplish the mission, and unusual in the critical role played by artillery in a battle.

In their retreat, Carrera Torres's men took the same route of egress that the Federals used in 1913 through La Herradura.

While the Battle of Ciudad Victoria was ongoing, General González had to inform the First Chief that he had been unable to undertake his Saltillo-San Luis Potosí movement because the pressure put up by Carrera Torres was continuous, and Caballero's division had no more than 1,500 men "commanded by inept jefes." Therefore, he did not feel secure in removing troops from the petroleum region. Then, on the final day of that battle, González had to inform the First Chief of the colossal defeat of Villarreal's Division of the Bravo at the hands of Conventionist General Felipe Ángeles.

Ramos Arizpe

In the third-class railcar *Zacatecas*, which doubled as a sleeper at night and a war room by day, General Ángeles spread a map before his generals on January 2. Citing the example of William T. Sherman in the U.S. Civil War, Ángeles proposed to leave a small holding force in Marte Station and cut loose from his base of operations. Taking his infantry division cross country to link up with General Madero's cavalry division, he would attack Villarreal's base of operations at Saltillo and Monterrey before Villarreal could do the same and capture Torreón from Ángeles.

Accordingly, two days later Ángeles began his march in a driving rain, linked up with Madero, and destroyed the Constitutionalist garrison in the town of General

Cepeda. In the meantime, Villarreal smashed Ángeles' holding force at Marte Station and reported a terrific victory to the First Chief, but his glee was premature. Ángeles sent reinforcements back to harass Villarreal, should he try to press onward, but the Constitutionalist general had already realized that he could not hope to capture Torreón before Ángeles took Saltillo since the loss of General Cepeda left Saltillo wide open to capture. Therefore, Villarreal immediately began backtracking to Saltillo, which Constitutionalist General Luis Gutiérrez abandoned in the utmost haste for Ramos Arizpe, a crossroads to the northeast.

General Maclovio Herrera and his staff were among the first to arrive at Ramos Arizpe to reinforce Gutiérrez and had some luck in pushing Ángeles' Conventionists, who had sliced through Saltillo, back to the outskirts of the state capital. That evening, General Ángeles held another council of war and issued a professional battle plan to his generals, which entailed the infantry advancing in the center under the protection of his artillery supported by Madero's cavalry on the right conducting a flanking movement. In the other camp, the Constitutionalist commanders agreed to nothing more than to attack the next day.

In the morning the Conventionists began making their evolutions in a dense fog, opening fire at 6 a.m. when they ran into the Constitutionalist lead elements commanded by Colonel Emilio Salinas. Constitutionalist officers came to get General Herrera, who left the trains never to return. General Gutiérrez waited in vain for orders before taking the initiative and joining the battle. By that time the Conventionists were entering the east side of the city with the train station as their objective and forcing the Constitutionalists to extend their right. Next, Madero's appearance mid-combat on the Constitutionalist left compelled them to extend their lines in that direction, thus weakening the center. In the confusion caused by the fog, panic entered into the Constitutionalist ranks, which broke and ran. The battle turned into a rout. Officers who tried to instill discipline into their charges were ignored or, worse, shot. Precisely in these moments, Villarreal arrived on the scene with the rest of his division. He managed to save the artillery that had not been unloaded by instructing the trains to back up to Monterrey, but the Villistas cut off the retreat of other trains. Villarreal then ordered a staff officer to set fire to the rolling stock and promptly fled the field of battle. There simply was not enough time to destroy the trains, however, and consequently a tremendous amount of booty—including the majority of the materiel that had been collected from the ex-Federals in Mexico City and subsequently sent north to Villarreal—fell into the hands of the Conventionists. The Constitutionalists dispersed in headlong flight, some headed for Paredón and then Monclova, others for Monterrey.

Villarreal then evacuated Monterrey without putting up a fight and marched to Los Ramones, in the direction of Matamoros. The Villistas moved in and occupied the capital of Nuevo León on January 11, supported on the left by General Rosalío

Hernández following the old Orozquista invasion route from Chihuahua into central and northern Coahuila. Hernández's mission was to destroy those Constitutionalists who had escaped the Battle of Ramos Arizpe and take possession of the state's coal fields.

Villarreal blamed the defeat at Ramos Arizpe on the fog, the betrayal of the railroad employees who intentionally derailed railcars (or so he claimed) thus blocking the retreat of the majority of his trains, the conduct of Colonel Emilio Salinas, whose troops initiated the panic that turned the battle into a rout, General Herrera, and the men of General Luis Gutiérrez—in sum, everyone except himself.

The First Chief was incredulous and demanded to know how Salinas, who commanded only three hundred men, could have been responsible for such a colossal defeat.

The true reasons for the Constitutionalist defeat were the lack of unity of command between Herrera and Gutiérrez, a leadership vacuum at the very top (Villarreal), the absence of a detailed plan of battle, and two dirty little secrets. First, the Division of the Bravo contained many ex-Federals from the Federal Division of the North that had been defeated at Ojinaga by Villa and imprisoned by the U.S. Army until repatriated. Those ex-Federals had reentered national territory through Piedras Negras and many joined the Division of the Bravo. Yet they proved less than worthless in the fog—just as they generally failed to perform in the dark. Second, and perhaps more disturbing, many Constitutionalists in leadership positions, including Villarreal himself, were secretly Gutierristas.

Finally, one must also recognize the skill of Ángeles and his Villista subordinates, not to mention the vagaries of war.

The Battle of Ramos Arizpe broke the Division of the Bravo and left the Constitutionalists at Tampico and the Gutierristas in San Luis Potosí vulnerable. No Constitutionalist position in the Northeast was secure, and so General Treviño began entrenching in a 180° line of defense around Tampico and created a five kilometer system of trenches around El Ébano as a precaution. To attenuate the psychological and publicity hit of such an enormous fiasco, Treviño also organized an operation to advance toward San Luis Potosí with small, fast-moving cavalry units cutting a wide swath to isolate the capital and capture it, using Ciudad Valles and later Cerritos, if need be, as the base of operations. That effort produced modest results but it would never advance much farther than Ciudad Valles.

FEBRUARY

By the end of January, the frenzy of the initial months settled down as the battle lines, so to speak, had been drawn. Obregón remained in Mexico City. Herrera and Villarreal were busy reorganizing the remaining 2,000 men of the Division of the Bravo, now called by its original and less grandiose designation: 1st Division of the Northeast. The Constitutionalists that González had brought with him from Pachuca (which became the new 3rd Division of the Northeast) continued battling the Cedillos along the Tampico-San Luis Potosí rail line while Caballero's 5th Division occupied key points of exit from the sierra, keeping Carrera Torres bottled up in the highlands and the Ciudad Victoria-Tampico line relatively secure. General Ángeles could not continue advancing on Tampico because after garrisoning the rails from Torreón to Monterrey, he had so whittled down his maneuver force of ten thousand men that he only had a few thousand remaining. Yet he faced Gutierristas to the south of Saltillo and the remaining Constitutionalists of the 1st Division of the Northeast to the north and east, each constituting a persistent threat to his lines of communication. Slowly closing in, by the end of January the Constitutionalists had initiated a loose siege on Monterrey and prepared to do battle.

Battle of Monterrey, February 6-7

On February 6, the Constitutionalists attacked Monterrey with infantry supported by cavalry and artillery. Ángeles responded by first pushing back Villarreal on the north side, then using interior lines, he pivoted those forces to demolish the Constitutionalists still engaged on the east side. The weakened state of the Constitutionalist horses caused by the long marches to get into position and a lack of fodder prevented the cavalry from fulfilling its role of protecting the infantry, which had started to run low on ammunition. The Conventionist counterattack put the Constitutionalist cavalry to flight, and machine guns firing from retreating trains in an attempt to stop the Conventionist push instead wound up ravaging friendly infantry, which was subsequently run down and decimated by the Villista dragoons. The attack on Monterrey had either been part of a strategy to distract Conventionist attentions from Tampico, northern Coahuila, and Obregón in Mexico City—as González's officers later claimed—or an egregiously miscalculated blunder.

Thereafter, the 1st Division of the Northeast was reduced to a mere skeleton of its former self. Whereas before, its brigades had contained at least one thousand men, by mid-February none had more than three hundred, "so that now an excess of generals and jefes and a deficiency of troops" obtained. Taking into account "the state of disorganization and demoralization in which the forces of General Villarreal find themselves," González proposed to position them "along the National, International, and Central rail lines and brigade them under a competent general." He would then send the excess officers to Veracruz in exchange for cavalry from the south to help take Monterrey in order "to establish there a base of future operations in Coahuila and Nuevo León...as well as to achieve complete extermination of the enemy on the border." The plea for cavalry echoed a similar request of January 5, when González had asked for men from Generals Cesáreo Castro or Fernando Dávila, which the First Chief had denied because Obregón needed them. Eventually, González got the men he needed—from the Gutierristas defeated at San Felipe Torres Mochas—whom he threw into the line at Monterrey. Villa's good news in Guanajuato turned into a curse for General Ángeles, who would have to endure a twenty day siege.

Villista Strategy

General Obregón's occupation of Mexico City on January 29 effectively split the Convention into two, North and South, a fait accompli enshrined in Villa's decree dated February 2 that formed his own government in the North. After attending to that and other administrative details, the Villistas were once again rolling. General Hernández was busy trying to capture the important coal region of northern Coahuila, where General Maclovio Herrera had assumed command of the remaining Constitutionalist forces. In Yucatán, the nominally Villista ex-Federal Colonel Abel Ortiz Argumedo initiated a rebellion that required the Constitutionalists to begin sealifting assets to Campeche in preparation for a campaign in the neighboring rebel state. General Manuel Chao was spearheading the general advance in San Luis Potosí toward Tampico. General Juan Cabral, leading an expedition from western Chihuahua, had arrived in eastern Sonora to cooperate with Governor Maytorena in destroying Constitutionalist General Plutarco Elías Calles. General Rafael Buelna was driving from Tepic toward Mazatlán with orders to continue up the coast until he linked up with Maytorena's Conventionists in Sonora, even as Constitutionalist General Ángel Flores's Expeditionary Column of Sinaloa had bogged down in southern Sonora on the way to rescue Calles. And Villa returned to his duties as a theater commander intending to invade Jalisco State and relieve pressure on Buelna that Diéguez and Murguía might apply.

At the time, General Murguía had lead elements as far as Yurécuaro and was moving eastward on orders from the First Chief to link up with Obregón in Mexico City. Villa pulled General Agustín Estrada back from San Felipe Torres Mochas to Querétaro as a blocking force against Obregón, then still in Mexico City, and asked the Suriano Conventionists to cut Obregón's lines in Tlaxcala and Hidalgo states. Villa then departed Irapuato traveling west with his infantry and artillery to attack

Murguía's Constitutionalists while his cavalry, reorganized and refitted after the debacle of El Cuatro, mounted up and headed south for Yurécuaro from Aguascalientes. As the Villistas moved into Jalisco, Murguía retreated and maneuvered out of the pincer between Villa's infantry and cavalry, which closed farther west than initially planned, at La Barca.

In La Barca, Villa generated a new campaign plan. He sent his cavalry around Lake Chapala to cut off the Constitutionalist withdrawal from Guadalajara at Atoyac while he continued over the rails to the state capital, essentially creating a new cavalry and infantry pincer to close at a different location farther south.

Once again Villa's infantry and cavalry linked up, at Atoyac, and once again the Constitutionalists had retreated out of the trap, but they were not home free.

Battle of Cuesta Sayula, February 17-18

Coming to the sierra at Cuesta Sayula the gradient was so steep and the trains so overloaded that the Constitutionalist infantry had to detrain to lighten the load. In these moments, Villa's cavalry caught up to the last trains on February 17 and attacked.

About every half hour, the Villista cavalry delivered charges in depth directed at the Constitutionalist flanks. Charging while shooting to within twenty or thirty meters of the Constitutionalist lines, they retreated, reformed, and charged again. In the center, after the requisite artillery preparation, the Villista infantry entered into action, deposited on the firing line by cavalry that carried them on the backs of their horses. The Constitutionalist infantry in the center waited behind a stone wall for the Villista infantry to advance to within fifty meters and then let loose a volley, dropping about seventy in an instant and forcing the rest to withdraw to a "regular fighting distance," probably meaning about three hundred meters. The only artillery fire returned by the Constitutionalists came from Marilerañela rockets, which caused more noise than destruction. At one point the Constitutionalists spied Villista artillery with its cavalry support still in column formation, and they sent concealed infantry to attack it but without being able to recover the piece.

Eventually, the Villista infantry in the center breached the Constitutionalist lines. The separated Constitutionalist wings on both sides made an obliquing movement to the rear that succeeded in reconnecting the firing line, but in those moments General Diéguez ordered Colonel Amado Aguirre to send word to unit commanders to retreat.

Colonel Aguirre actually directed the battle, not Generals Murguía and Diéguez, who were on site only briefly. Moreover, most of Murguía's division and much of Diéguez's division did not participate in the battle. The Villista advantage in numbers was an estimated 3:1 on the right flank and 5:1 in the center—the Constitutionalist left remained virtually unengaged throughout the battle.

The Constitutionalist position at Cuesta Sayula was very good, and with enough ammunition, Murguía and Diéguez might have dealt Villa his first defeat there, according to the Dorado Juan Vargas.

Colonel Aguirre, for his part, believed the absence of unity of command had been a greater problem for the Constitutionalists than a lack of ammunition, which Murguía blamed for the defeat. Operating on exterior lines of communication on a strategic level, the Constitutionalists' supply of arms and ammunition to Jalisco had to run from Veracruz through the Isthmus of Tehuantepec to Salina Cruz and from there to Manzanillo and then inland. The supply of arms and ammunition over that route had been sketchy, and without sufficient ammunition to resist Villa, General Murguía reported that he did not intend to offload the artillery pieces and shells due to arrive for fear they would just wind up in Villa's hands after another lost battle.

Villa pursued the Constitutionalists as far as Zapotlán, where he received a telegram from Ángeles asking him to return to the Northeast to join forces to defeat the Constitutionalists there definitively. Villa wanted to follow up on his Cuesta Sayula victory and destroy Diéguez's infantry once and for all, driving them into the mountains, denying them resupply via Manzanillo, and capturing all their rolling stock. Ángeles countered that occupying Manzanillo would accomplish nothing of importance except adding more railroad lines that needed to be garrisoned, and that Diéguez could easily slip into the mountains of Guerrero and Michoacán to avoid presenting battle, while Monterrey was the key to Torreón and Villa's supply lines. The Northeast was a much more worthy and pressing objective. By clearing the Rio Grande Valley of Constitutionalists, Villa and Ángeles could wheel south and attack Tampico in combination with Urbina. Just as against Huerta, the Norteños needed to control the entirety of the North and then turn their attentions to the Center and South. Ángeles was trying to create an overarching strategy that designated the Northeast as the first priority. Whether Ángeles needed Villa's troops to drive on Tampico remains a point of contention.

González's staff officers later explained that by properly distributing the 3,000 men surrounding Monterrey, they had convinced Ángeles that they had seven times more men than they really did, and this disposition of forces had been the genesis of Ángeles' request for Villa to come to his aid. Be that as it may, the forward momentum of Ángeles, Hernández, and Carrera Torres had stalled; only Chao in San Luis Potosí was experiencing any progress. And now Villa started pulling back to Guadalajara in preparation for a return to the Northeast.

Before leaving Jalisco, Villa sent General José I. Prieto, essentially in command of Teodoro Elizondo's former 3rd Division (renamed the Benito Artalejo Brigade), to invade Michoacán and fight the forces of Gertrudis Sánchez and Joaquín Amaro. Prieto took Morelia, the capital of Michoacán, at the end of February.

Mexico City, February-March

In early February, just after Obregón had occupied Mexico City and Chao entered San Luis Potosí during the brief campaign against the Gutierristas, the Zapatistas believed that Villa intended to push on to Tampico and then turn south against Veracruz and circle back through the states of Puebla and Mexico. Therefore, they finally started to make an effort to halt train traffic between Veracruz, Puebla, and Mexico, mainly focusing on the area around Otumba, Irolo, and Ometusco, and northern Tlaxcala just northeast of Mexico City. At the same time, the Zapatistas had to address Gertrudis Sánchez's Gutierristas driving eastward from Michoacán in an effort to capture El Oro, in Mexico State. In response the Zapatistas ordered General Inocencio Quintanilla to occupy El Oro and invade Michoacán.

This was the first time the Zapatistas had participated in anything resembling conventional war. General Emiliano Zapata's soldiers mainly came from the states of Mexico, Hidalgo, Puebla, Tlaxcala, and, of course, Morelos—precisely the states originally assigned to the newly-formed Rural Police Corps in 1857 to suppress banditry. The Rural Police Corps major units were "corps" of less than three hundred men, mirroring their quarry in unit strength, and dressed *de charro* like the bandits they hunted, most famously romanticized by the legendary gang *Los Plateados*. Zapatistas continued in the long Suriano tradition of civil unrest and guerrilla warfare that stretched back to General José María Morelos, the namesake of the home state of Zapatismo. They solemnly believed that they had played a decisive role in the defeat of Huerta's army and that they were every bit the equals of their Norteño foes and allies. That belief led them to prosecute a war of annihilation against the Constitutionalists in Mexico City and in Puebla, Tlaxcala, Hidalgo and Mexico states, but it was a strategy for which they were ill-suited.

The Zapatistas did not have the command and control necessary for conventional warfare and their troops were notoriously undisciplined. Many Zapatistas did not even know proper hierarchy of ranks, or the difference between corps, regiments, battalions, or brigades. Additionally, their units were top-heavy with officers, with officers sometimes outnumbering enlisted men, mainly because Zapatistas tended to appoint officers and then recruit the enlisted men as opposed to the Norteño tradition of forming units that then elected their officers. Zapatista commanders also adopted the bad habit from the Federals of inflating their rosters with phantom soldiers in order to pocket the excess pay. The Norteños who initially commanded the Conventionist War Department tried to rationalize the Zapatistas by placing a moratorium on all promotions and requiring inspections once a month. But the most pressing Zapatista defect was one of culture, which had adverse effects on command and control: Zapatista "generals" commanded only a few hundred men and exercised near total autonomy such that a commanding general often had to "recommend" that a subordinate general follow certain orders and ask near equals to "cooperate" in operations, and in other instances had to resort to threats to get them to comply. Moreover, the Zapatistas seemed to be much more prone to infighting than any other combat force, compounded by the natural animosity between North and South that played a much bigger factor in the Conventionist Army than among the Constitutionalists who had the steadying hand of the First Chief on the tiller. Simply put, there was not enough time to instill the discipline and organizational and institutional rigor required of conventional warfare into an army that according to Suriano tradition, mission, and campaigns of the last few years had been adapted to guerrilla warfare.

Nor did the Zapatistas have the resources—money, provisions, fuel, transportation, arms, and ammunition—to prosecute a modern war, because they lacked a substantial economic base, such as a cash crop (the sugar industry in Morelos had been destroyed) or an extractive industry that might generate hard currency, or a port through which to ship exports and receive war materiel. Additionally, they exacerbated their state of material want by selling and wasting ammunition and accomplished less with what they used because of their notoriously poor marksmanship. Without regular pay and logistical support, they resorted to taking arms and ammunition from each other—which increased the in-fighting— and arrogated what else they needed from the populace, further damaging the image of their cause.

Given their situation, the proper strategy for the Zapatistas was one of attrition: hitting soft targets and small garrisons along Obregón's supply lines would have afforded better chances of gaining battlefield salvage, and it would have served the larger strategy to debilitate Obregón and assist Villa. Instead, the Zapatistas pursued a strategy of annihilation, kicking off tertiary campaigns in the states of Guerrero and Michoacán—perhaps to gain access to a port of entry, which might have made some sense, but that purpose is not expressed in the extant documentation—while focusing their efforts on trying to recapture Puebla and Mexico cities. All of these mentioned objectives had the especially egregious effect of placing the largest concentrations of Zapatistas far from Obregón's umbilical, the indicated objective, while they tried their hand at conventional warfare in the nation's capital (the best-equipped Zapatista units were actually in Morelos State, idle).

Over six weeks, the Conventionists attempted several coordinated attacks to drive the Constitutionalists from Mexico City, but as Constitutionalist General Gavira stated:

They did it in such a way that it clearly revealed that there was not among them organization, or discipline, or unity of command, or anything. They are truly mobs that do not lend reciprocal aid except when in the mood. Soon it was seen, moreover, that they lacked ammunition and totally any money.

Villa suspected as much when he pondered whether the ineffectiveness of the Zapatistas was due to a "lack of organization, discipline, or lack of some elements, or some other causes of which I am unaware," but "the campaign it [the Liberator Army] has undertaken is not giving desirable results."

That is not to say the Zapatistas did not inflict hurt on Obregón. Indeed, the main casualty from combat in the nation's capital was a temporary slump in the Constitutionalists' morale as a result of sniping. After many years of enjoying the advantage of fifth columns in their assaults on plazas held by Federals in the North, the Constitutionalists were now on foreign soil and the shoe was on the other foot as Zapatista sympathizers shot at them from windows and rooftops. During his sojourn from hell in the nation's capital, Obregón's casualties averaged sixty soldiers per day, the most disconcerting for the Yaquis being the death of the beloved Lieutenant Colonel Tiburcio Morales.

In the larger picture, though, the Zapatistas gave Obregón six weeks to: wait for arms and ammunition to arrive from the United States, remove the powder from the national powder factory, retrieve the equipment for manufacturing cartridges from the Citadel, incorporate former Gutierristas who rejoined the Constitutionalist Army (including Joaquín Amaro, Miguel Acosta, and Alfredo Elizondo, among others), recruit thousands of new soldiers, and then finalize preparations to resume marching into the heartland.

The first step in that upcoming campaign had come on February 8 when Obregón sent about two thousand dragoons to occupy the capital of Hidalgo, Pachuca, in order to secure his supply lines. Two weeks later he put his vanguard into place with Colonel Eugenio Martínez's 1st Battalion of Sonora in San Juan del Río, Querétaro, backstopped by the 21st Battalion of Sonora at Huichápam, Hidalgo.

MARCH

Mexico City

At the beginning of March, the First Chief urged Obregón to destroy the National and Central Railroads east of Querétaro, evacuate Mexico City, and fall back on

Ometusco to receive his supplies. Obregón's policies (the specifics of which are beyond the scope of this writing) in the nation's capital had the expatriate community up in arms, figuratively speaking, and it was causing the First Chief all manner of diplomatic headaches. Obregón refused Carranza's request, saying that such a move would embolden Villa to destroy the Constitutionalist army corps in the West and Northeast in detail. Instead he argued that he should resume the offensive, make contact with Villa, and then decide whether to continue the offensive or go on the defensive. Carranza assented, ordering him to evacuate immediately and advance into the Bajío.

March 7 was a pivotal day for Obregón. He received a message from the U.S. State Department accusing him of starving the people of Mexico City, among other offenses, at the same time that the Villistas attacked his vanguard at San Juan del Río and the Zapatistas renewed their attack on Mexico City, attempting to block any Constitutionalist withdrawal.

By the end of the day, the vanguard had defeated the Villistas at San Juan del Río, which convinced Obregón he had the wherewithal to resume the offensive, and the last convoy bringing five thousand rifles from Veracruz had arrived. Obregón closed the day by directing the State Department to address its concerns to the First Chief. Then, sweeping aside the Zapatistas in his path, he began evacuating Mexico City on March 8. Almost simultaneously, Villa had started his return trip to the Northeast.

After spending several weeks in Guadalajara attending to administrative and diplomatic matters, Villa left for Monterrey at 3 p.m. on Saturday March 6, 1915. Using his interior lines of communication on a strategic level, he took his brigades to Monterrey, once again leaving Contreras and Fierro to defend Jalisco in cooperation with the regional forces of Governor Medina.

The Northeast

Upon his arrival at Monterrey, General Villa became instantaneously upset with Ángeles over his use of dismounted cavalry on the defensive instead of for offensive operations. It would appear that after Ramos Arizpe, Ángeles had grown too cautious surrounded by his ex-Federal compatriots—too concerned with keeping his supply lines secure, hesitant to deviate from the rails, in the old Federal Army manner of thinking. Even if Ángeles only had four thousand men for offensive operations after attending to garrison and security details, he should have been able to handily defeat Villarreal's weakened 1st Division after the Battle of Ramos Arizpe, "in which the brilliant Division of the Bravo was nearly annihilated, and its Regiments and Battalions reduced almost to squadrons and companies," as Constitutionalist staff officer Manuel González later lamented.

Civil War, September 1914 – May 1915

The arrival of Villa had the effect of further decreasing the morale of Villarreal and his men. The Constitutionalists surrounding Monterrey retreated to the north, east, and south and Villa sent three columns against each of those concentrations, initiating a campaign to capture Matamoros, Nuevo Laredo, and Ciudad Victoria on the way to Tampico. Villa personally led the column directed against Matamoros, defeating the Constitutionalists near Los Ramones on March 14, and sending the inept General Villarreal racing into exile.

After that scrap, Villa again chided Ángeles telling him "Decidedly, General Ángeles, sir, you have no military problem" here. And by calling Villa to the Northeast and away from the main Constitutionalist forces under Murguía, Diéguez, and Obregón, not to mention wearing out the horses during the five day trip, Ángeles committed what many Villistas later considered to be the key strategic blunder of the war—one that cost the Conventionist Army any chance of victory. Still others realized that Villa's excursion to the Northeast was symptomatic of a deeper problem: the lack of an overarching strategy.

With the reinvigorated Villista offensive in the Northeast, González asked the Constitutionalists in the Center to hurry their advance and again requested permission from the First Chief to negotiate a non-aggression pact with Gutiérrez and his supporters in southern Nuevo León.

Villa returned to Monterrey and left General José Rodríguez to direct the rest of the Matamoros campaign. Rodríguez caught up to and thrashed the Constitutionalists at Cerralvo on March 19, relieving them of their artillery. At the same time Villista Generals Máximo García and Severiano Ceniceros kicked off the campaign to take Tamaulipas State, surprising González and his Constitutionalists at Montemorelos and forcing them to continue pulling back toward Ciudad Victoria.

In Coahuila, Villista General Orestes Pereyra had already come up from Saltillo to support General Rosalío Hernández and renew the offensive in Coahuila in anticipation of Villa's arrival at Monterrey. General Maclovio Herrera spent weeks trying to outmaneuver the Villistas, sometimes surprising them and notching a victory and, at other times, being the one surprised and beaten. At the end of March, González ordered Herrera to abandon Coahuila for Nuevo Laredo to reinforce the garrison against the impending Villista attack on that plaza.

On March 19, González stopped in Ciudad Victoria and sent a message to the First Chief. He had lost many men in combat with Generals García and Ceniceros at Cadereyta, and with Villa and Ángeles together tackling the extremely weak and poorly led 1st and 5th Divisions of the Northeast, González could not be sure of a Constitutionalist victory. Therefore, he informed the First Chief that in the event of a Villista victory along the border he would evacuate Villarreal's infantry by sea to Veracruz in order to keep those soldiers from infecting the defenders of Tampico with their poor morale, and he would give Villarreal's remaining cavalry to General

Caballero to defend Ciudad Victoria. Should that city fall to the Villistas, then Caballero had instructions to break up the cavalry into guerrilla units and continue harassing the Villistas. González requested that transports be sent to Tampico to be prepared for Villista General Manuel Chao's capture of the port, in which case he would evacuate the 3rd Division's artillery and infantry by sea and send its cavalry to fight in the Huasteca. These measures, however, were only a worst case contingency that would go into effect "only in the event mentioned, that we lose Tampico and the Border, which, as I said...I will defend to the last moment."

Matamoros

General Rodríguez finally reached Matamoros on March 27 to face a formidable obstacle: "Breastworks high enough to hide a horse, with barbed wire and thorn-spiked brush in front, encircled the land side of Matamoros...the earthworks...broken by small lakes and marshes, impossible to cavalry or infantry." The only high-speed avenue of approach was the nearly half mile field of fire "cleared of brush and trees," beyond which were "woods and more lakes," according to newspaper accounts. The commander of the garrison, General Emiliano P. Nafarrate had ordered locks built utilizing the numerous irrigation ditches that surrounded the city to be able to flood the area known as "Las Matanzas de los Cárdenas," also known as the "Garita de Monterrey" to the front of the trenches. He only had maybe three hundred men, but just before the battle he had received a large shipment of carbines, cartridges, and machine guns that he was supposed to forward to Veracruz through the Bagdad Bar. He hurriedly took the machine guns out of their crates and placed them on the line of fire. The defenses were competently arranged, with the machine guns emplaced for grazing fire and trench works designed to accommodate Nafarrate's garrison forces, leaving the troopers from Villarreal's old 1st Division, who were soon to join in the defense, free to maneuver.

At 11 a.m. Rodríguez's Villistas divided into three columns and attacked, one on the left near the river, another in the center toward the Garita de Monterrey, and a third on the far right, due south of the city. There were three assaults, each one gaining in intensity, first with three hundred men, then six hundred, and finally two thousand. General Nafarrate was on the line of fire and saw what he thought were Villistas dismounting to continue fighting on foot, but he later learned that they had been knocked off their horses by gunfire. When they were about fifty paces from the Constitutionalist trenches, the machine guns let loose and mowed them down. It was a disaster for the Villistas who were taken apart by machine gun fire. Consequently the attack only lasted about four hours.

After the fighting of that first day the Villistas rested and waited for their artillery in the hope that it would be able to reduce the trenches to rubble. By all accounts, Rodríguez had suffered an incredible five hundred casualties.

On March 29, the troopers of the 1st Division of the Northeast under its new commander, General Ildefonso Vázquez, hit Rodríguez's rearguard and then entered Matamoros to reinforce Nafarrate. They took up position on the Constitutionalist left where they remained available for counterattacks while Nafarrate's men continued to man the trenches. Over the next two weeks the two sides continued to skirmish while Rodríguez brought up his artillery, an exercise complicated by the destruction the Constitutionalists had done to the rails.

El Ébano

Conventionist General Chao led the Villista push on Tampico, repairing the rails and bridges along the way to reach El Ébano on March 5, 1915. For the next three days the Constitutionalists fought Chao's men, eventually repulsing them. In the midst of the fighting, Constitutionalist Colonel Luis T. Navarro's column from the 1st Division of the East disembarked at Tampico to reinforce General Treviño, now in charge at El Ébano. Navarro requested that his own Morelos Battalion be forwarded to him, but those men had already been shipped off as part of a larger effort to put down the rebellion in Yucatán.

To participate in the Yucatán operation, the Ocampo Brigade of the Constitutionalist 1st Division of the East had boarded ships on March 10 in Puerto México and sailed to Campeche where it joined other units commanded by General Salvador Alvarado. The overriding mission was to reconquer Yucatán State and recover the valuable henequen fields from the nominal Villistas under Ortiz Argumedo. Driving up the coast, the Constitutionalists defeated the Argumedistas at Blanca Flor on March 14, and again at Halachó two days later. On March 19, the Constitutionalists entered Mérida after it had been evacuated by the Argumedistas and then began transporting units by sea to Tampico, where they disembarked and marched off to El Ébano to join the Constitutionalist 3rd Division of the Northeast and other units from the 1st Division of the East.

Meanwhile, General Jacinto Treviño oversaw the construction of fortifications at El Ébano. The central defenses consisted of earthworks in zigzag that provided interlocking fields of fire and enabled shooting while kneeling. By the later stages of the battle, the soldiers had deepened the trenches to permit standing, created large entrances and exits, and added subterranean rooms where the women cooked, sewed, and applied homespun cures. Machine guns had barbettes of compacted dirt near the gap in the line caused by the railroad track, and were placed at other high-speed avenues of approach. Hills dense with vegetation anchored the line on both flanks and on the left there was a tar pool and then La Pez Hill. Wire had been strung to the front of the trenches to impede cavalry charges, but it could not be electrified because of a lack of time and materials to do so. Mines placed to the front of the trenches near the railroad had been planned, but the Constitutionalists likewise ran out of time to make that improvement.

On March 21, Villista General Chao's centaurs renewed the battle, delivering a mounted cavalry charge in close order columns, with a reduced front, over open fields of fire and down the highway, against entrenched infantry in the center, and without any prior reconnaissance. The Villistas also hurled "against the trenches their numerous infantry," according to Treviño's artillery commander, Major Fernando Vázquez. They attempted to support the assault with artillery fire, but the shells landed too far to the rear, with high explosive impact type ordnance aimed at infrastructure instead of shrapnel directed against the Constitutionalist trenches. One explanation for why the shells may have landed too far away had to do with the condition of the guns and the quality of the artillery shells, which forced the gunners to go into battery close to the line of fire and engage in direct fire missions. After the assault had failed, the Villista artillery switched to counter battery missions.

The Constitutionalist artillery had been surprisingly accurate given that Captain Anselmo Brunicardi's guns were missing aiming apparatuses. Constitutionalist scouts sent to reconnoiter in the direction of La Pez Hill on the morning of March 25 reported seeing bloody remains that testified to the deadly effect of Brunicardi's artillery.

In subsequent days, the Villistas changed tactics from in-depth charges to probes and started to build a fort near the railroad bridge that threatened the Constitutionalist center; the Constitutionalists responded by sending a gondola loaded with dynamite down the rails to blow up the bridge near the fort and employed aimed shots ("*a salvo*")—as opposed to volleys—that decimated the Villista cavalry. Some Constitutionalists held their rifles aloft over the parapets to shoot without aiming while keeping their body covered. On that first day of battle the fear of legendary Villista cavalry charges had been dispelled, and the Constitutionalists learned to trust in their prepared defenses.

Reinforcements continued to pour into the area by sea for the Constitutionalists, including General Pedro Colorado and his brigade from Tabasco State and the 1st Red Battalion from Veracruz, which arrived at El Ébano on March 24 and at Tampico on March 25, respectively. The strategic significance of Tampico compelled the First Chief to give El Ébano the priority in receiving money, men, and materiel in March. Not only was the port city key to the petroleum industry but, if lost, would provide the Villistas with a second major port of entry, in addition to Ciudad Juárez. Then the Conventionists would be able to import war materiel by sea from New York and launch attacks up and down the Gulf littoral.

Meanwhile, after the March 23 deadline passed without General Chao being able to conquer El Ébano, Villista General Tomás Urbina decided to leave San Luis Potosí and personally take charge of the battle. From March 27 until Urbina arrived, the combat turned light and intermittent.

At the same time, Villista Generals Máximo García and Severiano Ceniceros were also making a push toward Tampico, having reached Garza Valdez just inside Tamaulipas state lines on March 26. González gave instructions to General Caballero to present battle at Santa Engracia and Caballeros Station, ahead of Ciudad Victoria, and those preparations were in the works.

The Center

The concerted Villista effort to conquer the Northeast cost them the state of Jalisco.

In the Battle of Tuxpan, waged March 21-25, Villista General Fierro launched suicidal open-field cavalry charges under the protection of his artillery. The Constitutionalist infantry, expertly handled, extended its front under fire toward the left, threatening the Villista right flank and forcing the regional Jalisciense forces on that end to break and run. Panic ensued, resulting in yet another disaster for the Villistas.

At the same time, Obregón was advancing through the heartland to take pressure off González under attack in the Northeast by Ángeles, Chao, and now Villa, but he had to move deliberately. He knew that Villa had him outnumbered while the Constitutionalists had to operate on external lines, with Diéguez and Murguía supplied via the Isthmus of Tehuantepec, not to mention the threat posed to Obregón's own supply lines by the Zapatistas.

On March 24 Obregón was in Cazaderos Station, where he expected to be attacked by the Villistas stationed in Querétaro. He took advantage of the opportunity to rehearse a disposition of combat similar to that which he would later employ at Celaya.

Back in Veracruz, and apparently concerned over the situation in the Northeast, the First Chief suggested that the strategy for Obregón should be to go through Querétaro, San Luis Potosí, and then Tampico, to avoid the arid north of San Luis Potosí and to be situated to receive supplies by sea. Because Obregón's intelligence suggested that he only faced Agustín Estrada's forces in Querétaro, and that Estrada would not receive reinforcements before Obregón planned to attack, Carranza scrapped his own suggestion and approved Obregón's earlier request to march toward Aguascalientes, link up with Diéguez and Murguía at Irapuato, and then proceed to San Luis Potosí and Tampico. When Obregón learned that Diéguez and

Murguía had defeated the Villistas at Tuxpan and that Estrada's Villistas had retreated from Querétaro, he moved up and occupied the latter plaza on March 31.

Estrada promptly telegraphed Villa about the Constitutionalist buildup of men and materiel around San Juan del Río. Villa ordered him to pull back to Celaya, told General Fierro to hold Guadalajara at all costs, and once again began concentrating troops at Irapuato, strategically located between Murguía and Diéguez to the west and Obregón to the east. He also tried to recall Rodríguez from Los Ramones, but General Ángeles informed him that those troops were engaged in a battle for Matamoros. General Ángeles sent Villa what forces he had available, about six thousand men, to Irapuato, and Villa again asked Zapata to focus his attentions on cutting Obregón's supply lines.

APRIL

In April Obregón's Army of Operations continued penetrating into the Bajío, with General Fortunato Maycotte's cavalry brigade pushing Villista General Agustín Estrada back to reach El Guaje. Another of Cesáreo Castro's cavalry brigades commanded by General Porfirio G. González headed north toward Dolores Hidalgo to destroy the rails from San Luis Potosí and thus secure the Constitutionalist right. The third cavalry brigade under Alejo G. González scouted down south toward Acámbaro because it was rumored that Zapata had kicked the Norteño Conventionists out of Mexico City and Obregón expected those Norteños to head north to link up with General Estrada. At the same time, Obregón continued to concentrate his infantry and artillery in Celaya and asked that the First Chief order Diéguez to make a forced march toward Irapuato in order to catch the Villistas from two directions. The communications between the First Chief in Veracruz and Diéguez in Jalisco, however, were interrupted at this point, cut by the Zapatistas in Guerrero State. In the last communique from Diéguez, dated March 30, he stated that he expected to be in Guadalajara around April 5, so Obregón decided to postpone his advance until April 10.

Once Villa learned that Fierro had abandoned Guadalajara and that Estrada was in retreat, he returned to the Bajío, arriving at Salamanca on April 5. He held a council of war in which he received intelligence reports about the forces under Obregón's command, which continued to grow with the arrival of every new train from Veracruz. Villa was particularly piqued to learn that Zapata had failed to execute or, worse, had ignored his requests to cut Obregón's supply lines. Villa decided to attack Obregón as soon as possible; he didn't even wait for his artillery or his infantry, which provided "powerful assistance in battles," before making contact, "habituated as he was to great clashes of arms being resolved through his famous cavalry charges," as the Dorado Juan Vargas explained.

While no one knows for certain how many men Villa had under his immediate command, twelve thousand would be a good estimate since he had on hand most of General Ángeles' infantry division, Estrada and Calixto Contreras' brigades, what remained of the Robles Brigade, and the regional forces of Guanajuato. Obregón had a similar number of men, but his cavalry was detached to vanguard and flanking duty, as previously described, and not at Celaya.

First Celaya, April 6-7

In the morning of April 6, Villa's cavalry smashed Obregón's vanguard at El Guaje and pushed through to Celaya where it crashed against "those one hundred machine guns of Álvaro Obregón, and those Yaqui Indians" on the firing line, protected by cavalry on the flanks and in reserve, and backed by artillery in the center to the rear of the line of fire. Obregón frantically tried to call in his cavalry detached to the north and south, but General Porfirio González was too far away to respond. General Alejo González, however, reached Celaya after a forced march and on April 7 executed a flanking maneuver around the right (north) in combination with a solid push in the center by the Constitutionalist infantry; simultaneously Maycotte's brigade, having recovered somewhat from the earlier drubbing at El Guaje, made a flanking attack on the left that put the Villistas to flight. The Constitutionalist cavalry, however, was too exhausted from fighting and riding over the previous two days and so could not carry out a proper pursuit.

Both sides suffered greatly in this first clash of arms between Obregón and Villa with the Constitutionalists probably losing approximately one thousand men and the Villistas about twice that number. The greatest casualty, however, had been the loss of Villa's aura of invincibility. After that first Celaya battle, the Constitutionalists could face Villa's centaurs without undue fear.

Villa received harsh criticism for sending his cavalry charging into territory not previously reconnoitered—something quite egregious for a command so heavily weighted in the mounted arm; nor did he employ his cavalry to hit Obregón's supply lines, which might have discovered the arrival of Alejo González's brigade and thrashed it before it had reached the safety of the Constitutionalist perimeter, much as his centaurs had done to Maycotte's brigade at El Guaje. Additionally, Villa did not maintain a reserve or even issue precise orders for the battle.

Nor was Villa's artillery arm up to the task. His guns were so dilapidated after four years of campaigning that the recoil systems sloshed fluid with each shot and had to be refilled with a mixture of glycerin and soapy water. Additionally, they were firing subpar shells (manufactured in Chihuahua City), which forced the pieces to get up dangerously close and conduct direct-fire missions in support of the infantry where Obregón's machine guns went to work on the gun crews. Other times, the Villista artillery fired high explosive shells at the city, which inconvenienced more than incapacitated those Constitutionalists inside the plaza.

Finally, one cannot help believing that if the Villista infantry had high crawled during the attack, it would have suffered much fewer casualties.

In more general terms, Villa's personal secretary, Colonel Pérez Rul, chalked up the defeat at First Celaya to a lack of preparation, rest, ammunition, and the necessary quantity of troops.

After extricating his troops from the battlefield, General Villa retreated to Irapuato and continued to collect stragglers and reorganize after the loss of so many soldiers. He also recalled minor units to join his immediate command: General Prieto arrived after quitting the Michoacán campaign as did more of the troops recently forced out of Jalisco; General Francisco Carrera Torres reported from the center of Guanajuato state, and even General Natera sent troops. Villa needed more, so he also telegraphed General Tomás Urbina and urged him to quickly conclude the Tampico campaign and report to Irapuato with his troops. The response from Urbina was a request for reinforcements, which Villa answered with nothing more than unreasoned advice: continue attacking. But Urbina faced a situation that was, if anything, worse than Villa's and not one that could be solved simply by throwing more resources at the problem, as Urbina quickly learned.

El Ébano, First Half of April

Urbina had arrived at El Ébano to assume command on April 1 and the next day greeted the Constitutionalists with a series of artillery salvos followed by a general assault. General Urbina's soldiers had never tasted defeat before, and after conquering Torreón and Zacatecas thought that with one forceful cavalry charge they could make short work of a little "ranch" in a few hours. General Treviño's Constitutionalists waited until the Villistas were almost on top of them and then opened fire, decimating Urbina's men just as they had done to Chao's earlier. Rather than volleys, the Constitutionalist soldiers fired at will ("*a salvo*"). Some Villistas made it past the wire, and in some sectors of the line hand-to-hand combat was reported, but those were the exceptions to the rule, according to which the Villistas reformed and delivered sequential charges, none of them successful.

One of the subsequent charges was noticeably feeble and in fact consisted of raw recruits, students, and policemen from San Luis Potosí pressed into service. Many of them dispersed or surrendered to the Constitutionalists. Evidence also suggested that ex-Federal officers had been employed in their previous roles to keep desertions to a minimum.

The next day Urbina made an attempt against the pumps on the Tamesí River, which anchored the Constitutionalist right and supplied water to the Constitutionalists. But that attack likewise failed.

Then Treviño employed a ruse to make the Villistas think he was receiving reinforcements by rail: infantry detrained at the station in full view of the Villistas and then, taking a concealed route to march to the rear, again boarded the trains and returned to the station, repeating the exercise.

In subsequent days, the Villistas continued to skirmish and relied on their superior artillery to constantly harass the Constitutionalists. Their earthworks consisted of three lines of defense designated the front, chain, and support, which represented a static form of the ternary system of echelons, which further evidenced the presence of ex-Federals. These lines were not slit trenches, but rather a series of individual fighting positions with loopholes and graduating depth to allow for shooting from the prone and kneeling positions. From those positions the Villistas delivered major pushes on April 7 and April 13, perhaps consciously in conjunction with the two battles at Celaya, but neither resulted in a victory for Urbina.

Second Celaya, April 13-15

Obregón continued to receive arms, ammunition, and reinforcements bringing his total effectives up to about fifteen thousand men, almost evenly divided between infantry and cavalry. But communications between Veracruz and Diéguez still had not been reestablished after the Battle of First Celaya, so Obregón did not know the status and disposition of the Constitutionalists coming from Jalisco. In the meantime Obregón prepared to face Villa alone, creating formidable defenses. He had infantry in foxholes on the west and north side of the city about one kilometer from the suburbs—the direction of the expected attack—and cavalry on the east and south sides with machine guns distributed along the perimeter. His artillery crews had measured distances to where Villistas guns had gone into battery in the previous battle and their pieces remained arrayed on a line parallel to and behind the infantry on the west side. The perimeter was divided into three sectors, each with its own reserves and headquarters linked by telephone or telegraph to Obregón's headquarters.

Castro's cavalry was posted thirteen kilometers south of the city and hidden in a thicket. At the appropriate movement it would move up to within seven kilometers of the city and be prepared to deliver a flanking movement in support of the infantry counterattack, a textbook plan of defense for a plaza according to Federal Army doctrine. Heliograph and flag signals were to be used to signal the attack instead of the insecure telephone and telegraph. Most of the trains had been removed with the impedimenta to Querétaro—except for the hospital trains, those of the "victory convoy" that arrived on April 12 with another one million rifle cartridges and three hundred .30-30 rifles, and the ones transporting General Gavira's brigade that rolled into the plaza in the early morning hours of April 13.

About an hour after Gavira's arrival, the Villistas began departing Salamanca for Crespo where their infantry detrained. They encircled Obregón's perimeter and initiated combat. But their attacks spread out, weakened, and broke before they could fall upon the Constitutionalist line. General Gavira reported the effects of the Villista cavalry charges on the north side:

> They started out hell for leather and then little by little decreased their speed until coming to a stop. When this happened the distance that separated them from our trenches was from 200 to 300 meters, that is, in effective range of our Mausers. Our soldiers then were shooting their weapons, aiming while entrenched and began to kill them, which started them little by little turning their horses, presenting a better target, and soon turning completely around, leaving on the field a multitude of dead, without having caused but little damage.

Villista Colonel Gustavo Durón González seemed to agree with Gavira's characterization of the assaults, stating later that "from where we were [south, on the Villista right], our artillery opened fire furiously without any infantry or cavalry ostensibly advancing," although he did allow that "it is probable that in some other sector it [spectacular assaults] may have happened." The entire Villista effort seemed to have no coordination, no effective command and control between the various units, the staff, or with Villa, who rode up and down the line, issuing one-off commands. At one point, the commanding general ordered Durón to send one cannon to the east side to assist the assault taking place there, but Durón properly observed that "one cannon is a useless thing, it can be said it is good for nothing. Ángeles had taught us how to use them massed. So...the three batteries in my command shot together at a determined objective to make their employment effective." Durón succeeded in getting permission for an entire battery to go.

After two days of combat, Obregón sprang the trap. His infantry in the center made a forceful push forward and the cavalry attacked on Villa's flanks. At first Villa thought the cavalry might be the arrival of his Zapatista allies since he had sent a request through Convention President Roque González Garza for Zapata to attack the Constitutionalist rearguard, but no. It was Constitutionalist cavalry.

As a result, Villa suffered even greater losses than in the first battle, probably some 3,500 men and untold materiel. The Constitutionalist casualties were much lower, officially just over four hundred. The adverse effects of the surprise

Constitutionalist counterattack were compounded for the Villistas by jefes of the infantry support who did not let their artillery counterparts retreat until it was too late, with the result that all their guns were lost to the Constitutionalists.

Still, one cannot discount the "brilliance" of Obregón in simply arranging his defense of the plaza of Celaya according to tactical doctrine published almost twenty years earlier: paying due attention to communications and stationing troops outside the perimeter for a flanking counterattack to be executed in combination with a push from inside the plaza. Obregón also utilized interior lines on a tactical level to rush reinforcements, namely General Gavira's infantry, to various vulnerable points on the line, especially those manned by the untested Red Battalions.

In the final analysis, though, both of Villa's defeats at Celaya were self-inflicted. He attacked not once but twice without having the requisite men and ammunition to achieve victory. Similar to General Robert E. Lee and the Confederacy, which had the advantage of interior lines on a strategic level and suffered crushing defeats at Sharpsburg and Gettysburg, General Villa failed at Celaya fighting on unfamiliar ground without the previously enjoyed advantage of intelligence and close proximity to his supply lines. Villa's Conventionists, like Lee's Confederates, shared a boldness borne of tradition and necessity. Additionally, Villa did not retain a reserve in this or, really, any battles. On the other hand, Obregón did maintain reserves—consistently. And Villa, if he were truly a student of the art, would have known of this practice by Obregón and never for a second would have confused a flanking cavalry counterattack with imagined Zapatistas acting according to a coherent overarching strategy.

His Suriano allies were simply unwilling and incapable of acting in concert with the Villistas. As Venustiano Carranza's chief of staff, Juan Barragán, stressed: "For further clarification, in the campaign plan arranged with Villa, the Zapatistas had *no other mission* [emphasis added] than to interrupt communications between Obregón's army and the port of Veracruz. Villa trusted that his allies, at least, would lend this cooperation." Moreover, "if there had been any difficulty for the passage of General Norzagaray's ["victory convoy"] train, or better said if the Zapatistas had destroyed a short section of the railway between Ometusco and Tula, surely the army of General Obregón would have been destroyed in the battle that was about to be waged." Numerous historians have tried to exonerate the Zapatistas for their inexcusable parochialism, but the facts are incontrovertible. The Surianos failed their Norteño allies egregiously and unforgivably.

Officially, Villa blamed his defeat on a lack of ammunition, a condition for which he implicated the United States government. While no one denies that he was short of ammunition from the outset of the battles, his case against the Americans is devoid of any proof and defies all logic. If any grand narrative properly and succinctly describes the battles of Celaya, it is not to be found in alleged covert machinations by the U.S. government, but rather Villa's gross incompetence and the superiority—in 1914—of entrenched infantry and machine guns over Villa's favored cavalry-artillery combination.

Northeast

The fighting in the Northeast was no more successful for the Villistas than battles waged at El Ébano or in the Bajío. By the end of March the Constitutionalists in Nuevo Laredo were preparing to defend against a pincer attack consisting of Villista General Rosalío Hernández advancing down along the Rio Grande from Piedras Negras, while General Orestes Pereyra's Villistas marched northward from Monterrey. The Americans were concerned about another battle on the border, and General Robert R. Evans informed General Maclovio Herrera, the Constitutionalist commander, that no Mexican combatants would be allowed to enter the United States, meaning to seek medical care or sanctuary. Most likely because of diplomatic sensitivities, Herrera moved south to Huizachito Station where he deployed and waited for Pereyra. On April 12, Herrera defeated Pereyra and chased his fleeing Villistas all the way to Candela. Hernández did not continue the campaign and slowly retreated back north to Piedras Negras.

The day after Herrera's victory at Huizachito, and after almost three weeks of encirclement, the Constitutionalists inside Matamoros made a frontal assault in the center, pushing back the Villista outposts while General Vázquez led his troopers from the 1st Division in a sweep from inside the plaza. The 1st Division soldiers attacked General Rodríguez's headquarters at Rancho San José de las Rusias, recovering booty and killing Rodríguez's second-in-command, General Saúl Navarro, before retreating back to the safety of the Constitutionalist lines. The operation resulted in no change in position and surprisingly few casualties, with the total killed on both sides estimated at less than fifty. Three days later, the Villistas announced to U.S. Army authorities that they were lifting the siege and began pulling back to Monterrey, and ultimately to the Bajío.

Villista sympathizers have tended to credit General Nafarrate's numerous machine guns for the Constitutionalist victory, somehow implying that the United States bore some responsibility for this imbalance of forces. Yet Villa could have purchased machine guns from the Americans just as easily as the Constitutionalists had. Most probably he did not invest in them because there was no doctrine that explained how to employ machine guns with mounted troops, nor would there be one until the advent of armored cavalry. Villa was much more enamored of field artillery that complemented his force composition overwhelmingly weighted in dragoons.

A second factor in Rodríguez's failure, mentioned by Villa's private secretary, was that he did not have enough men to accomplish the task.

El Ébano, Second Half of April

After repulsing Urbina's reinforced probe of April 13, General Treviño agreed to General Lárraga's request to execute a rearguard action against the Villistas. Lárraga led the southern part of a pincer movement that was to be complemented by General César López de Lara of the 5th Division on the Constitutionalist right. It is probable that General López de Lara's commanding general, Luis Caballero, overruled his participation because López de Lara never showed and subsequently González complained to the First Chief that "the operations of General Caballero's forces have not been favorable, by virtue of the fact that he effects all his movements in complete disaccord with the orders dictated by this headquarters." Caballero had more immediate concerns, namely, defending Ciudad Victoria against Villista Generals García and Ceniceros, who defeated López de Lara at Puerto del Aire on April 16. Two days later Caballero abandoned Ciudad Victoria to the Villistas. Without López de Lara to form the right side of the pincer, Lárraga moved south into the Huasteca, a region swarming with Conventionists.

Indeed, on April 14, the Constitutionalists had been compelled to land five hundred men and numerous rifles to counter the attack directed by Conventionist General Manuel Peláez against Tuxpan, a potential sea entrepôt for the Conventionists and a threat to the southern flank of those Constitutionalists at El Ébano.

With Peláez's offensive thwarted, the danger to General Treviño's position at El Ébano from the north, likewise, eventually diminished. The Villistas of Generals García, Ceniceros, and Carrera Torres could not press forward against Tampico because their operations against Caballero's division had taken too long to develop. By the third week of April the Villista campaigns to capture Nuevo Laredo and Matamoros had ended in failure, and the Constitutionalists had returned to put pressure on Monterrey, forcing García and Ceniceros to leave Ciudad Victoria in the hands of Carrera Torres and begin pulling back toward the capital city of Nuevo León to avoid being cut off.

Meanwhile, El Ébano had become impregnable. The Constitutionalists continued to perfect their trench works and General Treviño ordered improvements to the telephone lines connecting his headquarters to the various sector commanders. He also posted two searchlights on El Ébano Hill to light up the battlefield at night in response to requests from the trenches, and Major Alberto Salinas Carranza and his aviation "fleet" that consisted of two planes arrived at El Ébano from Yucatán. Initially, the airplanes dropped propaganda leaflets and when the Villistas came out of their trenches to shoot at the planes, the Constitutionalist infantry on the firing line picked them off. Later, the airplanes dropped bombs on the Conventionist artillery, trenches, and headquarters as an armored gondola loaded with infantry attacked down the railway supported by ground forces in a form of proto-blitzkrieg.

With these developments in mind, General Urbina wisely started to consider Treviño's flanks in earnest. The Villistas hacked out a breach through the brush with machetes and used the new trails to slip into Treviño's rear and create mayhem. But the path was too narrow for a large-scale penetration, and the Constitutionalists repulsed the raiders and implemented countermeasures to prevent any recurrence.

A few days later Constitutionalists in and around the town of Pánuco reported "persistent rumors" that a combined-arms column of two thousand Villistas was on the march to attack the linchpin plaza. Treviño sent a brigade to reinforce Pánuco but his aerial reconnaissance also reported the continuous arrival of trains bringing men and materiel to General Urbina which redirected his attentions to the front. Then on April 29, the Villistas unleashed a slow cadence artillery barrage against the Constitutionalist defenders of El Ébano. The Constitutionalist guns on the left could not respond because their recoil systems had been damaged. Urbina then launched a general assault to start what would become five of the most brutal days of combat. It was a diversion that allowed Urbina's flanking column to take Pánuco on that first day, April 29. But for unknown reasons, the Villista column simply remained there without advancing any farther. General Pablo González sent two columns into the Huasteca that finally forced the Villistas out of Pánuco on May 6.

A New Battle in the Bajío

After the Battle of Second Celaya, the Villistas pulled back to Irapuato and, subsequently realizing he had to permanently foresake the states of Jalisco and Michoacán, Villa moved even farther north to Aguascalientes, leaving lead elements behind in León. The Constitutionalists cleaned up the battlefield, incinerated the dead, and slowly began moving northward. Obregón's Army of Operations stopped in Irapuato to wait for the arrival of Generals Diéguez and Murguía from Jalisco.

At the time, Villa controlled the National Railroad between San Luis Potosí and San Miguel de Allende with General Pánfilo Natera's men along the Dolores Hidalgo-Guanajuato line threatening Obregón's right and causing the Constitutionalist general to approach León very cautiously. Obregón sent General Porfirio González and his brigade to Guanajuato to safeguard his flank, while Maycotte's cavalry brigade in the vanguard skirmished with probes sent by Villa around Los Sauces Station and Hacienda Nápoles.

Around this time, General Felipe Ángeles joined Villa in the Center and recommended a "serious and ordered reorganization" of the army and a lengthy period to recuperate from the loss of men and materiel and its attendant impact on morale. He suggested pulling back to Chihuahua, maybe even as far as Sonora, but at least as far as Torreón. Villa would not listen because he thought that to retreat would be to admit defeat, and giving up so much territory might lead the United States to grant recognition to Carranza's government. In that case, Ángeles urged Villa to pull back to Aguascalientes, at a minimum, dig in and force Obregón into a tactical and strategic offensive. He reminded Villa that the same aggressive tactics employed against the Federals in 1914 would not work against the Constitutionalists who were highly skilled, motivated, and numerous. Villa again demurred. The Bajío was Conventionist in sentiment and he could not just abandon its people to Constitutionalism. Moreover, digging in and going on the defensive after such a loss as suffered at Celaya would further impair, not repair, morale. Villa then asked Ángeles to scout the area around León and give his assessment of a possible pitched battle in the area. The report was discouraging. There were no natural defenses, no terrain features upon which to anchor one's flanks, and therefore the front would naturally be extensive and result in the army being widely dispersed. Once again Villa discarded Ángeles' appreciations and went ahead with his plans. He reasoned he could use the Arroyo de Asperas as a natural trench and La Cruz Hill and Mirador as anchors and observation points on the left.

On April 25, General Obregón went to meet Generals Murguía and Diéguez in Pénjamo and La Piedad, respectively. The long-awaited link-up was now complete. After reviewing the troops, the commanding general ordered Murguía's cavalry to advance to La Romita on horseback while the infantry rode the rails to Silao and disembarked to form a large defensive perimeter around the town, similar to what had been done at Celaya.

Four days after the linkup, on April 29, the Constitutionalists made a push to occupy Trinidad Station. After a brief engagement with Villista cavalry, Obregón's infantry detrained and started to dig in around Trinidad Station while he took a scout train to probe toward León. Villista cavalry suddenly charged out of León like a swarm of angry hornets headed for the scout train and the five hundred Constitutionalist cavalry that had disobeyed orders to remain at Trinidad. The scout train had to pull back slowly in order to cover the retreat of friendly cavalry who, after the day's travel and fighting, were exhausted and more a vulnerability than an asset. At length the Constitutionalists regained the safety of Trinidad and fighting for the day came to a close. The next day, however, Villista cavalry, far superior in numbers and quality than any possessed by the Constitutionalists, smashed Murguía's 2nd Cavalry Division on the Constitutionalist left, forcing him to retreat to La Romita where his men rallied and dug in. The Villistas returned to León and the next day Obregón sent two regiments and a small brigade to reinforce Murguía, who complained bitterly that Obregón did not come to his rescue during the fighting because he did not want to disrupt his "little figure," meaning his infantry square. Obregón countered that Murguía had exceeded his authority, advanced too far toward León, and allowed his troops to be forced into fighting a battle on terrain of the Villistas' choosing. Either way, the episode only added to the estrangement between the two generals.

Villa continued to probe the Constitutionalist lines, sometimes personally accompanying his Dorados, trying to ascertain the particulars of the disposition of forces. He also sent an airplane to scout the battlefield as shipments of arms and ammunition as well as reinforcements, including the Cedillos pulled in from El Ébano, continued to arrive.

MAY

Trinidad Station, May – June 5

By May 7, General Obregón had created an infantry square centered on Trinidad Station, a grand implementation of the concept, but not Mexican Army doctrine as some have claimed. The Constitutionalist cavalry remained on the flanks and to the rear, outside the infantry square for various reasons among which the lack of sufficient water for such a large force in such a constrained area figured prominently. Villa's infantry moved in and occupied lines on three sides of the square, which actually more resembled a rectangle, opposite the Constitutionalist front and laterals, on external lines of communication.

Obregón's generals suggested an attack against Villa while he was still in the process of receiving reinforcements and before he had concentrated all his forces, but the Constitutionalist commander wanted to wait until Villa had put his infantry and artillery into place because the Villista cavalry could maneuver too easily out of defeat and Obregón wanted to destroy those less mobile arms.

During this time, the Constitutionalist 3rd Brigade of the 1st Infantry Division had to leave Trinidad for Colima to protect the rail line to Manzanillo which had come under threat by Conventionist marauders. Obregón needed to keep that line open because if the Zapatistas succeeded in severing his direct lines of communication to Veracruz, he would need to be supplied from the West Coast via the Isthmus of Tehuantepec.

On May 11, the Constitutionalists performed reconnaissance by fire to locate weak points on the Villista line, and the next day General Obregón ordered a push against the Villistas occupying a series of hills opposite the Constitutionalist right. In response, Villista cavalry exited León and counterattacked. In a deep line of foragers the Villistas started at a trot, and then between four and five hundred meters from the Constitutionalist infantry square they fired off all five rounds of their Mauser carbines, sheathed them, and then filling both hands with pistols, reached a full gallop. Only the first rank riding hell for leather could be discerned in the approaching onslaught that forced the Constitutionalist skirmishers to retreat to the safety of the square, at which point those on the firing line unleashed volleys at the Villista cavalry. The Villistas returned fire and then making a half turn sought a weak point in the defenses and, finding none, returned toward León. General Obregón was so impressed with the charge that day that he later wrote, "In none of the campaigns that I have found myself—against Orozco, against Victoriano Huerta, and against Villa—did I witness a cavalry charge so brutally delivered as that of the Villistas on that day."

It had made an impression on Villa, too, one of disgust. The attack had been the brainchild of General Fierro, who had led that charge with General Villa's Dorados because apparently he was upset that so many had joined the escort, and he wanted to thin their ranks saying, essentially, "I see many new Dorados for the escort of General Villa...! So many are not needed; I see the escort now very large, you have to send it into combat to thin it a bit!" The loss of so many of his Dorados infuriated Villa who ordered Fierro to take his own Fierro Brigade and charge against La Cruz Hill, one in the series of hills that Obregón's Constitutionalists had taken on May 12.

Two days later Fierro retook La Cruz Hill, but an aide later recounted: "It was a triumph with enormous casualties for the enemy, but to get there and remove them from their trenches cost many valuable lives of our own and many wounded, among them General Fierro himself, who continued fighting right up to the edge of the stone walls with his grand roan horse with white hooves, to kill Yaquis with his pistol."

By mid-May, the Battle of Trinidad Station had settled into trench warfare when Obregón received news that an attack on Celaya, to his rear, was imminent. Therefore, he ordered Generals Murguía and Martín Triana to take their forces and smash the Villista column at Dolores Hidalgo and then proceed to destroy the rails between Aguascalientes and San Luis Potosí farther to the north. The day Murguía left to begin the Dolores Hidalgo mission, May 22, just so happened to coincide with the Villista evacuation of Monterrey, and Villa's next big push at Trinidad Station.

In the May 22 assault, the infantry on the Villista right advanced to within one hundred meters of the Constitutionalist firing line, trying to get into range to lob grenades into their trenches. At the same time, Villista artillery hammered the Constitutionalist left at Hacienda Santa Ana del Conde, killing several Constitutionalist gunners and mules. Constitutionalist General Gavira remarked:

It is probable that more intense and better-directed shelling will never again rain on us like that of the 22nd. The shells exploded over our heads with mathematical precision, inflicting little damage upon us, nonetheless. One shell falling four meters in front of me but without exploding filled the jar of milk that I was drinking with earth. Another fell on the 'Oasis House' [his field accommodations] and all very close. In truth, the artillery very necessary for us in certain cases only produces an effect on morale when troops occupy positions as we were doing, without bunching up.

After that preparatory artillery barrage, the cavalry on the Villista right charged. The Constitutionalists repulsed the first two assaults, and afterward, the Villista infantry entered into action riding on the backs of the cavalry, then dismounted at three hundred meters distance to assault the trenches. The Constitutionalists had to withstand fourteen such assaults. Few Villistas who actually made it into the Constitutionalist trenches escaped alive.

When the frontal assaults failed, the Villista cavalry made a flanking maneuver around the Constitutionalist right to attack their rearguard near Hacienda Los Sauces seeking to cut their lines of communication until Murguía's division—recalled from its Dolores Hidalgo mission at the outset of the Villista attack—reinforced Castro's cavalry and together repulsed the column.

At the end of the day's fighting, dead Villista infantry were lying in rows as if they had been electrocuted at once, according to Obregón, most likely complete echelons of assault killed at a time. General Gavira reported "that our soldiers were entertaining themselves in checking the pockets of the dead, finding many wealthy Jefes. All were carrying well-manufactured hand grenades that they intended to use to attack our trenches." The Constitutionalist artillery fired 1,800 shells in repulsing the attack, and Obregón estimated that Ángeles' guns had at least doubled that number. After that push, there was very little in the way of combat around Trinidad until the end of the month. However, Villista activity around Guadalajara started to pick up and once again Obregón had to pull more men, this time two battalions, off the perimeter and send them to reinforce the chief of operations in Jalisco, General Enrique Estrada.

After weeks of trying to follow the advice of Ángeles to remain on the tactical defensive, Villa had an idea he wanted to run past the ex-Federal; he wanted to attack Obregón's rear with all his cavalry. Ángeles was concerned that removing artillery support and reserves for the maneuver would leave the artillery and infantry vulnerable. If Villa wanted to deliver his surprise attack, he would have to conclude the attempt quickly and return to again support the infantry. Villa agreed.

In the early morning hours of June 1, the Villistas made a surprise attack on the Constitutionalist left, to the west. Simultaneously, after making a roundabout detour behind the hills to the east, the bulk of Villa's centaurs emerged out of the morning sun and attacked Hacienda Nápoles, hitting Castro's Constitutionalist cavalry so hard that it was sent reeling. Castro's troopers raced around the corner of the infantry square to join Murguía on the west side, where the day's action had started. Villa then assaulted the back side of the square hoping to capture or destroy the military trains. Some of his men reached the third echelon of trenches before being taken prisoner but most never made it that far before being killed or turned back. Soon the Constitutionalists noticed smoke billowing up from Silao where a smaller Villista column under General Chao, recently pulled from El Ébano, had captured the Constitutionalist supply trains, hospitals, and—even worse—the Yaqui women and children, the Achilles' heel of that bellicose tribe. Once the Constitutionalists realized they were cut off and their wounded and families killed or captured, morale started to slump. Villa had finally done what Zapata had been unwilling and unable to do, sever Obregón's lines.

After Villa's bold rear attack of June 1 at Trinidad, the Villista leadership realized that the lack of reserves for their principal front left them imperiled.

Ángeles ordered an inspection of the lines that returned a woeful report.

> There are a great many unburied cadavers and the stench was unbearable. After our "misguided brothers," the Carrancistas, our worst enemies are the flies, lice, and rats. The flies are precious, green peacocks, and there are thousands that, from the mouths and eyes of the cadavers fly to take up residence in our food. The rats are so voracious that in spite of having bellies full from the flesh of the dead, they go to take bites of our meager provisions. The lice, large as grains of rice, scarcely can move they are so engorged and produce sores, although by now we are well vaccinated. Two or three days after bathing we are once again lice infested. After eight days without removing our clothes we can barely sleep, at times, in the trenches themselves.

General Murguía finally tired of the stalemate and informed Obregón of his decision to attack on the morning of June 3. Obregón agreed but requested a one-day delay so that the entire army could coordinate an assault. In the meantime, Obregón got runners to make it south and tell General Joaquín Amaro in Celaya to move up to La Romita and at the appropriate time attack Silao. The support troops and office personnel attached to headquarters then created a smaller square inside the larger disposition with three hundred meter laterals around the station as a contingency.

That afternoon Obregón was wounded by Villista artillery and lost part of his right arm; he was out of action.

On June 4, General Hill convened a council of war comprised of the four division commanders. It had previously been arranged for Hill to assume command should Obregón become incapacitated. Hill laid out the general situation for all to consider: the commanding general was wounded and in the hospital, the supply lines were cut, and stockpiles of ammunition and morale were on the decline. Generals Hill and Diéguez, commanders of the infantry divisions, wanted to retreat to Irapuato and receive resupply via Guadalajara. Murguía strenuously objected: the infantry could simply load up on trains and pull out of battle, but the cavalry would have to cover their retreat and against the superior Villista cavalry would incur tremendous casualties.

Murguía then said that the next morning he was going to push back the Villista right wing and attack León and that General Alejo G. González, who commanded one of Castro's brigades, had agreed with him and had already left with 2,500 men to cut the Villista lines of retreat between San Francisco del Rincón and Lagos, to the north of León. With that information, the infantry commanders had to back Murguía's move and prepared to support the cavalry's maneuvers, but it was evident from the exchange that Hill and Diéguez did not fully support the plan, and their reticence was borne out in succeeding day's combat.

Murguía did attack the next day, supported by a single brigade of Constitutionalist infantry that left the security of the square to assault and reach León. However, the Villistas reformed their lines and the rest of the Constitutionalist infantry remained in their trenches. Villa did not become aware of the breakout until mid-day, at which time he called in his reserves from the south, hidden behind Hacienda Nápoles by the hills and mesquite. It would seem that Villa had in mind a repeat of the road to Guadalupe massacre during the Battle of Zacatacas, with a strong contingent of troops aligned along the presumed direction of retreat that the fleeing Constitutionalists would have to take. What he got instead was a reprise of the Battle of Second Celaya. When his troops saw the large dust cloud of their fellow Villistas coming around the flank from down south to push the Constitutionalists out of León, they assumed the cavalry was Constitutionalist cavalry. Just as at Celaya, they figured this was a counterattack by Obregón's reserve, and they fled into the mountains. Had it not been for that lack of communication—if Villa had only sent word to his subordinate infantry generals that large numbers of Villista cavalry would be passing behind their lines—the entire debacle might have been avoided. Moreover, if General Murguía had not imposed himself on his fellow Constitutionalist generals, it is quite possible the result might have been a disaster of epic proportions for the Constitutionalists. In sum, it is highly conceivable that Villa might have won the battle.

It was only after the Villista infantry opposite Diéguez became disoriented and fled that the rest of the Constitutionalist infantry became enlivened and pushed forward to put the Villista army to headlong flight. The Constitutionalists captured a paltry six cannons, thirteen machine guns, and about three thousand firearms with some ammunition. The booty might have been much greater, but Villa's trains in León had been protected by the city's garrison long enough for them to escape so that Villa did not lose one train, and General Alejo González got lost and failed to cut off the Villista retreat. The following day General Amaro and other Constitutionalists chased the remaining Villistas out of Silao.

Villa reached Lagos de Moreno and dispatched orders for his men to concentrate around Aguascalientes for the next great battle.

The Northeast

After the battles of Celaya, Villa had begun drawing down the troops in the Northeast, emptying garrisons and returning key units, such as his own Villa Brigades, to the Bajío. El Ébano was no exception; it was estimated that he had removed eight thousand men from under Urbina's command, including those belonging to Carrera Torres and the Cedillos.

By the second week of May, the Villistas besieging El Ébano tended to rely more on artillery, with duels between guns becoming the norm. Urbina had access to shells manufactured in Chihuahua while the Constitutionalists did not have a shell production facility. At one low point during the battle, the batteries on the Constitutionalist right had been reduced to ten shells on hand.

On May 14, the Constitutionalists went on the offensive, captured a portion of the forward line of Villista defenses, and then incorporated that section into their system of defenses. This sequence of events upset the integrity of the Villista chain of shooters, making the support ineffective, and forcing the Villistas to rely solely on the reserve, speaking in terms of the three echelons of defensive works.

Thereafter, General Pablo González handed over command of the Army Corps of the Northeast to General Treviño and set sail with his staff and a few select troops for Veracruz in order to receive his new assignment as commander of the Constitutionalist Army Corps of the East.

The Battle of El Ébano finally came to a close on May 31. An armored train bearing machine guns and infantry supported by artillery fire pierced the Villista center, which isolated the Villista left and right wings. The Constitutionalist infantry then made a push on the left and right with the modest goal of breaking up the integrity of the Villista lines and pushing back their left: the maneuver exceeded all expectations and turned into a full-fledged rout. By breaking through the Villista center, the two wings remained in danger of being cut off and defeated in detail and therefore had fled the field.

Urbina pulled back to San Luis Potosí—which he promptly sacked—and then joined Villa at Aguascalientes.

By the end of May the Constitutionalists had captured all three capitals in the Northeast: Monterrey, Saltillo (aided by the Gutierristas), and Ciudad Victoria, all as a result of Villa's forces being weakened by numerous defeats and the removal of key units to the Bajío for his great gamble at Trinidad.

6. CIVIL WAR, JUNE – DECEMBER 1915

In the last week of May, it looked as if Villismo was collapsing in the Northeast: the Constitutionalists held Monclova and Saltillo in Coahuila, all of Tamaulipas, and most of Nuevo León. Villista General Raúl Madero was in Paredón, a crucial crossroads regarded by many as the key to Coahuila, where he was hemmed in from three sides, north, east, and south. Going on offense, Madero recaptured Monclova on May 26 and began moving toward Monterrey. At Icamole, he attacked the Constitutionalist 1st Division of the Northeast on June 6 and again four days later. The ensuing Battle of Icamole would exceed El Ébano in duration and involve a reversal of roles inasmuch as the Villistas at Icamole blocked the advance of Treviño's Constitutionalists, thereby denying them access to the coal and cotton of northern Coahuila and La Laguna, respectively.

JUNE

Treviño and Obregón

The commander of the 1st Division of the Northeast, Poncho Vázquez, was mortally wounded in the June 10 attack on Icamole and died a few days later. The Constitutionalists in the Northeast had never recovered after the debacle of Ramos Arizpe, and now Vázquez had been killed and Maclovio Herrera was dead from a tragic friendly fire incident in April. Carranza needed a strong commander in the area, so he decided to take the San Luis Potosí objective away from Treviño and reassign it to Obregón. In turn, the First Chief tasked the hero of El Ébano with the defense of Monterrey and promised to send him plenty of cartridges. Treviño was to establish advance positions either at Icamole or Paredón and keep the rails between Piedras Negras and Monclova interrupted in order to deny coal to Villa's army. Upon receiving this new assignment, Treviño immediately ordered the various commands of the 1st and 5th Divisions to begin concentrating in Monterrey for the upcoming campaign.

Meanwhile, on June 13, the Villistas under Madero recaptured Saltillo without a fight, most likely taking advantage of the disorganization between the Gutierristas and Constitutionalists inside the city. President Eulalio Gutiérrez had officially resigned as President of the Convention on June 2, 1915, in Ciénaga del Toro, Nuevo León, and had gone into exile. With Saltillo the Villistas now controlled the line all the way to Piedras Negras and had an estimated four thousand soldiers remaining in the capital of San Luis Potosí. The Constitutionalists feared that the four thousand might join with Pánfilo Natera's two thousand Conventionists in Dolores Hidalgo and advance toward Celaya and Querétaro, cut Obregón's lines of communication, and link up with the Zapatistas in Mexico City. That chain of events would force the Constitutionalists in the Bajío to have to receive their supplies on exterior lines, through the Isthmus of Tehuantepec to Salina Cruz, on the water, and then inland via Manzanillo.

Instead, General Tomás Urbina left San Luis Potosí with most of the Villistas and headed north through Coahuila—facilitated by Madero's recapture of Saltillo—and joined General Villa in Aguascalientes, where the Villistas were now determined to assume a defensive posture and let the Constitutionalists attack.

Since there were no longer any sizeable Conventionist units in Jalisco that might present a strategic threat, the First Chief urged Obregón to hurry on his way to Aguascalientes, capture that city, advance with the bulk of his army to San Luis Potosí, and then link up with the Constitutionalists repairing the rails from Tampico. Once the railroad between San Luis Potosí and Saltillo had been repaired, he would march north and conquer Saltillo in cooperation with General Jacinto B. Treviño, if the latter had not already done so. The combined forces of Obregón and Treviño would then clear out northern and central Coahuila and attack Torreón, where the First Chief expected Villa to make his final stand. A smaller column from Obregón's Army of Operations would march north from Aguascalientes to Zacatecas and act as a blocking force to keep Villa from trying to drive southward, and then at the appropriate moment join in the attack on Torreón, or drive into Durango and capture that state in combination with the Arrietas. Carranza suggested that General Murguía should lead the Zacatecas column since he knew the area from the time of the Orozco Insurgency. Finally, he recommended sending a 2,000-man cavalry force from Coahuila through Cuatro Ciénagas and into Chihuahua to cut Villa's lines of communication and block his retreat from Torreón.

Map 7 Constitutionalist Strategy and Movements, June – October, 1915

On June 20 General Treviño arrived at Monterrey with the main body of his forces and quickly organized all the various leaders now under his immediate command in preparation for driving on Saltillo, where the Villistas were concentrating "a great number of elements." He also began forming a "strong defense in Villa García to defend Monterrey." That same day, Obregón's headquarters reached Encarnación de Díaz. According to what was now becoming unwritten Constitutionalist tactical doctrine, Obregón ordered General Hill's infantry to form a square around the city and pulled Murguía's cavalry inside the protected zone and waited for the next convoy of supplies to arrive from Veracruz.

After "meditating" on the First Chief's proposed strategy and discussing it with several jefes familiar with the region—and guessing that Saltillo had been recaptured by the Villistas based on "the tenor of your message"—Obregón first asked the First Chief to keep him constantly apprised of such developments in the North. He went on to explain that if Saltillo had indeed been recaptured then the

Villistas "could establish railroad communication with San Luis [Potosí] and we would be in very difficult conditions if we [did not] succeed in occupying San Luis before said communication were established." Therefore, Obregón recommended the Constitutionalists around El Ébano make haste to capture San Luis Potosí while he prepared a cavalry force under General Joaquín Amaro to march against the same objective from the south.

At this time, General Amaro's "stripers" (so-called because they wore prisoners' uniforms acquired from the jail in Celaya) provided protection for Obregón's supply lines, keeping an eye on the sierra in northern Guanajuato for any Villistas leaving León in the aftermath of the Battle of Trinidad Station. Amaro specifically monitored San Felipe Torres Mochas and San Miguel de Allende, where it was rumored that the Cedillos were about to evacuate, although it was not believed that they would try to move south to Querétaro. Indeed only minor advance forces were reported in Dolores Hidalgo, San Miguel de Allende, and San Diego, none more than one or two squadrons in strength, to provide security to those Villistas remaining in San Luis Potosí.

Carranza responded to Obregón's strategy counter proposal by informing him that the Villistas in Saltillo could not repair the railroad to San Luis Potosí because it was completely destroyed, and that General Herminio Alvarez was marching on San Luis Potosí, and with General Luis Gutiérrez would block any Villistas trying to drive southward from Saltillo. It must be pointed out that none of those factors had impeded Urbina's retreat through northern San Luis Potosí.

Nevertheless, after receiving assurances from Carranza that the Villistas would not be able to advance down from Saltillo to San Luis Potosí, Obregón agreed to Carranza's strategy. He only needed to wait for the convoy that left Veracruz on June 21 to arrive with one million pesos before marching on Aguascalientes.

The convoy in question reached Pachuca without incident, and at that point received an escort train consisting of General Abundio Gómez's "stripers." The convoy was arranged in a particular sequence and with categorical orders, to wit: the scout train, which doubled as the repair train, at the front, with the resupply convoy in the middle, and then followed by Gómez's escort train. The trains had orders not to lose sight of each other except around curves, and if a train lost sight of the others, the engineers were to stop their trains immediately.

During this time Villista activity continued between Dolores Hidalgo and San Miguel de Allende, so Obregón ordered garrisons established to the south to protect his rearguard. He also directed Diéguez to continue to León from Irapuato instead of going to Celaya, which Diéguez did that very day. Finally, Obregón tired of the continual threat to his rearguard, and had Diéguez send a detachment from General Gonzalo Novoa's brigade commanded by Colonel Natividad Sánchez to destroy the rails north of Dolores Hidalgo and if possible to clear the Villistas out of San Felipe

Torres Mochas. Sánchez accomplished the mission on June 27. That development forced the Villistas to pull back to San Luis Potosí. The next day five hundred troopers from General Amaro's brigade occupied Dolores Hidalgo and San Miguel de Allende and reestablished telegraphic communications to Celaya.

The most harrowing news for Obregón, in the midst of all this activity, came from General Cesáreo Castro's scouts who reported a strong Villista column marching to the south from Aguascalientes through the Plains of Tecuán. As a precaution, Diéguez moved up to Lagos to protect the convoy from Veracruz bearing the pesos and ammunition that was expected to arrive soon. In the early morning hours of June 30 the line to Lagos went dead as the 3,000-man raiding party led by Villista Generals Canuto Reyes and Rodolfo Fierro attacked. The fighting was fierce and Diéguez was hit by a soft-nosed bullet that almost caused him to lose his arm, but doctors managed to save it. The Villista column broke contact and continued with its mission. Hours later the Veracruz convoy safely entered Lagos to witness the Constitutionalist trains riddled with bullets, their dead piled on both sides of rails. The convoy had been virtually unmolested in its passage through "Zapatista territory," an event worthy of note because, in the words of Obregón, "That train, captured or destroyed by the enemy would have changed the whole complexion of the Revolution in an instant."

As a matter of fact, it was precisely because the Zapatistas had done little or nothing to cut Obregón's lines of communication that Villa had finally decided to send the Reyes-Fierro column to do the job. He had become so thoroughly disgusted with the Suriano Conventionists that he had actually assigned a tertiary mission to the column to punish those who had failed—for whatever reason—to execute the plan to cut the Constitutionalist umbilical linking the Constitutionalist Army of Operations to Veracruz. The column's two main objectives had been to cut Obregón's lines of communication and then escort those Norteño members of the Convention who had fallen out with the Suriano members, such as Juan Banderas and Roque González Garza, back north. The intention, then, was to preserve a sense of legitimacy for Villa's movement by bringing members of the Convention government to his side, disrupt Obregón's communications, and to make the Zapatistas pay for not committing to the Convention's strategy.

After the scrape at Lagos, the Reyes-Fierro column continued to penetrate deep into Obregón's rearguard and, employing a ruse similar to the one used to capture Ciudad Juárez in 1913, its commanders sent fake telegrams under Obregón's name ordering Constitutionalist garrisons to abandon the plazas in their path. Then, quite unlike the Zapatistas who never took up more than a few kilometers of rails at a time, the Villistas proceeded to destroy the rails for a stretch of 171 kilometers. As Constitutionalist General Gavira later remarked, "This is called knowing how to destroy." Cut off from his base of operations, Obregón would have to achieve victory to his front, or rely on Pablo González's Army Corps of the East to his rear to re-establish his communications with Veracruz.

González

During the month of June, the First Chief had once again reordered priorities. Obregón and Treviño were now subordinated to General Pablo González's mission to conquer Mexico City, which had taken precedence since the miserable living conditions inflicted on the civilian population by the Convention's gross mismanagement had turned into an international and diplomatic disgrace. González's mission, therefore, was first and foremost humanitarian in nature and military only as a means to that end. At the same time, the Zapatistas concentrating in Mexico City played into Carranza's strategy to "gather a greater number of reactionary elements there, with the purpose of delivering a definitive blow" to those "disposed to defending the city."

González arrived in Puebla at the end of May and began organizing his new Army Corps of the East, composed mostly of divisions from his old Army Corps of the Northeast and the 2nd Division of the Center—especially the Hidalgo Brigade and Coss's Division—and troops from Cándido Aguilar's 1st Division of the East, perhaps ten thousand men total.

González next ordered his director of engineering to make improvements to plazas on the lines of communication so that a minimal number of troops would be needed to garrison and guard them. He also implemented a scouting and mapping service in an ongoing effort to gain situational awareness. Then he began slowly moving in for the attack. González led the center column up the Mexican Railroad through Mexico State with the 4th Division and the main grouping of artillery, skirmishing along the way through Apizaco and Ometusco Station. In that last location he established his headquarters, on June 14, and installed telegraphic communications to all the axes of advance along the main rail lines: Mexican, Interoceanic, Pachuca branch, and Central Railroads.

As the center column advanced, it was protected on the left by General Coss's 2nd Division, which was the largest and best organized. Leaving behind two brigades to cover southern Puebla, Coss advanced up the Interoceanic through western Puebla and Tlaxcala and captured Calpulalpam, the headquarters of Zapatista General Porfirio Bonilla, who was killed and his command destroyed in the event. Afterward, Coss's men pushed the Zapatistas out of Texcoco.

The right flank of the Constitutionalist advance consisted of General Agustín Millán advancing due south toward Mexico City from Tula, supported on his immediate left by Alfredo Machuca's brigade advancing diagonally from Pachuca toward Lechería.

With the Constitutionalist offensive underway, Zapatista headquarters in Mexico City did not hesitate to load up and evacuate the Hotel Sanz on June 15, headed for points south.

Still, as the Constitutionalists closed in on the city and solidified their lines along an arc running from the north around Tlalnepantla to the southeast at Texcoco and Los Reyes, the Zapatistas contested the ground. As Constitutionalist staff officer Manuel W. González later wrote, the Conventionist Army was "not inactive, and as it still possessed regular contingents of troops...it undertook partial attacks, especially on places that had scant garrison or that it considered it could conquer." On June 21, the Norteño and Suriano Conventionists attacked the Constitutionalists in an attempt to dislodge them from the north of Tlalnepantla and to the south of Barrientos Canyon. In response, the Constitutionalists launched a máquina loca against the Conventionist artillery that knocked the guns into ditches on the roadside. The Conventionists then counterattacked and drove Millán and Machuca back to Lechería. The counterattack imperiled the Constitutionalist disposition, and Millán was threatened with court martial for having pushed too far forward in the first place. But when the Constitutionalists recovered and continued to close in, Pablo González forgave the error.

On June 23, González moved his headquarters up to Tepexpan. The key to the Conventionist position was the Great Drainage Canal, which functioned similar to a large moat that was overlooked by Cerro Gordo, where the Conventionists had placed numerous cannons, including some naval guns. Under the protection of these guns the Zapatistas sallied forth in an attempt to push back the Constitutionalists, especially Coss's men around Texcoco, but without success. González's sappers dug individual fighting positions and trenches opposite Cerro Gordo and Santa Clara for the infantry since the desiccated lakebed of Texcoco offered no cover. Then González simply bided his time and waited for a large shipment of .30-30 ammunition to arrive. Meanwhile, the Zapatistas in Mexico City started to disperse, compromising Carranza's plans to quash the large concentration in one blow and also subjecting González's rear to possible harassment, especially around Pachuca. Yet the dispersal of so many Zapatistas was also symptomatic of their growing sense of hopelessness amid quickly evaporating possibilities of an ultimate victory.

JULY

Mexico City

At the beginning of July the skirmishing around Mexico City started to increase in intensity, leading the populace to believe that by July 4 it would fall to the Constitutionalists at any moment. After his resupply train from Veracruz arrived with the much needed .30-30 ammunition, González initiated the attack, led by General Juan Lechuga, "the Maclovio Herrera of the South," assaulting in the center. Lechuga's objective was the San Cristóbal Bridge, where the Conventionists had a blockhouse and two machine guns emplaced. As Lechuga's men pressed forward, supporting small-arms and machine-gun fire opened up, soon followed by the field artillery. On the Constitutionalist left, Coss's division likewise attacked at La Magdalena and Los Reyes. After two hours of combat, Lechuga's men successfully employed machine guns to dislodge the opposing forces and, supported by a regiment of cavalry, swiftly crossed the bridge. Once the bridgehead had been established, the Conventionists started to abandon Cerro Gordo and fled through Villa de Guadalupe, where they jettisoned five Saint Chamond field pieces, and entered the city of Mexico. They used any means available to skedaddle: horses, trolley cars, automobiles brimming with soldiers, anything. It was a rout. General Coss's 2nd Division pushed through Los Reyes and attempted to cut off the retreat of the Zapatistas headed south, but the Constitutionalists were poorly situated to block the retreat of the Norteño Conventionists, who escaped west to Toluca.

González's victory has often been discounted because observers never thought much of the Zapatistas as soldiers, as one anonymous observer wrote:

> The Conventionist Army that occupied Mexico City lacked organization, discipline, morale, and unity of command; the majority of the soldiers, many barefoot and almost all with tattered uniforms that consisted of a shirt and trousers made of bleached cotton, received scant wages, and were seen during the day covering the large arteries of the metropolis humbly asking businesses or individuals for some money to buy their day's sustenance.

Nevertheless, as Obregón had learned, the Zapatistas could be lethal and were not to be trifled with. Due care had to be exercised in any campaign against them, and if nothing else González managed the affair quite capably.

The next day the Constitutionalists reconnoitered the city and then moved in to occupy the capital. It was July10, the same day the Battle of Aguascalientes concluded.

Aguascalientes

On July 3, General Obregón held a council of war with his generals that opened with a recapitulation of their condition: with fuel to power the trains for just four hours and five days of provisions, the Constitutionalist Army of Operations was cut off from its base of operations by the Reyes-Fierro column and threated to the front by Villa's army, which was entrenched in Aguascalientes. After distributing the ammunition from the last and final convoy, the army's reserve stood at a paltry 100,000 cartridges, or roughly five rounds per soldier, which would be consumed in the first engagement. They had to make do with what had been sent them because the First Chief had done everything possible to keep the Army of Operations supplied, but González's army corps now had priority claim on all available war materiel for the very necessary campaign to conquer Mexico City. Obregón's plan was to "burn our ships like Cortés," meaning send the trains back to the safety of Lagos and strike out overland to the north of Aguascalientes to turn the tables on Villa. Threatening his rear would pull Villa out of the safety of his defenses and force him to fight on open ground since the Constitutionalists could not carry off prolonged frontal attacks or implement a siege given their supply situation. After much discussion, his generals accepted Obregón's plan. The meeting closed with a joke about the pope.

In the following days, Diéguez left Encarnación de Díaz for Lagos, taking most of his division, the trains, and excess impedimenta. As he retreated, Diéguez lifted the track at Las Salas in order to frustrate any plans the Villistas might have to capture the trains, attack Obregón's rear, or press onward in the event of a Constitutionalists defeat.

The rest of the army set out from Encarnación on the morning of July 6. The first major encounter took place at Los Gallos Hill, where the Villistas held the high ground overlooking *el camino real*, the route being traveled by the Constitutionalists. Obregón gave orders to his generals: infantry would attack in the center and against Los Gallos Hill on the Villista right, while the artillery laid down fire on San Bartolo Hill on the Villista left; Castro and Murguía would charge through the valley against the Villistas just to the left of San Bartolo Hill. Obregón then went to a clump of mesquite to observe the battle. Shortly after the battle opened, the Villistas strengthened and extended their right, threatening to outflank the Constitutionalist left; from the distance of his position Obregón could not issue correcting orders, but exercising initiative his experienced generals on the line knew to extend their left and created a hammer that allowed them to avoid defeat and win the battle. The Constitutionalists were soon underway again.

Reaching Hacienda San Sebastián near the barranca of Calvillo, Obregón ordered his infantry to form a six by four kilometer infantry square, where he waited for Villa to attack him. Villa did not disappoint. His temperament simply would not permit him to wait in his defenses at Aguascalientes, as General Ángeles had urged him to do. Villa once again lined up around Obregón's square and attacked, and again, once the battle had reached its tipping point and with their own ammunition running low, the Constitutionalist infantry made a push in concert with a flanking movement by the cavalry, Murguía's on the right and Castro's on the left. The result was another rout, only this time Murguía's cavalry succeeded in reaching Chicalote Station to cut off the Villista retreat and capture valuable materiel, including four million rounds of ammunition.

After his crushing defeat at Aguascalientes, Villa retreated to Torreón, where he regrouped and called in General Raúl Madero, who was holding the line at Paredón, to confer. Apparently, he planned to make a stand in the hills of Jimulco, just a few kilometers south of the enormous Picardías Bridge, which he could blow in case of a fourth consecutive defeat against Obregón. By July 15 San Luis Potosí had been abandoned and the Villistas were in the process of evacuating Zacatecas, after which they destroyed a large stretch of rails between Torreón and Zacatecas to slow the Constitutionalist advance from the south and buy some time. Villa also recognized that his supply lines back to Chihuahua were under constant threat by troops from Treviño's Army Corps of the Northeast operating in northern Coahuila and Nuevo León. Therefore, he directed Madero to lead an offensive in those areas to forestall any invasion of Chihuahua via Cuatro Ciénagas and to reopen the coal fields in northern Coahuila.

Obregón and González, July 11-23

Meanwhile, Obregón put Carranza's strategy into action. He sent General Murguía north to Zacatecas and General Gavira east to San Luis Potosí. Murguía reached Zacatecas on July 17 and Gavira entered San Luis Potosí a day later to link up with Constitutionalist General Herminio Alvarez, who had entered the state capital two days before. General Castro's 1st Cavalry Division followed after Gavira to San Luis Potosí and some of the Constitutionalists began moving northward toward Saltillo. Before Obregón could link up with General Treviño to roll up the Northeast, however, he had to deal with the Reyes-Fierro column threatening both his own rearguard—in the event that it tried to return to the North—and General González in Mexico City. González, in turn, not only had to deal with Reyes and Fierro pushing toward Pachuca, but a Zapatista offensive from the south toward Puebla. If the

Reyes-Fierro column captured Pachuca and then drove south along the Mexican Railroad and the Zapatista captured Puebla and then drove north, the two Conventionist forces might link up and cut González's lines of communication. Under the circumstances, González held a very protracted telegraphic conference with the First Chief before abandoning Mexico City on July 18. González did, however, leave a sizeable force under General Lechuga at Cerro Gordo, because he did not want to have to reconquer that position, and Coss remained at Texcoco. Those two commands would be well positioned to effect a pincer attack against Mexico City just as before.

Over the next ten days, the Zapatistas continued to pour assets into western Puebla, and on July 27 General Emiliano Zapata personally directed the attacks on Atlixco and Los Frailes, preparatory to an assault on Puebla. The Zapatista campaign ended rather abruptly, however, when General Coss's brigade commanders left in charge of defending southern Puebla handily defeated Zapata in a single day. Lechuga and Coss also repulsed Zapatista attacks on their positions around Mexico City, which freed up González to direct a three-pronged attack against Pachuca on July 28. González's Constitutionalists swiftly recaptured the capital of Hidalgo State because General Reyes had detached only a minor command of Surianos to Pachuca. The bulk of the Reyes-Fierro column had never advanced farther to the east than Tula. And on the day González evacuated Mexico City the bulk of the Reyes-Fierro column, joined by the Banderas and González Garza brigades, had already pulled back to San Juan del Río, to the west. González then did an about face and easily retook Mexico City from the Zapatistas on August 1. He left Reyes and Fierro to Obregón.

Obregón arrived in Celaya on July 23 to meet Generals Diéguez and Amaro, who had once again formed a large square around the city just in case the cavalry of Reyes and Fierro attacked. The Constitutionalists spent four days preparing for the campaign and then began moving eastward toward the large Villista raiding party. The first engagement took place at Mariscala Station (eighteen kilometers west of Querétaro) on July 28, which coincided with González's attack on Pachuca farther to the east. Obregón's Constitutionalists easily defeated the Conventionist infantry, most of which probably consisted of Surianos who quickly dispersed, while the Villista cavalry fled to the south. From that point on, it became clear that Reyes and Fierro were attempting to return to the North, and it became a guessing game as to which route they might take. Obregón's cavalry and scout trains patrolled to the south of Celaya, eventually catching the Reyes-Fierro column at Jerécuaro and again at Valle de Santiago, each time thinning the ranks of the Villistas. The last encounter occurred on August 1, after which the Villistas exited the area. The Reyes-Fierro column, now heavily weighted in officers, reached Torreón on August 18.

AUGUST

General Villa spent much of August in Chihuahua attending to administrative matters, but toward the end of the month he returned to Torreón. General Arrieta's Constitutionalists had finally succeeded in recapturing Durango City on August 12, an occurrence that many blamed on the weakening of the state's forces due to Villa's many requests for manpower during the Battle of Aguascalientes and Murguía's subsequent march into Zacatecas. Regardless, Villa could not allow Arrieta to threaten his flank if he were to defend Torreón from the Constitutionalists approaching to the south and east. So he loaded up his cavalry and rode the rails to El Chorro Station where he detrained and then closed in on Durango City. Deployed on a long line that extended past the limits of the city, in the early morning hours of August 23, 1915, Villa's centaurs charged. The Arrietas could not hope to put up any meaningful resistance and once again fled into the sierra after losing copious amounts of men and materiel. They would not return until late October.

For the Constitutionalists, events tended to happen in a rather lock-step fashion according to Carranza's strategy after Mexico City had been secured. In August, the cavalry of Generals Amaro and Maycotte continued driving toward Tula, Hidalgo, repairing the rail lines and defeating the Zapatistas where they found them. General Diéguez took his 2nd Infantry Division to Jalisco, and after being assigned to the position of "Chief of Operations in the States of Sonora and Sinaloa and the Territory of Tepic" began moving up the West Coast toward Sonora. Diéguez first sent General Enrique Estrada with his cavalry into Tepic, forcing the Villista General Rafael Buelna into the Sierra of Huajicori in the northern part of the territory, while he accompanied the infantry on ships to Mazatlán. Constitutionalist General Iturbe turned over his 3rd Division of the Northwest to Diéguez and went to Manzanillo to take charge of operations to pacify Colima and Jalisco states, where Villista General Julián Medina still commanded quite a large contingent of regional forces. Also, Constitutionalist General Luis Herrera put the strategy he had developed with his brother the year before into action and invaded southern Chihuahua. He didn't get far. Herrera was checked at Creel Station on August 28 by Villista Generals Julio Acosta and Julián Granados who formed what would become the vanguard of a column to try to recapture northern Sinaloa for the Convention. At the same time, Obregón sent minor brigades to continue pacifying eastern San Luis Potosí and began moving Hill's 1st Infantry Division and Castro's 1st Cavalry Division up through northern part of the state to cooperate with General Treviño's Army Corps of the Northeast in securing southeastern Coahuila.

Icamole

Ever since June, General Luis Gutiérrez had been operating out of the Sierra of Arteaga, harassing the Villistas in Saltillo, while General Treviño faced off against the Villistas of General Rosalío Hernández and Raúl Madero at Icamole. The fighting around Icamole and Villa García had been nearly constant, and on several occasions had flared up into heated combat. One such scrape took place on July 6. In a very detailed plan of attack that involved a behind-the-lines assault on Monclova, an attempt on Saltillo, and a three-pronged attack at Icamole, the Constitutionalists initially enjoyed some success, pushing the Villistas back. But the battle-hardened Villistas soon regained their composure, and calling in reinforcements from Paredón, mounted a fierce defense at La Llorona Hill. The Constitutionalists finally succeeded in winning the day and forcing the Villistas to retreat to Paredón, but their horses and men were too dehydrated to pursue.

Five days later, on July 11, the Villistas counterattacked and Generals Carlos Osuna and Emiliano Nafarrate telephoned Monterrey from Villa García to ask Treviño for permission to clear the Villistas from the valley. Treviño agreed and the two generals attacked with their cavalry on the flanks and infantry in the center, a competent disposition (except that no reserves had been established) that nevertheless resulted in disaster for the Constitutionalists. The Villistas put their cavalry to flight and the Constitutionalist infantry, now accustomed to victory, continued to advance to the point of being cut off and captured in large numbers. Constitutionalist General Manuel García Vigil's artillery had never even had a chance to go into battery. After recovering from that debacle, which Treviño blamed on his subordinates and their inattentiveness to his instructions, Treviño assumed a defensive posture and established a defense in depth consisting of trenches, wire, and emplacements for cannons and machine guns.

Another Villista push came on August 10. General Hernández had his buglers blow *diana* as his cavalry and infantry charged the Constitutionalists' improved defenses. Mounted Villistas charged right up to the fortified positions, the fire from their weapons characterized as "ineffective and sterile." After almost two days of combat, Treviño decided to send General Ignacio Ramos through Nacataz Pass on the Villista right, which forced Hernández to retreat to protect his rearguard. Treviño faulted Hernández's failure to properly reconnoiter the Constitutionalist lines before ordering the attack as the main cause of his defeat. Others cited the Villistas' lack of ammunition and rolling stock.

SEPTEMBER

At the beginning of September, Obregón prepared his final orders to conquer southeastern Coahuila. Generals Hill and Castro would attack the Villistas entrenched at La Angostura, south of Saltillo, while General Gutiérrez slipped in behind Saltillo to cut off their retreat. Simultaneously, General Treviño would make a concerted general attack against Icamole. The operation came off without a hitch on September 4. Generals Hill and Castro encountered limited resistance while Gutiérrez's blocking brigade suffered the brunt of the Villista effort and came away much the worse for wear. General Treviño also had little trouble routing the Villistas at Icamole and then Paredón, although a body of evidence exists to show that he did not attack until September 6, and by then the Villistas were already in headlong retreat.

Shortly thereafter both Generals Hernández and Madero sued for amnesty. General Urbina was suspected of trying to do the same and executed by Rodolfo Fierro. General Buelna left Sinaloa with his troops for Durango where he then abandoned them and went into exile in the United States. Hernández and Madero suggested that Villa should do likewise and leave the country, but he was not done fighting.

General Treviño's command now came under the umbrella of Obregón's headquarters and the two forces began moving toward Torreón, the infantry taking the northern route through Hipólito and repairing the rails while the cavalry moved along the southern road through Parras. Simultaneously, General Fortunato Zuazua's Constitutionalists cleared northern Coahuila of remaining resistance, forcing the Villistas—including some former Constitutionalists who had been compelled to switch sides or die after Villarreal's disastrous campaign in January—to surrender or return to Chihuahua via Cuatro Ciénagas and Sierra Mojada.

As Obregón and Treviño moved cautiously and deliberately toward Torreón hoping to catch Villa in a pincer, they learned that the Villistas had already started to abandon Torreón, which General Murguía occupied on September 26 after briefly engaging the Villista rearguard. In essence, Villa had decided not to make a stand at Torreón, but rather to invade Sonora from western Chihuahua, link up with Governor Maytorena, and then start moving down the West Coast and ultimately attack Mexico City.

Constitutionalist General Calles, who had not gone operational until July—when the First Chief's nephew, Carlos Carranza, arrived on the scene and shamed him into action—had accomplished little in the way of campaigning in Sonora, even though he had substantially more money, men, and materiel than Governor Maytorena. Upon learning of the impending arrival of Villa from Chihuahua, Calles immediately began backtracking from the environs of Nogales toward Agua Prieta and started entrenching to await the onslaught of Villa's centaurs.

General Ángel Flores's Expeditionary Column of Sinaloa, which had spent the entire year in a virtual state of siege in southern Sonora trying to push northward and link up with General Calles, finally finished reconnecting its lines of communication to San Blas and began driving northward from Navojoa in September. The Sinaloans reached Esperanza Station and halted in order to bring up supplies and assets left behind at Navojoa during the push northward and to prepare for a general offensive.

Governor Maytorena did not wait to see what might develop, but rather went back into exile on October 1.

OCTOBER

At the beginning of October, Diéguez left his cavalry in Sinaloa and sailed from Mazatlán with two brigades of infantry. He disembarked at Guaymas, easily brushing aside the Conventionist forces in the area, and linked up with General Flores's Sinaloan column. Over the next few weeks, Diéguez waited for the rest of his command, namely the infantry of General Gabriel Gavira and the cavalry of Miguel M. Acosta, to arrive from San Luis Potosí via Manzanillo. He also requested replacement parts for the wireless that the Conventionists destroyed upon evacuating Guaymas, inflicting de facto radio silence on his column. For the duration he would have to communicate with the Constitutionalists in the eastern part of the state via the consul in San Francisco, California, Ramón P. Denegri. Meanwhile, Diéguez's infantry, now under the direct tactical command of General Flores, began pushing toward Hermosillo at the end of October, advancing as far as Torres Station and skirmishing with Maytorenistas along the way.

The first half of the Villista pincer into the Northwest got started with a column commanded by the Sinaloan Juan Banderas leaving Chihuahua City on October 6 to invade northern Sinaloa. The column included Orestes Pereyra and his brigade from Durango and other units from Chihuahua. The movement over the sierra took quite a while, and by November 1 the Banderas Division had only reached as far as Choix.

General Villa was to lead the part of the pincer driving into northern Sonora, but he spent weeks at Casas Grandes preparing for the campaign and receiving a succession of bad news. First, he learned that his loyal General Rodolfo Fierro had died. Fierro had killed Villa's long-time friend, Tomás Urbina, and now Villa had the services of neither general. Worse still, Urbina's men, upset over the unwarranted murder of their leader, refused to report to Casas Grandes for the Sonora campaign and instead joined the Constitutionalists. Then, an explosion of ordnance killed a great many men in the camp. Finally, and most disturbing for Villa, on October 19, the U.S. government recognized Carranza's Constitutionalist Army as the de facto government of Mexico. That meant that Villa would no longer be able to legally import war materiel, and the Constitutionalists could apply for permission with the U.S. government to transport troops via American territory. Undeterred, Villa invaded Sonora.

General Obregón wasted no time in filing an official request with the U.S. government to send General Hill's infantry from La Laguna to Sonora. The 1st Infantry Division entrained on October 20 and left for Piedras Negras, passed through the United States, and began arriving at Agua Prieta in the final days of October.

At the end of the month, General Murguía also began the campaign to secure Durango, advancing from Torreón toward Durango City, catching the Villistas in a pincer with the Arrietas, who crossed the sierra into western Durango. Murguía defeated Calixto Contreras on October 26 and the Arrietas did the same to General Máximo García at the beginning of November. Murguía entered Durango City on November 4.

NOVEMBER

Agua Prieta

Villa reached northeastern Sonora at the end of October and on November 1 attacked Agua Prieta. He encountered an opposing force highly motivated, with superior assets, and entrenched with elaborate defensive works that featured a field of mines that could be remotely detonated, a six-feet-deep system of wires to maintain any assaulting force in the kill zone, and numerous machine gun nests.

This battle, played up by Villistas and Constitutionalists alike, was actually much ado about nothing. According to U.S. Army officers who witnessed the battle from Douglas, Arizona, Villa only "nibbled" at Agua Prieta before deciding that it was too well-fortified.

On November 3, Villa left part of his force under General José Rodríguez to hold the Constitutionalists at Agua Prieta in check while he took the remainder west and then south to unite with the Maytorenistas and attack Hermosillo. Most likely, he realized that he had too much impedimenta to move swiftly, and he needed to act fast to close the pincer with Banderas's Sinaloa invasion column.

El Fuerte

The speed of Villa's advance would not matter, however, because Diéguez's cavalry defeated the Banderas Division at El Fuerte soon after the Battle of Agua Prieta. Constitutionalist General Jesús Madrigal formed a perimeter around El Fuerte divided into sectors with field works according to the "tactics employed by you [Obregón] in the defense of Celaya and the march to Trinidad." In the morning of November 6, Banderas' Villistas deployed in two lines about three kilometers long anchored on the river and occupying two hills to the south and attacked El Fuerte, shooting sparingly "in the Zapatista style." General Enrique Estrada, commander of Diéguez's cavalry division, arrived after noon in the middle of the battle and the next day, November 7, Madrigal executed a dismounted flanking attack in combination with a push in the center—the same tactic employed time and again by the Constitutionalists. The maneuver broke the Banderas Division and effectively ended any Villista hopes of a strategic Northwestern pincer.

Back in Sonora, Villa was once again operating on interior lines against the Constitutionalists who were in the northeast and southwest parts of Sonora, operating on exterior lines, with incomplete intelligence and delayed communications that had to go through the United States.

Repairing the rails as they went along, General Diéguez's Constitutionalists entered Hermosillo without a fight on November 6. Diéguez only had about two hundred mounted men under Colonel Fermín Lugo that could be used for reconnaissance and he was short on .30-30 ammunition. Obregón, who by now occupied Naco and had pushed Rodríguez back to Cananea, told Diéguez not to present battle unless he could be sure of victory. The Maytorenistas commanded by General Francisco Urbalejo were at Zamora Station and Villa was on his way to join them.

Nevertheless, on November 17 Diéguez left Hermosillo and began marching northward to encounter the Maytorenistas. He actually had wanted to retreat as far back as the Mayo River but he moved forward because of one or more possible reasons. Either Flores (supported by other brigade commanders) convinced the commanding general to press onward, or the Constitutionalists had grown weary of waiting for the Conventionists to attack Hermosillo, or Diéguez believed that Villista General Manuel Madinaveytia was on his way from Ures with reinforcements and so he needed to defeat the Maytorenistas before the Villistas joined them. No matter the reason, the Constitutionalists left Hermosillo on November 17 and attacked the Maytorenistas at El Alamito the next day, just as Villa was arriving with his troops.

Strategy and Tactics of the Mexican Revolution

The Constitutionalists thoroughly trounced the Conventionists, even going so far as to capture the women and children of Villa's Yaqui allies.

The day after the Battle of El Alamito, Flores concentrated his forces at Zamora Station and began forming an infantry square. General Gavira arrived from Hermosillo with his infantry and assumed the front of the square, placing the 27th and 28th Battalions on either side of the railroad with two companies of the 32nd Battalion on each flank and in hammer. General Melitón Albáñez's 13th and 14th Battalions formed the right and left laterals, respectively.

The previous day's defeat did not sit well with Villa, and on November 20, the fifth anniversary of the official start of the Mexican Revolution, Villa returned and attacked the Constitutionalists at Zamora Station. It was believed that Villa had lost the previous battle because of poor communications through the hills and brush, so using a system of smoke from burning palms and shrubs—which the Constitutionalists initially mistook for the arrival of more trains—the Conventionists maintained a dressed line of advance. Diéguez lost his nerve and decided to retreat to Hermosillo at the last moment, leaving General Gavira's infantry to bear the brunt of the Yaqui rage over the capture of their loved ones. After being thoroughly decimated, Gavira's Constitutionalists arrived at Hermosillo, covered in their retreat only by Flores' chief of staff, Colonel Guillermo Nelson, who rounded up the few mounted Constitutionalists and counterattacked against Villa's centaurs.

At Hermosillo, the Constitutionalists occupied their previously prepared field works that encircled the city. Diéguez, again, wanted to retreat all the way to Guaymas, but news that General Madinaveytia had reached Torres Station, effectively cutting off the route of egress, forced his decision to stay and fight. After fighting for two days, on November 22 the Constitutionalists counterattacked. Albáñez's brigade hit the left flank of the Villistas, while Flores, who was opposite the Villista left wing, ordered his four pieces and eight machine guns to train fire on the hill that anchored the enemy position until it was covered in dust. With the defenders effectively blinded, Albáñez and his men moved to within rifle range at which point Flores ordered his men to charge the hill. The two Constitutionalist commands effectively created a hammer strike that forced the Villistas to flee. Albáñez and Flores then turned west and rolled up the Yaquis, who held the right wing of the Conventionist line.

Villa ordered his troops back to Gándara and El Alamito stations for a rest. Diéguez could not pursue because he lacked a worthy mounted contingent. Moreover, Diéguez was still considering whether or not he should retreat. He sent a party to the south to scout and begin repairing the rail lines. Several days later Villa's Yaqui allies accepted amnesty in exchange for the return of their families and Villa, ruminating on the destruction of Banderas's column, called an end to the western campaign, abandoned his artillery, and retreated over the sierra back to Chihuahua.

General Flores continued north toward Nogales repairing the rails, and Diéguez finally and happily returned to Guaymas.

DECEMBER

Although Villa's Northwest campaign had ended, there was still the issue of General Rodríguez in northeastern Sonora. The Constitutionalists knew he would be retreating to Chihuahua as soon as he learned that Villa had left the state. The idea was to catch him as he crossed the railroad to the south of Agua Prieta. For that purpose, Obregón ordered General Flores to bring his column to Nogales and then, entering the United States, come out at Agua Prieta and begin moving down the rails.

General Flores spotted Rodríguez's Villistas near Fronteras in the evening of December 8 and positioned his men in a defensive circle around the convoy, with his artillery in the middle, and established reserves.

The next morning what became known as the Battle of Fronteras began as, supported by dozens of their guns, the Villistas attacked. Yet once again artillery was not a factor in the battle's outcome. General Flores held his ground and Rodríguez abandoned his artillery and drifted back into Chihuahua.

While Rodríguez was fighting General Flores, Villa was arriving at San Pedro Madera, Chihuahua, after allegedly sacking Dolores, where he telegraphed Chihuahua City to get caught up on events. He learned that General Treviño was in the process of invading the state.

After spending the month of November making slow progress from Torreón while repairing the rails, General Treviño decided to leave behind his infantry and, taking his cavalry, invaded Chihuahua. It is notable that Treviño, a Northeasterner, took his cavalry to invade Chihuahua and left behind his infantry while Diéguez, a product of campaigning under Obregón, did the opposite when entering Sonora, taking infantry and leaving his cavalry behind in Sinaloa.

Treviño linked up with General Herrera, who had come up from Sinaloa in July and most recently had captured Parral, at Jiménez. Then, Villista General Rosalío Hernández joined Treviño's army corps and after a series of minor scrapes, on December 21 the Constitutionalists entered Chihuahua City, which had been evacuated by the Villistas the day before. Villa retreated to Bustillos where he disbanded the remainder of his army and told his men to be prepared to be recalled to service at a later date.

General Gavira took his brigade from Hermosillo to Naco, Sonora, continued through U.S. territory and reentered Mexico at Ciudad Juárez, where his men began mustering out Villistas in northern Chihuahua and sending them home. The revolution was not over, but it had been won.

CONCLUSION

TACTICS AND WAYS OF WAR

T he tactics employed during the Mexican Revolution of 1910-1915 were a function of time and space. As time went on, the territory experiencing conventional warfare expanded from Chihuahua to other northern states during the first three years, then to the Center with the revolution against the Huerta administration, and finally in the civil war of 1915 engulfed the entirety of the national territory, even to the point of requiring sealifts. Consequently, each new twist and turn over the years not only brought new sections into the fight, but each region also contributed its own special fighting traditions, or "ways of war," to the mix of tactics. Finally, legacy technology from the Federal Army played a role as well. Different armies inherited a disproportionate share of the Federal Army's arsenal and adopted Federal Army tactical doctrine—itself specifically crafted in the late nineteenth century to the peculiarities of future combat expected in Mexico—to varying degrees. Hence, in terms of tactical fingerprints, we can identify at least six unique ways of war: Eighteenth-Century European (Federal Army), Sonoran, Chihuahuan (military colonies), Northeasterner, Lagunero (Colorados), and Zapatista. The main characteristics of each of these are summarized below along with two additional sections to cover unconventional tactics as broadly practiced by Norteños and select cases of stratagems witnessed over the course of the revolution.

Federals

The Federal Army infantry was organized on a base of three's, which corresponded to the number of echelons prescribed for combat according to tactical doctrine under the ternary system. Yet there are few mentions in the historical record of Federal Army units deployed for combat in echelon because the ternary system required motivated and intelligent soldiers and officers, which the Federal Army did not have. Additionally, to perfect the very complicated and dangerous coordination of three echelons of combat demanded rigorous training and practice, which the Federal Army did not do. Instead of emulating professional armies of the world, with patriotic citizens and an officer corps drawn largely from the middle class, the Federal Army still resembled those of the eighteenth century. The enlisted men were "sad Indians, gloomy, indifferent, who did not care if they died or killed," and criminals, who "had to be always watched over in order to avoid, as much as possible, incessant desertions. The officers and noncoms were true jailers." Additionally, most officers lacked the most basic technical knowledge and the intelligence to apprehend tactical concepts, such that they were resigned to using superior courage to will victory, even though bravery and force of will are not tactics, or at least not effective ones. Finally, "command in the Federal Army was regularly held by mature, and at times not just mature but old, men," incapable of enduring the rigors of campaign, devoid of martial spirit, and experts only in garrison life and regulations. Therefore, it would appear that the Federals rarely assumed an echeloned disposition, one of the few instances being the crack 29th Battalion's performance in the Battle of First Rellano. And yet in that battle, a journalist accused the 20th Battalion of firing on the soldiers of the 29th, perhaps because of the confusion of the echeloned disposition. Another reason for the friendly fire may have been that after the chief of staff countermanded the 29th's flanking movement and ordered it to assault in an oblique direction, the 29th came into the 20th's line of fire.

Another example of a complex order of battle occurred in the same campaign against Orozco's Insurgency. In the Battle of Conejos, in 1912, General Huerta's Federal Division of the North deployed in an oblique order of battle in which the left, suspended, wing comprised the reserve. Mid-battle it advanced in echelon to close with and kill the Orozquistas. That campaign of 1912 against Pascual Orozco occurred at the Federal Army's high-water mark.

In the campaigns of 1913 and 1914, when the Federal Army had to quickly and greatly increase its numbers, the press gangs went into high gear, filling out the ranks with more recalcitrant and subpar specimens of soldiers. By presidential decree, the school of the soldier took place in garrison and was to be rudimentary in nature. As such, the Federals dared not leave the rails and major population centers and penetrate into the countryside for fear that the enlisted might seek the nearest opportunity to desert. For the same reason, the Federals could not initiate battles at night. Their duties became largely constabulary and increasingly they fought only in defense of towns, haciendas, rails, and other high value targets. Yet even in the

Conclusion

relatively simple combat disposition of a defense, they committed the same errors with amazing consistency. Most significantly, protocol dictated that defensive works for a plaza were to be constructed to accommodate the personnel of the garrison; any operations brigades or flying columns that might join the garrison to defend a plaza would only have recourse to hasty field fortifications. The Federals, however, routinely established perimeters excessively long in relation to the number of defenders or, almost as bad, did not construct any fortifications at all because Federal officers tended to believe, as expressed in Federal General Jacinto Guerra's report on the defenses at Monterrey, that the "care of works increases the fatigue of the troops and the individual and collective valor of the same decreases when they are frequently forced to utilize fortifications and fight under their protection." Federals also dispensed with nighttime illumination on a regular basis and neglected communications in defense of plazas—the Battle of Chihuahua City in 1913 being a rare exception—even though doctrine specifically called attention to the need for signals in these types of battles because of the compartmentalized nature of urban fighting. Federal commanders therefore exploded their magazines to signal a retreat or, worse, simply abandoned their men in a most cowardly fashion. Further contributing to the lack of a coherent plan of defense in many battles were frictions between garrison and operations commanders, on the one hand, and regulars and irregulars, on the other, which weakened the unity of command; the atomization of Federal Army battalions and regiments, a condition caused by weeks and months of being detached to some locations for short-term assignments and then hurriedly concentrated in others for battle; and an overreliance on civilian volunteers, who deserted at the first opportunity or, in the worst of cases, joined the revolutionaries.

The one prescription that seemed to receive compliance was the designation of a redoubt in the plaza, usually the barracks or bank or church, of sturdy construction, for a last stand in "hold at all costs" battles. But all too often, the Federals crowned the top of these structures with machine guns to provide indirect fire in defilade, a faulty disposition.

The state of the Federal cavalry was even worse than that of the infantry, if such were possible. Soldiers of the U.S. Army in the Mexican-American War had been terrified of Mexico's lancers, but by 1910 the Federal cavalry had turned into a joke, literally. Because of the high degree of corruption, the cavalry officers were said to "eat more hay than their horses," meaning they arrogated resources intended for their mounts. Consequently, the cavalry was few in numbers, its horses in a poor state, and its subpar and apathetic officers ridiculed by Norteño horsemen. The only record of the Federal cavalry being used according to doctrine was during the campaign against Orozco. At First Rellano, the commanding general placed the cavalry on the flanks, as support for artillery, and to the rear to guard the trains. During the subsequent retreat, the cavalry properly covered the egress of the artillery and infantry. Still, the cavalry had performed its scouting mission poorly during the movement to contact, denying the commanding general crucial situational awareness. Only under Huerta's competent leadership in the advance to Conejos and Second Rellano did the artillery and cavalry working together in the vanguard attempt to secure favorable terrain for the follow-on infantry, as per doctrine, although the Federals had to rely overwhelmingly on mounted Maderista irregulars to fulfill the role of cavalry. More typical was General Landa's lethargic movement to Ojinaga in September 1912, where he ultimately wound up relying on Maderista irregulars to carry out the pursuit of the Colorados after the battle even though his Federal regular cavalry had endured no more hardship than the irregulars during the campaign.

Once Huerta usurped the presidency, the Maderista irregulars joined the Constitutionalist Army en masse, depriving the Federal Army of its most competent mounted contingents. The regular cavalry was too few in numbers to fulfill its mission, and several regiments had been completely wiped out in Chihuahua by the beginning of 1914. Therefore, the Federals employed scout trains and relied on irregulars for any dangerous or skilled missions away from the rails, which in turn led to friction between the freewheeling irregulars and Federal Army martinets who could not help nursing inferiority complexes. In sum, Federal Army cavalry was a virtual nonentity.

Federal artillery, too, played a largely insignificant role in the Mexican Revolution for two main reasons: the contents of the arsenal and the competence of the personnel. The Federal Army cannons consisted almost solely of medium size guns, a smattering of mortars, and no howitzers. The flat-trajectory shrapnel fired by field guns would have performed well against enemy infantry advancing in closed ranks over open terrain, but it was ill-suited against properly dispersed and entrenched enemy combatants. A plethora of eyewitnesses testified to the ineffectiveness of Rubio's guns at Bachimba including the American, Edwin Emerson. And as the American military attaché to Mexico remarked, "the present mountain gun of the Mexican Army is described by officers as being a play-thing."

Indeed, among the key reasons why the Federals failed to subdue the rebels and save Madero's government during the Tragic Ten Days, as recounted by General Rubio, were: a lack of howitzers and mortars and reduced-charge shells for high arched shots over buildings, guns of small caliber that could not punch through thick masonry, and a preponderance of shrapnel (as required by regulations) instead of high explosive shells. To these limitations, the general also added command and control deficiencies, such as a failure to designate a particular sector for attack, bombard in coordination with a major ground assault, and establish proper telephone communications and observation points.

Conclusion

Although Sánchez Lamego lauded the employment of a rolling barrage to support an assault at Conejos—the first time employed by the army in Mexico—he identified the main defect as one of mismanagement in the allocation of guns. The Federal Army failed to mass artillery for indirect fire, the one exception being Chihuahua City, 1913, where a great number of the artillery batteries held over from the Orozquista campaign of 1912 were still on station. More typically, the Federals flouted doctrine by detaching single batteries, sections, or even lone guns and employed them in direct fire missions from exposed positions and on, or near, the line of fire where the crews were savaged by enemy rifle and machine-gun fire and their pieces were subject to capture.

Sonora

While the culture of the Federal Army prevented it from reaching its aspirational effectiveness, it had developed tactical doctrine that was not just well-suited to combat in Mexico, but contained some of the best recommendations for its time. Constitutionalist General Álvaro Obregón studied the ternary system and put it into practice to create the dominant fighting force of the Revolution. His infantry, in which Yaqui warriors predominated, fought in echelon, could extend the line under fire and go into hammer (at El Cuatro, January 1915; Los Gallos, July 1915; Flores and Albáñez at Hermosillo; and Gavira at Zamora Station, November 1915) as appropriate, and made effective use of hasty field fortifications (known as *loberos*) without becoming ensconced and immobilized in the relative safety of those confines.

At the end of 1914 the operating environment vastly changed and Obregón adapted accordingly. The Federals had been relatively immobile and chose to fight large-scale battles in defense of plazas. The generals of revolutionary origin who prosecuted the war in 1915, by contrast, campaigned over long distances and fought wherever they encountered opposing forces—often at chokepoints or crossroads—which resulted in large open field battles.

In the wide open space of the Bajío, where Constitutionalists from the Northwest came up against the powerful Villista cavalry for the first time, Obregón resorted to enormous infantry squares, which, in spite of popular belief, ran contrary to doctrine. Given the tremendous lethality of firepower in modern warfare, the defensive-minded square significantly increased morale in the short-term, but it could produce a negative siege mentality if the disposition dragged on too long. The first such army-corps-sized square was formed around Trinidad Station, essentially modeled on the defense of the city of Celaya. From then on it became a go-to disposition and unwritten tactical doctrine for Constitutionalist infantry against Villistas: in June around Encarnación and later during the Battle of Aguascalientes;

in July when Diéguez occupied Celaya to face the returning Reyes-Fierro column; Estrada at the Battle of El Fuerte; Diéguez, again, at Zamora Station after the Battle of El Alamito and subsequently during the Battle of Hermosillo; and Flores at Fronteras. Fighting in the square until reaching the tipping point, and then counterattacking with cavalry on the flanks while pushing forward with infantry or dismounted cavalry in the center, concluded the signature Constitutionalist tactic.

For defense of a plaza, official doctrine recommended the stationing of cavalry outside of the plaza in anticipation of using them for a counterattack, a tactic never employed by the Federals or, for that matter, by anyone else besides Obregón. Federal Army doctrine also placed a high priority on communications in defense of a plaza, precisely because it could be difficult to deliver orders to numerous pockets of soldiers stationed in buildings. At Celaya, Obregón had telephone and/or telegraph connections from his headquarters to all brigade headquarters. In order to communicate with forces stationed outside the perimeter he was prepared to use heliographs, which he considered more secure than transmissions over wire. Later, he developed a system of train whistles. Constitutionalist General Gavira used smoke for signals in Mexico City, 1915, and the Conventionists also used smoke in Sonora in 1915 to avoid friendly fire and keep their lines dressed.

In other theaters, the battles of 1915 (for example, Naco, Agua Prieta, El Ébano, and Icamole) resembled more closely the static fighting and complex field works typically seen in Flanders, with nighttime illumination, cleared fields of fire, crenulated trenches in echelon with covered communication trenches, wire to keep attacking forces outside of grenade (or dynamite) range, mines, and machine gun nests.

For offensive operations during the Mexican Revolution, infantry relied on railroads for any sense of meaningful mobility, and in this regard the Sonorans were the first to insert combat troops railborne, first in search and destroy missions against Colorados in western Chihuahua, and then directly into combat against Federals in the Battle of Culiacán. There is also at least one story—which may or may not be true—of a railborne Villista assault at Celaya. According to a newspaper article, at Second Celaya the Villista General Manuel Banda and his staff rode a train into combat, stopped about thirty feet from the Constitutionalists trenches, assaulted through directed small-arms and artillery fire to reach a building which they mined and blew up and then retreated.

The other main application of railroad assets in combat was, of course, the máquina loca: runaway trains loaded with explosives sent barreling down the tracks toward the enemy. Examples of the employment of this weaponized rolling stock include its use by Sonorans in 1913 at Naco, the Colorados in 1912 at First Rellano, and Constitutionalist Generals Millán and Machuca against the Zapatistas at Tlalnepantla in June 1915.

Conclusion

Regarding artillery, Obregón and his generals attached no particular importance to artillery, and never seem to have employed more than a few batteries. Even when the Army of Operations captured dozens of pieces from Villa during the battles of the Bajío, it did not incorporate those assets. Rather, as the newspapers reported, "not needing them in his campaign," Obregón sent the guns back to Veracruz, far removed from the battlefield.

Military Colonies

Unlike the Sonorans, the revolutionaries from Chihuahua employed copious cavalry and artillery to the point of almost completely neglecting the infantry arm. Culture accounted for much of this bias, since Villa's men instantly discounted any foot soldier, but the vagaries of history resulted in the Constitutionalist Division of the North attaching an outsized importance to artillery. General Villa had received his practical education in conventional warfare under the Federals during the campaign against Orozco when he and his irregular cavalry had been in the vanguard with the artillery during movements to contact. In those operations, he observed how the two arms worked hand in glove to fix the location of the enemy and then secure the best ground. Later, in the battle proper, he had seen how the artillery walked in shells ahead of the mounted forces during assaults that concluded with close-quarters fighting with pistols, not sabers.

Precisely because of the campaign against Orozco and Villa's participation in that campaign, the marriage between the artillery and cavalry had been fixed upon his psyche, and the largest concentration of artillery in the Federal Army was still stationed within the confines of Chihuahua when Villa mobilized the military colonies in the summer of 1913. After battles at San Andrés, Torreón, Tierra Blanca, and the capture of Ciudad Juárez and Ojinaga, by the beginning of 1914, Villa possessed a large number of artillery assets formerly owned by the Federals.

Thereafter, Villa's main battle tactic could best be described as "surround and pound": encircle and fully deploy—foregoing any reserves—with superior numbers of artillery pieces and soldiers and then attack, attack, and attack again, until the defender capitulated. In those assaults, Villa employed his cavalry along the hybrid model, which is to say in both mounted and dismounted. Hence in the campaigns against Huerta, the Division of the North sometimes fought dismounted and at night, such as at Torreón, September 1913; and Lerdo and Gómez Palacio, March 1914. Alternatively, Villa's dragoons also deployed in a line of foragers across an extended front and mounted daytime charges: ten kilometers wide at Gómez Palacio, 1914; four kilometers wide at Paredón, 1914; and later at Durango City, 1915. The mounted assault at Gómez Palacio failed, but Paredón, like Beersheba in 1917, proved that in the age of modern weaponry such an attack could succeed if done quickly, unexpectedly, and over the right terrain—flat with few obstacles. Durango City, a surprise attack, likewise succeeded.

The main problem with fighting dismounted was the immediate decrease in firepower since one trooper in six or ten, according to Norteño tradition, or one in four, per Federal cavalry doctrine, had to hold the horses of the others. Perhaps for that reason, during the campaigns of 1915 the Villistas tended to reject the hybrid model and used cavalry almost strictly for mounted assaults. However, in 1915 the Division of the North was no longer fighting incompetent Federals on the arid plains of the North, nor could it count on fifth columns in attacks on plazas. The Villista cavalry charged entrenched Constitutionalist infantry—among the best in the world at the time—and, at El Ébano, in closed order over the highway and hilly, vegetated terrain with the result that it failed miserably. At Celaya, the cavalry also tried to deliver mounted assaults over unfavorable ground, rain-soaked fields crisscrossed by ditches and canals, with disastrous results. Charging to within about three hundred meters of the Constitutionalist lines the Villistas stopped, fired off their weapons, and then slowly turned around and retreated, offering their backs as targets.

Conversely, Villa failed to employ his cavalry in more orthodox roles for scouting, reconnaissance, and raiding until too late, meaning the Reyes-Fierro column. Consequently, in one of the great ironies of military history an army overweighted in cavalry suffered from a lack of situational awareness. And since he did not have an overwhelming advantage in troop numbers and failed to use reserves, Villa was consistently caught by surprise and in an overextended disposition of troops with little or no opportunity to respond. On those occasions, his infantry and artillery paid the price.

Of all the major factions in the Revolution, the Division of the North under the supervision of General Felipe Ángeles handled artillery to the greatest effect, which is to say massed and hidden (Fourth Torreón and Zacatecas) or, alternatively, by using speed to unexpectedly go into battery and shoot at targets of opportunity such as at Trinidad where Obregón lost his arm. However, the artillery continued to decrease in effectiveness during the campaigns of 1915 for several reasons.

First, with World War I raging across the ocean, the Mexicans had to rely on defective artillery shells produced in Chihuahua or Cananea that sometimes damaged the guns. And since the manufacture of timed fuses could not be mastered, almost all the shells were high explosive—as opposed to anti-personnel shrapnel—designed to explode on impact. That meant that now, even more than before, the artillery was limited to raining down harassing fire on the plaza, which accomplished little, or had to go into battery extremely close (within five or six hundred meters) and engage in direct fire missions at the enemy's trenches. As a result, at Celaya the Villista gunners experienced the same punishing return fire

Conclusion

from Obregón's machine guns that the Federal artillerymen had endured earlier at Santa Rosa in 1913.

Second, after normal wear and tear caused by years of campaigning, the use of defective ammunition, and roughed up by constantly loading and unloading from trains, many of the guns had damaged aiming devices and rifling, not to mention defective recoil systems that sloshed fluids. Some pieces could only fire one shot before having to be refilled with liquid. Indeed, Constitutionalist General Calles later complained that he had to place his artillery in the open on the plains because damage to the recoil mechanisms shortened the effective range, and hence options for placement. Yet he still maintained a proper distance to reach the Maytorenista trenches and machine-gun positions, or so he claimed.

The combination of damaged guns and faulty ammunition added to the tactical challenges that the weapon systems themselves had presented to their previous owners, the Federals, meaning that Villa, in particular, trusted in the artillery to support his infantry and cavalry in combat at a time when it was increasingly unable to do so. Mortars and howitzers, not guns, were needed for trench warfare. As proof, General Gavira confirmed that the artillery barrage of May 22, 1915, at Trinidad Station was both well-directed and intense, but nevertheless was not very effective in the end. He concluded that artillery was generally only good for its effect on morale and not particularly consequential, especially when troops on the receiving end were dug in and spread out.

In the final analysis, Villa relied heavily on mounted troops (which as Obregón pointed out had the added disadvantage of being very expensive) and placed undue faith in the power of artillery. Conversely, the Constitutionalists scarcely had any artillery to speak of, employed cavalry almost solely in its doctrinal roles—in combat during movements to contact and for counterattacks, and in ancillary roles for forage, scouting, reconnaissance, and security—and trusted in the constancy of infantry. Villa, for his part, had little regard for infantry and relied on ex-Federals for that combat arm. A signature tactic consisted of Villa's cavalry delivering infantry into combat on the back of their horses, such as at the battles of Tierra Blanca and Trinidad. In the latter instance, the infantry endeavored, unsuccessfully, to get within hand grenade range of Constitutionalist trenches.

In sum, the pressures of culture and the influence of experience tipped Villa's force composition out of all normative proportions, heavy in cavalry and artillery, when infantry and machine guns were much more apt for the sustained actions of 1915.

Northeastern

Like the revolutionaries of Chihuahua, those of the Northeast counted many ranchers among their ranks and consequently almost all their forces went mounted. However, unlike the Division of the North, the Army Corps of the Northeast was outfitted with an overwhelming preponderance of .30-30 hunting carbines, with maximum effective ranges of some two hundred meters. Therefore, the evidence points to their fighting almost always dismounted and even attacking at a crawl, such as at Tampico, in order to get within range. On the defense, in the trenches at El Ébano, the Constitutionalists fired in volleys and even "at will" with aimed shots. But they also employed the old Federal tactic of holding rifles aloft over their head and shooting.

Unlike the Division of the North, again, the Constitutionalists of the east never acquired many artillery pieces, mainly because the Federals in their section of the country did not have a large number of guns and set-piece battles where the revolutionaries might have the opportunity to take cannons away from the Federals seldom took place in the region. Consequently, Pablo González's army corps never did develop this combat arm to any extent. Ironically, however, the 1914 Battle of Ciudad Guerrero provided the best demonstration of the different fire missions during the different phases of combat: *con precisión* at a single target in counterbattery, *segadora* at the deployed Federal infantry, and *progresivo* against the Federals retreating along a single axis.

The one area where the Northeasterners did excel was in the development of their aerial service, mainly because Carranza's nephews were aviation enthusiasts and pilots. While the Villistas delivered messages and scouted with airplanes and the Sonorans tried their hand at, possibly, the first ever attempt at aerial naval bombing, the Northeasterners innovated the most. They employed their air service in dropping pamphlets—and later bombs—in psychological warfare, in conducting aerial reconnaissance, and, finally, in practicing a form of proto-blitzkrieg at El Ébano, with armored trains attacking in combination with aerial bombing and infantry support.

Colorados

The agricultural laborers and industrial workers of La Laguna who joined the Colorados became, in the opinion of many, the best soldiers of the revolution, surpassing even Villa's centaurs. The American Edwin Emerson, a journalist and U.S. Army informant, reported to General Leonard Wood that "from my observations within the last three years and from what I now witnessed at Gómez Palacio, Torreón, and San Pedro de las Colonias I consider the Federal Rurales and the

Federal Irregulars commanded by Benjamín Argumedo better individual fighters than any of Villa's men."

In the battles mentioned by Emerson, which took place in March and April 1914, the Federal commander, General José Refugio Velasco, employed these irregulars in the execution of difficult and bold tasks: under Colonel Reyna to deliver the counterattack at Gómez Palacio; under Argumedo to screen the evacuation of Gómez Palacio, deliver counterattacks at Torreón, provide the delaying force that allowed the Federal regulars to abandon Torreón, scout ahead and link up with the Federal relief force at San Pedro de las Colonias, return with sorely needed ammunition for the Division of the Nazas, and then guide the division back to complete the link up; to carry out (with Juan Andrew Almazán's Rurales) the diversion/counterattack to facilitate the Federal retreat at San Pedro, cover the retreat at Saltillo in May, 1914, and finally, travel cross-country in June 1914 and reinforce the Federal garrison under attack at Zacatecas. As Velasco famously commented, "None of the generals who were in Torreón and San Pedro de las Colonias are worth anything. Argumedo is better than all the rest put together. He alone got out a million cartridges on one thousand horses, that is, one crate for each horse" so that the Division of the Nazas could confidently complete the linkup at San Pedro. There was no other force in the entire annals of the revolution that consistently received the toughest assignments and executed them with more élan and success than the Colorados of La Laguna.

After Huerta's war, many of the Colorados traveled south to join the Zapatistas and continue fighting the Constitutionalists, but there simply were not enough of them to tip the balance in favor of their new allies.

Zapatistas

After four years of fighting guerrilla warfare and with absolutely no tutelage fighting alongside the Federal Army, the Zapatistas attempted to fight a conventional war in the winter and spring of 1915. They were, of course, not organized to direct such an effort. In the Liberating Army of the South there was no rhyme or reason to what constituted a squadron, regiment, battalion, corps, brigade, or division. Nor did the officer corps have any proportion in its rank structure, with officers sometimes outnumbering the enlisted in certain units.

Moreover, the Suriano Conventionists were ill-prepared for conventional combat, all of which explains why Constitutionalists and impartial observers, alike, described the Zapatista assaults as uncoordinated and resembling more of a mob riot. Additionally, all recognized the Surianos as poor marksmen. And because they lacked sufficient quantities of arms and ammunition they typically shot "in the Zapatista style," meaning sparingly, as described of Banderas' assault on El Fuerte. Attempting aimed shots that missed the mark rather than volleys muted the lethality of Zapatista arms. Sniping, an example of asymmetrical warfare, carried out by Zapatista sympathizers in Mexico City proved their most effective tactic.

Norteño Tricks and Traps

In addition to the foregoing tactics, Norteños employed tricks and traps redolent of nineteenth century warfare on the frontier, for example:

- Orozco's men tunneling under the Federal barracks to set dynamite at Ciudad Guerrero in 1910;
- the Maderistas at Cerro Prieto, trying to lasso Federal machine guns (not an uncommon event) and then shooting and moving in retreat—in textbook fashion—until slung up on the back of a comrade's horse;
- Toribio Ortega's men using lowered telegraph lines at night to unhorse Federal cavalry near Cuchillo Parado;
- Maderista ambushes of Federals at Boquilla de la Laguna and Malpaso, where Orozco's men placed their hats on bushes meters away from their actual location as a decoy and to make their numbers look deceptively larger than in reality;
- Maderistas burrowing through buildings to avoid exposure to Federal fire in the streets of Ciudad Juárez in 1911;
- Maderista "guides" in the extreme vanguard riding hard in a line of foragers to kick up enough dust to make the Colorados at Conejos in 1912 think that infantry was present;
- General Jesús Carranza's soldiers dragging branches tied to the tails of their horses (*echar las rastras*) to kick up a large dust cloud and make the Federals think reinforcements were arriving at the Battle of Ciudad Guerrero in 1914;
- General Campa's Colorados sending a stampede of horses and mules to overrun Colonel Ricardo Peña's Federals at Huarichic Station;
- Maytorenistas stampeding cattle toward the Constitutionalist lines at Naco in 1914 and then attempting to tunnel under the Constitutionalist defensive works;
- Yaquis quietly slithering in the dark in an attempt to enter Constitutionalist lines unnoticed at Naco in October 1914; and,
- Calles' sentinels employing reconnaissance by fire to try to get the Yaquis to reveal their location.

Stratagems and Asymmetry

Finally, no army was completely above employing stratagems, some wily and others quite scurrilous; some worked while others did not. Among the most dishonorable tricks employed by the Federals was to wave a white flag of parley and then open fire when the envoys approach to talk. In another instance, in 1912, General Blanquet at Avilés started loading troops on trains in view of Campa's Colorados as a show of vulnerability to entice them to attack—it produced the desired results. Obregón at Naco in 1913 sent his trains back to Nogales empty to make Federal General Ojeda think the bulk of the Constitutionalists had left the area and trick him into attacking—it didn't work. Constitutionalist General Treviño employed a ruse at El Ébano with the opposite objective; he used trains to deliver the same infantry over and over. The troops detrained in the station in full view of the Villistas to make them think that the Constitutionalists were receiving numerous reinforcements.

The Maytorenistas at Naco in 1914 attempted a cocktail of subterfuge that included: feigning to be wounded and making pathetic cries to elicit aid from the Constitutionalist defenders only to kill any would-be rescuers; attempting to suborn Constitutionalist officers; and firing at the American side to try to provoke an armed intervention.

Of course, the most famous stratagem of the revolution was Villa's Trojan horse: the "Federal" train that returned to Ciudad Juárez because it was unable to deliver coal to Chihuahua City only to disgorge numerous Constitutionalists who easily captured the border town.

In a similar move, the Reyes-Fierro column in 1915 sent telegrams in Obregón's name to Constitutionalists in command of the plazas in its path ordering them to evacuate the towns so the Villistas could march down Obregón's lines of communication unopposed, destroying railroad tracks and telegraph lines.

STRATEGY

As the revolution progressed from its opening shots to the climactic battles of 1915, the Scope, Scale, Sophistication, and Speed of operations continued to increase until reaching the point of the largest land battles on the North American continent, aside from the U.S. Civil War. What follows is a recapitulation of the many strategies proposed and executed through the various phases of the revolution in a condensed narrative to give the reader a better appreciation of the complexity and interrelatedness of actions and consequences that might have been missed in the more detailed text.

Madero's Rebellion

At the start of the revolution, officially set to begin November 20, 1910, Madero's supporters in Chihuahua proposed to conquer the axis running from Ciudad Guerrero to Chihuahua City. The Federals first sent a company-sized group of soldiers from Chihuahua City to reinforce the garrison at Ciudad Guerrero, but the Maderistas shot the train to pieces at San Andrés and then Pascual Orozco's Maderistas destroyed the remainder of the column at Pedernales. Next, the Federals sent an "operations brigade" under General Navarro to reinforce Ciudad Guerrero, but it got a late start.

Navarro soon realized that the rebels controlled the rails and so he set out cross country. He managed to defeat Orozco and Pancho Villa at Cerro Prieto, but his operations brigade lacked the mobility, numbers of men, and proper supplies to continue forward and he soon found himself cut off from his base of operations. The Federals sent a relief train to bring reinforcements and ammunition under Colonel Guzmán but the rebels ambushed and decimated the column at Malpaso. Thereafter, the Federals had to send successive relief columns with the proper gear and sufficient numbers of men to skirt Malpaso and rescue Navarro. By the time Navarro reached Ciudad Guerrero the garrison had long since surrendered and the rebels were preparing to leave the area to head north.

The Maderistas moved toward the border in two columns, Orozco from Guerrero District toward Ciudad Juárez to meet with leading Maderistas gathering at the border, and Blanco from Bachíniva headed for northwestern Chihuahua. In response, the Federals started sending columns from Chihuahua City to Casas Grandes, one via Ciudad Juárez, and a second marching only as far north as Gallegos Station and then turning to the west. The second column fought a bloody but indecisive battle against Orozco at Sierra Mojina while Blanco fought and destroyed smaller Federal garrisons in western Chihuahua. As Orozco approached Ciudad Juárez the Federals had to concentrate their forces in Casas Grandes and forward a column under Colonel Rábago to reinforce Ciudad Juárez. Orozco ambushed Rábago at Bauche Station and left him escaping on foot to Ciudad Juárez, where he was soon joined by Navarro's Operations Brigade.

At this point, Madero entered Chihuahua with his foreign advisors and began moving southward. The Federals gave chase from Ciudad Juárez but could not catch him before he made an ill-advised and poorly-conducted attack on Casas Grandes, which failed. Thereupon, Madero continued south to Bustillos, west of Chihuahua City, where he concentrated his forces under Orozco and Villa. The Federals assumed that his intentions were to attack the state capital and abandoned Casas Grandes, the key to northwestern Chihuahua, and started pulling in troops to reinforce Chihuahua City. At this point, the Maderistas wheeled back to the north through the wide-open west, and isolated and laid siege to Ciudad Juárez. With overwhelming numbers, they crushed the Federal defenders inside the city and forced the resignation of President Díaz.

Orozco's Insurgency

According to the Treaty of Ciudad Juárez, the Federal Army had to evacuate the state of Chihuahua, leaving General Pascual Orozco in command of four corps of Rurales as the only national security force. In February, however, it appeared as though Orozco was planning to join Emiliano Zapata in rebellion and mobilize troops to march on Mexico City. Therefore, the Federal commander in Torreón sent Major Ramírez with a company-sized column to advance up the Central Railroad and remain in overwatch. Governor Abraham González, too, reactivated Maderista officers who had been discharged from service, most notably Pancho Villa, to protect the rails and be ready for what might develop. Then, at the end of February, adherents of the PLM captured Ciudad Juárez and a few days later Orozco resigned his commission and declared himself in rebellion. Supported by Rurales from his former command, his new PLM allies, Vazquistas (Vázquez Gómez), and a host of enthusiastic adventurers, Orozco began moving down from Chihuahua City, sweeping along Pancho Villa and destroying Ramírez's command.

Meanwhile, the Federal Division of the North moved up to intercept Orozco, attempting a pincer movement designed to close at Jiménez. Orozco's Colorados roundly defeated the main body of the Federals at Rellano on the Central Railroad, but had to reverse trajectory to deal with the second prong of the pincer commanded by General Trucy Aubert. The Colorados subsequently trounced Trucy Aubert and forced Pancho Villa out of Parral, but they lost precious time and forward momentum in doing so.

Orozco then sent General Salazar to invade neighboring Coahuila and ordered General Campa to join Argumedo in eastern Durango to cut the rails south of Torreón toward Durango and Zacatecas. Salazar was a poor choice to lead such a logistically difficult mission across the desert of western Coahuila and, using interior lines, the new commander of the Federal Division of the North, Huerta, sent Trucy Aubert to reinforce the state forces of Coahuila and defeat Salazar west of Monclova. Afterward, Trucy Aubert returned to Torreón to rejoin the division.

The Federals attempted to catch Campa in a vice, with one group of Federals traveling southward from Torreón toward Durango City, and another headed in the opposite direction from that latter city. Campa however, defeated the two Federal commands in detail. In the meantime, General Blanquet returned from convalescing in Mexico City and, using Torreón as a pivot point, employed his 29th Battalion in successively defeating Campa and Argumedo in engagements at points on the railroads to the southwest and southeast of Torreón. At the same time, and after having gathered superior firepower and sufficient numbers, Huerta drove into Chihuahua and garnered one victory after another against Orozco. Whereupon, the Colorados initiated unsuccessful campaigns to invade Coahuila from the north and Sonora from three points on its eastern border, which spurred the mobilization of paramilitaries in those states.

Orozco's rebellion was significant for several reasons. First, for the first time the number of combatants engaged in battles ascended into the thousands for both sides. Second, the war was transformed from one of politically-engaged activists who had fought for Madero to a more populist-inspired army, with an attendant decrease in the average age of the combatants. Finally, maneuver warfare burst from its original confines in Chihuahua to include other states, such as Coahuila, Durango, and Sonora.

The Constitutionalist Revolution in 1913

After former and active-duty officers of the Federal Army overthrew the government of Madero in February 1913, the state best prepared to offer resistance was Sonora. The highest ranking commander, Colonel Obregón, succeeded in gathering the major militia commands in Hermosillo, in the middle of the state, so that once the legislature decided to rebel, he was well-positioned to operate on interior lines. While Obregón marched north with superior numbers and destroyed the Federal garrisons along the border, his cousin Colonel Hill moved through the south collecting enough volunteers to conquer Alamos. By May 1913, the Federals in Sonora only held Guaymas and Torín.

On orders from Mexico City, the Federals attempted to march from Guaymas and retake Hermosillo, but they did not have the requisite men, equipment, enthusiasm, skill, or an adequate logistical system to undertake such an operation. Consequently, they suffered a serious defeat at Santa Rosa that forced them to abandon Torín. An even worse loss at Santa María in June relegated the Federals in Sonora to Guaymas and neighboring Empalme for the rest of the war.

Afterward, the Federal Division of the Yaqui turned its attentions on subduing the Constitutionalist in northern Sinaloa, landing troops at Topolobampo to cooperate in a pincer with troops marching up from Sinaloa de Leyva to attack San Blas. That operation failed miserably and a second force sent to Topolobampo fared no better. After that, the Federals had to evacuate Topolobampo by sea to reinforce Mazatlán and the Constitutionalists marched down with superior numbers, conquering Sinaloa de Leyva and Culiacán so that by year end they controlled all of Sonora and Sinaloa except for Guaymas and Mazatlán, which they ultimately invested and bypassed.

Conclusion

In Durango and southern Chihuahua, state-based paramilitaries and former Maderista commanders mobilized their troops and recalled their former soldiers to oppose the Federal Army in those states. Employing overwhelming numbers, they destroyed garrisons on the Central Railroad, such as Camargo and Jiménez, which forced the evacuation of Parral, and cleared central Durango of Federal troops while taking up track between Durango City and La Laguna. Once Durango City had been sufficiently isolated, the Constitutionalists attacked with dominant numbers and scattered the Federal defenders. By June 1913 the Federals were reduced to operating in Chihuahua along the Ciudad Juárez-Chihuahua City axis and to possession of the northeastern corner of Durango.

Once Pancho Villa had raised a brigade and had sufficiently trained, organized, and equipped it, he marched down from northwestern Chihuahua and joining the Constitutionalists from Durango captured Gómez Palacio and Torreón, and with them badly needed resources. Returning to the north, he attacked Chihuahua City, without success, and then with cunning surprised and took Ciudad Juárez. In the subsequent Battle of Tierra Blanca, Villa's Division of the North beat a combined force of Federals and Colorados, which compelled the Federals to abandon Chihuahua City for Ojinaga. At the beginning of 1914, Villa's men attacked Ojinaga and forced the Federals to cross over into the United States, where they were detained. The Constitutionalists held all of Chihuahua and most of Durango.

In the Northeast, the Constitutionalists were not as numerous, as well-equipped, nor did they have the same unity of command enjoyed by the revolutionaries of Chihuahua and Sonora. Moreover, the Federal Army had a distinct interest in maintaining control of one of the wealthiest sections of the country where Blanco's Constitutionalists had captured Matamoros, causing quite a stir. Accordingly, in June 1913, the Federals launched dual offensives against the Constitutionalists of Coahuila, one driving north from Saltillo and another doing the same from Monterrey. At the time, the Constitutionalists held Monclova, the First Chief was in Piedras Negras, and Jesús Carranza with Pablo González operated out of Candela in far eastern Coahuila where they harassed the rails leading to Nuevo Laredo. When the Federals captured Candela, the First Chief ordered forces in Monclova to retake the town. However, the movement of troops to retake Candela weakened Monclova to such a degree that the Federals moving up from Saltillo captured the city and forced the Constitutionalists back to Las Hermanas and ultimately out of the state by the beginning of October.

At that point General González decided to attempt a *coup de main* on Monterrey, which failed to take the town but resulted in the capture of invaluable supplies. With these, and leaving the Blanco Brigade to keep the Federals in Monterrey from making an attempt on Matamoros, General González began his Tamaulipas campaign. He took Ciudad Victoria in November, and then attacked Tampico in

December and Nuevo Laredo on New Year's, at opposite ends of the state. González's Constitutionalists won neither battle, but by year's end, he controlled most of the state of Tamaulipas.

The Constitutionalist Revolution in 1914

By 1914 the Constitutionalists controlled a fair amount of territory in the North. As the First Chief's senior commanders debated strategy, it was clear that for the first time all had inter-theater cooperation on their minds. Because Sonora had no direct rail links to the east and was somewhat isolated by the Sierra Madre, Generals González and Villa had the greater opportunity to operate together. They devised a plan whereby González's brigades would isolate Nuevo Laredo, try to capture Monclova, if practicable, cut the rails to the west of Saltillo, and then drive in the garrisons surround Monterrey and attack that plaza in conjunction with the Division of the North's attack on Gómez Palacio and Torreón on or about March 23, 1914.

Villa completed his part of the mission, attacking and taking Torreón by April 2, but González's brigades, ranging over a vast amount of territory did not fare so well. First, a Federal operation designed to recapture González's base of operations at Matamoros in February and March tied up his 1st and 3rd Brigades and other smaller columns. Then, his 2nd and 4th Brigades were defeated at Monclova and again at San Buenaventura in mid-March, such that the 4th was in no condition for battle and the 2nd remained out of commission waiting for supplies for almost a month. The 5th and 6th Brigades were down south in San Luis Potosí and around Tampico, and General Jesús Agustín Castro's 8th Brigade simply ignored orders to report to Monterrey. The 7th Brigade had done its best to cut the rails west of Saltillo, but by itself could not contain the combined-arms columns of Federals flowing through Saltillo to reinforce Torreón. Therefore, after weeks of campaigning, Villa's Division of the North had to battle three Federal divisions at San Pedro de las Colonias in mid-April. Shortly thereafter, aided by the geopolitical shift caused by the American invasion of Veracruz, González finally conquered Monterrey. Threat of an American offensive through the Northeast also caused the Federals to evacuate Nuevo Laredo and Piedras Negras to begin a scorched earth campaign, thus permitting González's refitted 2nd Brigade to conquer northern Coahuila.

With the bulk of his brigades, González then moved down south and took Tampico—assisted indirectly by the 1st Division of the East attacking Tamiahua— while Villa, making good on his promise to help the Army Corps of the Northeast, destroyed the large concentration of Federals at Paredón in southeastern Coahuila and occupied Saltillo, which the Federals evacuated in haste for San Luis Potosí.

Villa's Division of the North left Saltillo almost as fast as it had entered, returning to Torreón to prepare for the Zacatecas campaign. The First Chief, however, had other plans. He was becoming suspicious of Villa and Ángeles increasingly acting independent of the plans and wishes of the First Chieftaincy, so Don Venustiano urged Obregón to quicken his campaign—which had bogged down in trying to take Mazatlán—and he assigned the objective of Zacatecas to General Natera. Natera's 1st Division of the Center, however, proved unequal to the task. Therefore, disobeying orders, Villa took his entire division and with Natera defeated the Federals at Zacatecas. In retaliation, the First Chief cut off Villa's access to coal and arms and ammunition due to arrive in Tampico, forcing him to retreat back to the North. This was the beginning of the schism within the Constitutionalist Army.

With the Federal Army freed of any threat from Villa, it could direct its attentions solely toward Obregón, so with an added sense of urgency the Constitutionalist general hastened for Guadalajara, which he occupied subsequent to winning a battle of reverse fronts at Orendáin. A week after the loss of that key city, Huerta resigned and Velasco took over command of the Federal Army. He ordered all Federals to abandon San Luis Potosí, the Bajío, and Guaymas and concentrate in Mexico City for a decisive battle. The Division of the Yaqui landed at Manzanillo and temporarily threatened Obregón's rearguard before he forced it to board its ships and sail away. The Army Corps of the Northwest and Army Corps of the Northeast then linked up at Querétaro, and the Federal Army realized that it could not hope to fight a successful culminating battle and surrendered.

Civil War

The main reason for the Villista defeat was a strategy inferior to that employed by the Constitutionalists, notwithstanding the fact that the Conventionist Army enjoyed interior lines on a strategic level. The Conventionist Army executed strategy as a collection of theaters, mainly at the insistence of Emiliano Zapata, instead of employing a holistic approach. The lack of a coherent effort was made more permanent by the defection of President Eulalio Gutiérrez, which had the effect of compelling both a de facto and de jure split in the Conventionist Army that already existed between its units from the North and South. The Conventionists continued to fail to set priorities and by March were conducting offensive operations on all fronts: Sonora, Tepic, Coahuila, Tamaulipas, and San Luis Potosí, everywhere except Jalisco where Villa had been convinced by General Ángeles to suspend operations to redirect his energies to the Northeast. When Villa returned to fight in the Bajío, exhausted by the trip, the Zapatistas refused to support the stated strategy by systematically destroying Obregón's lines of communication. Instead they tried to win heads-up battles for Mexico City and Puebla with a force entirely incapable of winning such battles.

Strategy and Tactics of the Mexican Revolution

As Constitutionalist General Treviño later stated, "the principal error of the Villista high command consisted of forming two simultaneous fronts of attack, one in the vicinity of Celaya and the other around Ébano, without the possibility of triumphing or imposing themselves in either place. The logical result of this error was that we could defeat them in detail." Villa was also spread too thin, having personally assumed responsibility for the government, judiciary, and military field operations.

In contrast the Constitutionalists did set priorities, first in February and March at El Ébano, to the detriment of Diéguez and Murguía in Jalisco and with Obregón having to wait for arms and ammunition in Mexico City for months. Priority again shifted in June and July to González's mission to retake Mexico City, once again leaving Obregón in Encarnación to his own devices. In sum, the Constitutionalists managed the principles of mass and economy of force better than the Conventionists.

The Convention also suffered from systemic defects related to the organization of forces. Zapata commanded a guerrilla force woefully lacking in organization, training, resources, and command and control that was ill-suited to the mission that he tried to assign it: to annihilate the Constitutionalists. A strategy of attrition in the South would have yieded better results for the overall Conventionist strategy. The Villistas, on the other hand, were top-heavy in mounted forces and artillery and deficient in infantry and machine guns for reasons, just as with the Zapatistas, related to regional traditions in fighting, or in modern parlance, "ways of war."

Thus, poor strategy formulation and execution and systemic defects in force composition and capabilities—in addition to numerous tactical errors and the vagaries of war—explained the Convention's defeat.

* * *

In conclusion, we can affirm that as the revolution spread to other states from Chihuahua in 1910, it increased the reach, complexity, and interrelatedness of campaigns and drew in different ways of war, which had different capabilities and traditions as expressed in the tactics employed. Those tactics, in turn, impacted battlefield outcomes and determined the Revolution's ultimate victors. In spite of popular lore, the Mexican Revolution was not just another guerrilla or "small war."

APPENDIX A: KEY BATTLES, 1910–1915

Madero's Rebellion

Cerro Prieto, Chih. (December 11, 1910) 850 Federals defeat 500 Maderistas

Malpaso Canyon, Chih. (December 18, 1910) 550 Maderistas defeat 620 Federals

Casas Grandes, Chih. (March 6, 1911) 1,060 Federals defeat 400–800 Maderistas

Ciudad Juárez, Chih. (May 8–10, 1911) 3,000 Maderistas defeat 817 Federals

Orozco's Rebellion

First Rellano, Chih. (March 24, 1912) 4,500 Orozquistas defeat 1,791 Federals

Conejos, Dgo. (May 12, 1912) 5,000 Federals defeat 4,500 Orozquistas

Second Rellano, Chih. (May 23, 1912) 5,200 Federals defeat 6,000 Orozquistas

Bachimba, Chih. (July 3–4, 1912) 4,800 Federals defeat 5,600 Orozquistas

The Constitutionalist Revolution

The Citadel, D.F. (February 9–18, 1913) 4,000 Felicistas and praetorian Federals defeat a similar number of pro-government Maderistas and Federals

Naco, Son. (April 8–13, 1913) 3,000 Constitutionalists defeat 450 Federals

Hacienda Santa Rosa, Son. (May 9–11, 1913) 3,100 Constitutionalists defeat 1,800 Federals

Matamoros, Tamps. (June 3–4, 1913) 900 Constitutionalists defeat 450 Federals

Zacatecas, Zac. (June 5–6, 1913) 1,200 Constitutionalists defeat 400 Federals

Durango City, Dgo. (June 17–18, 1913) 6,000 Constitutionalists defeat 1,300 Federals

Hacienda Santa María, Son. (June 19–26, 1913) 3,700 Constitutionalists defeat 2,650 Federals

Monclova, Coa. (July 10, 1913) 1,800 Federals defeat 1,500 Constitutionalists

First Torreón, Coa. (July 22–31, 1913) 4,000 Federals repulse 9,000 Constitutionalists

Las Hermanas, Coa. (August 16–17, 1913) 1,500 Federals defeat 1,000 Constitutionalists

San Andrés, Chih. (August 26, 1913) 1,025 Constitutionalists defeat an estimated 400 to 980 Federals

Second Torreón, Coa. (September 22 – October 1, 1913) 8,000 Constitutionalists defeat 3,500 Federals

First Monterrey, N.L. (October 23-24, 1913) 1,700 Federals repulse 1,800 Constitutionalists

Chihuahua City, Chih. (November 5-9, 1913) 6,400 Federals repulse 5,000 Constitutionalists

Culiacán, Sin. (November 5-14, 1913) 3,000 Constitutionalists defeat 1,250 Federals

Ciudad Victoria, Tamps. (November 16-18, 1913) 2,500 Constitutionalists defeat 950 Federals

Tierra Blanca, Chih. (November 23-25, 1913) 6,200 Constitutionalists defeat 2,300 Federals

Nuevo Laredo, Tamps. (January 1-2, 1914) 1,200 Federals repulse 2,000 Constitutionalists

Ojinaga, Chih. (January 1-10, 1914) 3,500 Constitutionalists defeat 4,500 Federals

Fourth Torreón, Coa. (March 22 – April 2, 1914) 15,700 Constitutionalists defeat 7,650 Federals

San Pedro de las Colonias, Coa. (April 5-14, 1914) 14-16,000 Constitutionalists defeat 9,000 Federals

Second Monterrey, N.L. (April 18-24, 1914) 4,000 Constitutionalists defeat 3,000 Federals

Allende, Coa. (April 26-7, 1914) 600 Constitutionalists defeat 500 Federals

Tampico, Tamps. (May 9-13, 1914) 4,000 Constitutionalists defeat 3,000 Federals

Zacatecas, Zac. (June 21-3, 1914) 20,000 Constitutionalists defeat 10,000 Federals

Orendáin, Jal. (Guadalajara) (July 6-7, 1914) 8,000 Constitutionalists defeat 2,000 Federals

El Castillo, Jal. (July 8, 1914) 4,000 Constitutionalists defeat 1,500 Federals

The Civil War

Naco (October 4, 1914 to January 15, 1915) 3,000 Maytorenistas fight 2,500 Constitutionalists to a negotiated truce

Puebla (December 29, 1914 to January 5, 1915) 12,000 Constitutionalists defeat 4,000+ Conventionists

Ciudad Victoria (January 5-9, 1915) 1,500 Constitutionalists repulse 2,000 Conventionists

Ramos Arizpe (January 8, 1915) 11,500 Conventionists defeat similar number of Constitutionalists

El Cuatro (January 17-18, 1915) 10,000 Constitutionalists defeat an unknown number of Conventionists, perhaps 5-8,000

Appendix A: Key Battles, 1910–1915

San Felipe Torres Mochas (January 28-30, 1915) 4-5,000 Conventionists defeat 5-6,000 Gutierristas

Mexico City (January 28 – March 11, 1915) 10,000 Constitutionalists defeat 13,000 Conventionists

Monterrey (February 6-7, 1915) 8,000 Conventionists repulse 4,000 Constitutionalists

Cuesta de Sayula (February 17-18, 1915) 10,000 Conventionists defeat 5,000 Constitutionalists

Blanca Flor (March 14, 1915) 5,100 Constitutionalists defeat 4,000 Argumedistas

Halachó (March 16, 1915) 5,100 Constitutionalists defeat 1,300 Argumedistas

El Ébano (March 5 – May 31, 1915) 6,000 Constitutionalists repulse 10-15,000 Conventionists

Tuxpan, Jalisco (March 22-25, 1915) 6-7,000 Constitutionalists defeat 5,000 Conventionists

Huizachito/Jarita Station (April 12, 1915) 1,800 Constitutionalists defeat 2,000 Conventionists

Matamoros (March 27 – April 16, 1915) 300 to 1,200 Constitutionalists repulse 2,000 Conventionists

First Celaya (April 6-7, 1915) 11,000 Constitutionalists repulse 10-12,000 Conventionists

Second Celaya (April 13-15, 1915) 15,000 Constitutionalists repulse 10-12,000 Conventionists

Trinidad (April 27 – June 5, 1915) 25,000 Constitutionalists defeat 25,000 Conventionists

Aguascalientes (July 6-10, 1915) 20,000 Constitutionalists defeat 8-10,000 Conventionists

Icamole (May 23 – September 4, 1915) 5,000 Constitutionalists defeat 5,000 Conventionists

Agua Prieta (November 1-3, 1915) 7,000 Constitutionalists repulse 8-12,000 Conventionists

El Alamito/Hermosillo (November 18-22, 1915) 5,000 Constitutionalists repulse 4-7,000 Conventionists

APPENDIX B: WEAPON SYSTEMS

SMALL ARMS

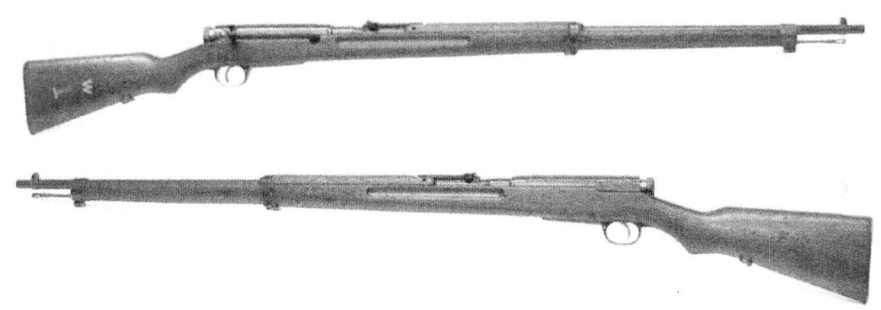

Figure 2 The Mauser Rifle

The 7-mm. Mauser rifle and carbine each took a five-round clip and had open sights for aiming at targets up to three hundred meters away and pop-up rear sight plates that accommodated a maximum effective range of 2,000 and 1,400 meters, respectively. The cavalry used the carbine, which had a turned-down bolt handle so that it could be sheathed. It weighed 3.5 kilograms, and measured 95.25 centimeters in length, compared to the 4.0 kilograms and total length of 123.2 centimeters for the rifle. War planners anticipated that combat would take place at between four hundred and six hundred meters distance, but in practice it often took place at a slightly closer range, perhaps owing to the poor marksmanship of the Federal soldiers or the predominance of Winchester .30-30 carbines favored by the revolutionaries. These U.S.-manufactured .30-30's tended to weigh slightly less than the Mauser and could carry from six to eleven cartridges in the magazine tube versus only five for the slightly heavier Mauser carbine, but the latter could be loaded much more rapidly using clips. Additionally, since the Winchester carbines were intended for hunting instead of as military-grade weapons they often overheated and broke down under combat conditions and had a much shorter effective range of some two hundred meters. The main advantages to the Winchester carbines were, of course, that they were relatively cheap and readily available.

Figure 3 Winchester .30-30 Carbines. Top: Winchester 1894 Saddle Ring carbine manufactured in 1914. Bottom: Winchester 1894 saddle ring .30 WCF (Winchester Center Fire) caliber carbine manufactured in 1908 with 20" barrel.

In 1893 the 8-mm. Pieper, manufactured by the H. Pieper firm in Liege, Belgium, became the regulation sidearm per presidential decree. The Federal Army ordered 5,000 of these. During the revolution, the M1902 Colt .38 automatic replaced the Pieper as the official sidearm and it was said that Captain Federico Montes used one of these automatics to defend President Madero during the palace coup of February 1913.

Figure 4 Regulation 8-mm. Pieper Sidearm

The Federals also purchased small lots of single- and double-action Colt .45s. Revolutionaries, of course, used whatever pistols they could acquire. Pancho Villa's Division of the North experienced some difficulty in getting pistols because its main suppliers were located in El Paso, Texas, and contrary to popular belief, throughout the twentieth century the State of Texas had some of the strictest handgun laws in the United States. Among the pistols mentioned in the literature are Colt .45s in the service of officers of the Division of the North and unspecified automatics and revolvers among the rank and file of González's Army Corps of the Northeast.

Figure 5 Regulation Colt .38 Automatic

HAND GRENADES

The Federal Army did not have hand grenades, but rather used Marten Hale rifle grenades, which were first tested in November 1910 and made a debut on the battlefield at Bachimba in July 1912. The rifle grenades failed to impress Captain Jacinto B. Treviño, who observed them in use at Bachimba and later said that their effect on enemy morale was greater than their ability to inflict physical damage. The Constitutionalists, on the other hand, did not have hand or rifle grenades, but rather created "hand bombs" with dynamite arrogated from the many mining operations in the North. There can be no doubt but that those hand bombs were quite effective in combat. There was limited mention of hand grenades in connection with the campaigns of 1915.

MACHINE GUNS

The rapid-fire rifles most commonly used by the Federals were Rexers and the Madsen M1911. Rexers were a knock-off of the Madsen manufactured in England by the Rexer Arms Company until court-ordered to cease production due to patent infringement. The Colt M1895 "potato digger" also received frequent mention, although typically in connection with revolutionaries. Other machine guns with fairly similar attributes also made an appearance on the battlefield: 7 millimeters, with a cyclic rate of four to five hundred rounds per minute able to cover a front just over one hundred meters wide at one thousand meters and a maximum range of three kilometers.

ARTILLERY

The workhorses of the Federal Army were the 75-mm. Saint Chamond-Mondragón (muzzle velocity of 500 meters per second, range of 100 to 4,000 meters) and Schneider Canet (range of 100 to 3,000 meters) field guns. At the beginning of 1913 the Federal Army had forty-five and thirty-two of each, respectively. The arsenal also included fifty-eight light and powerful type 80-mm. Mondragón field guns and twenty-one Mondragón mountain guns. Later, the Federals acquired twelve 75-mm. Vickers-Maxim mountain guns. Obregón's Army Corps of the Northwest and Villa's Division of the North came into possession of the field and mountain guns as they defeated their Federal foes, but General González's Army Corps of the Northeast did not have the opportunity to acquire as many guns simply because the Federals in the Northeast did not have as many as the other theaters and neither did his army corps have as many monumental victories. Ergo, for quite some time the Constitutionalist Army Corps of the Northeast used homemade cannons manufactured in the railroad workshops of the North.

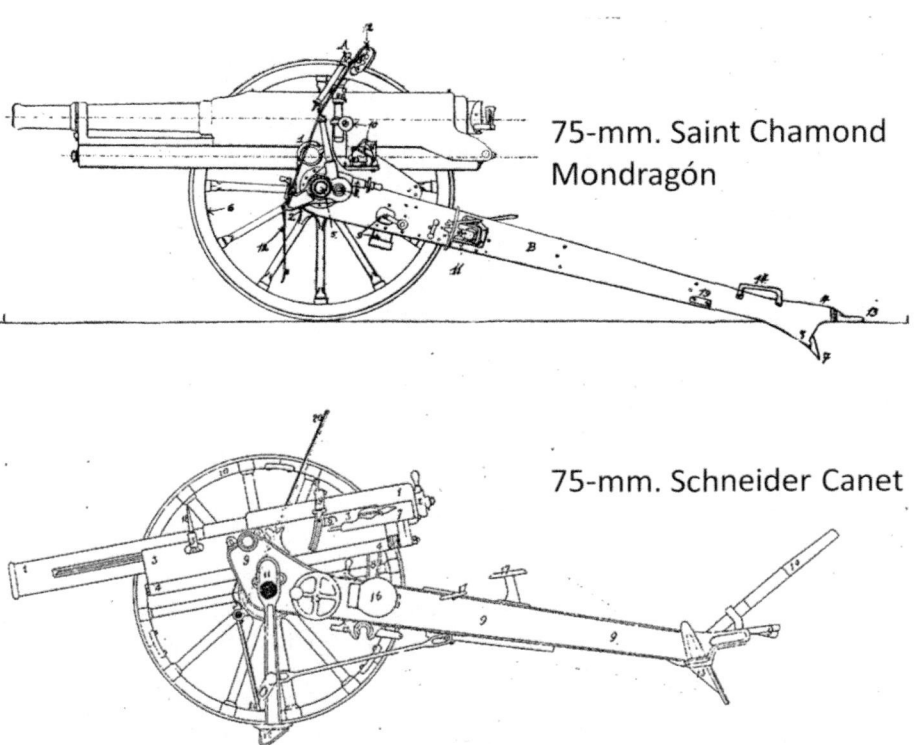

75-mm. Saint Chamond Mondragón

75-mm. Schneider Canet

Figure 6 Field Artillery Pieces

ARCHIVAL
ABBREVIATIONS

AGN-CDLR—Archivo General de la Nación: Colección Revolución
AGN-CONV—Archivo General de la Nación: Convención Revolucionaria
AGN-CR—Archivo General de la Nación: Cuerpos Rurales
AGN-EZ—Archivo General de la Nación: Emiliano Zapata

AHSDN—Archivo Histórico de la Secretaría de Defensa Nacional

AJBT—Biblioteca y Hemeroteca Nacional at Universidad Nacional Autónoma de México: Archivo Jacinto B. Treviño

AREM—Archivo de Relaciones Exteriores de México

BLAC-CPGMR—Benson Latin American Collection, General Libraries, University of Texas at Austin: Collection relating to Pablo González and the Mexican Revolution, 1911-1969
BLAC-LGC—Lázaro de la Garza Collection

CEHM-GRN—Centro de Estudios de Historia de México, Fondo DLXXIII Guillermo Rubio Navarrete
CEHM-HS—Centro de Estudios de Historia de México. Fondo XXI-1 Hojas de Servicio. Primer Jefe del Ejército Constitucionalista Venustiano Carranza
CEHM-JA—Centro de Estudios de Historia de México. Fondo VIII-3 Archivo del General Jenaro Amezcua.
CEHM-MWG—Centro de Estudios de Historia de México. Fondo LXVIII-1 Manuscritos del General Manuel Willars González.
CEHM-VC—Centro de Estudios de Historia de México. Fondo XXI Primer Jefe del Ejército Constitucionalista Venustiano Carranza.

Archival Abbreviations

NARA-AGOMI—National Archives and Records Administration. U.S. Adjutant General's Office, Record Group 94, Entry 25, "Mexican Intervention" File 2149991

NARA-AGOVR—National Archives and Records Administration. U.S. Adjutant General's Office, Record Group 94, Entry 25, "Villa's Revolution," File 2212358

NARA-GSS—National Archives and Records Administration. Records of the War Department General and Special Staffs, Record Group 165

NARA-USDS—National Archives and Records Administration. U.S. Department of State, 1910-29, Record Group 59, Military and Navy (812.20-30)

SOURCES CONSULTED

Archives

<u>México</u>:
Archivo General de la Nación
      Colección Revolución
      Convención Revolucionaria
      Cuerpos Rurales
      Emiliano Zapata
Archivo de Relaciones Exteriores de Mexico
Archivo Histórico de la Secretaría de Defensa Nacional
Archivo Histórico del Centro de Estudios de Historia de México
      Fondo DLXXIII Guillermo Rubio Navarrete
      Fondo LXVIII-1 Manuscritos del General Manuel Willars González.
      Fondo XXI Primer Jefe del Ejército Constitucionalista Venustiano Carranza.
      Fondo XXI-1 Hojas de Servicio. Primer Jefe del Ejército Constitucionalista Venustiano Carranza
      Fondo VIII-3 Archivo del General Jenaro Amezcua.
Biblioteca y Hemeroteca Nacional at Universidad Nacional Autónoma de México
      Mexican Newspapers, 1912-1915
        El Demócrata
        El Imparcial
        El Pueblo
        El Tiempo
      Archivo Jacinto B. Treviño

<u>United States</u>:
National Archives and Records Administration, Washington, D.C. and College Park, Maryland
      Adjutant General's Office Records, Record Group 94
      General Records of the Department of State, Record Group 59
      Records of the War Department General and Special Staffs, Record Group 165
Benson Latin American Collection, General Libraries, University of Texas at Austin

Sources Consulted

Collection relating to Pablo González and the Mexican Revolution, 1911–1969

Lázaro de la Garza Collection

New York Times Online

NewspaperArchive.com

Laredo Weekly Times

Nevada State Journal

The Abilene Daily Reporter

The Atlanta Constitution

The Galveston Daily News

The Salt Lake Tribune

The San Antonio Light

The Washington Post

Manuscripts, Benson Latin American Collection, General Libraries, University of Texas at Austin

Anonymous. "Heroicos Cuerpos de Ejército del Noreste y Oriente del General Pablo González."

González, Pablo Jr. "El Centinela Fiel del Constitucionalismo."

Archival Guides

Archivo General de la Nación. *Documentos inéditos sobre Emiliano Zapata y el Cuartel General, seleccionados del archivo de Genovevo de la O, que conserva el Archivo General de la Nación.* México: Comisión para la Conmemoración del Centenario del Natalicio del General Emiliano Zapata, 1979.

Espejel Lopez, Laura. *El Cuartel General Zapatista, 1914-1915: Documentos Del Fondo Emiliano Zapata Del Archivo General De La Nación.* 2 Vols. México: Instituto Nacional de Antropología e Historia, 1995.

Hernández y Lazo, Begoña. *Catálogo del Archivo Jacinto B. Treviño. Ramo: Ejército Constitucional. Subramo: Operaciones Militares.* México: Universidad Nacional Autónoma de México, 1984.

Moguer Flores, Josefina. *Guía e Índeces del Archivo del Primer Jefe del Ejército Constitucionalista, 1889-1920.* México: Centro de Estudios de Historia de México Condumex, 1994.

Muro, Luis, and Ulloa, Berta. *Guía del Ramo Revolución Mexicana, 1910-1920, Del Archivo Histórico de la Defensa Nacional y de Otros Repositorios del Gabinete de Manuscritos de la Biblioteca Nacional de México.* México: El Colegio de México, 1997.

Ulloa, Berta. *Revolución Mexicana, 1910-1920: Archivo Histórico Diplomático Mexicano, Guías para la Historia Diplomática de México*. México: Secretaría de Relaciones Exteriores, 1963.

Military Theory, Manuals, and Reference

Comisión para La Formación del Reglamento para el Servicio de las Piezas de 75 milímetros S. Schneider-Canet, Tipo Ligero. *Reglamento Provisional para el Servicio de los Cañones de 75mm S. Schneider-Canet Tipo Ligero, Aprobado por La Secretaría de Guerra*. México: Tipografía del Departamento de Estado Mayor, 1904.

Hughes, James B., Jr. *Mexican Military Arms: The Cartridge Period, 1866-1967*. Houston, TX: Deep River Armory, Inc., 1968.

Jomini, Baron Antoine Henri de. *The Art of War*. Introduction by Charles Messenger. First published 1838. This translation published by J.B. Lippincott & Co., Philadelphia, 1862. Novato, CA: Presidio Press, 1992.

Liddell Hart, B.H. *Strategy*. New York: Meridian, 1954, [1991].

Limmenkohl, Hans. *Vom Einzelshuss Zur Feuerwalze: Der Wettlauf Zwischen Technik und Taktik im Ersten Weltkrieg*. Bonn, Germany: Bernard & Graefe Verlag, 1996.

Martynov, A. S. *Enseñanzas de la Guerra Ruso-Japonesa*. Translated by Alfredo Escontria. México: Talleres del Departamento de Estado Mayor, 1908.

Paz, Eduardo. *A Donde Debemos Llegar: Estudio Sociológico Militar*. México: Tipografía Mercantil, 5a Ayuntamiento 100, 1910.

Paz, Eduardo. *Reseña Histórica del Estado Mayor Mexicano*. 2 vols. México: Talleres del Departamento de Estado Mayor, 1907. Vol. 1: 1821-1860.

———. *Temas Tácticos Aplicados a Pequeñas Unidades de Infantería*. México: Talleres Del Departamento de Estado Mayor, 1908.

Secretaría de Guerra y Marina. *Instrucción Práctica Para el Servicio de Infantería en Campaña*. México: Imprenta del Gobierno en el Ex-Arzobispado, 1898.

———. *Reglamento de Maniobras de Infantería*. México: Imprenta y Fototipia de la Secretaría de Industria y Comercio, 1914.

———. *Reglamento para El Ejercicio y Maniobras de la Caballería*. 2 vols. México: Librería de la Vda. De Ch. Bouret, 1921.

———. *Reglamento para el Servicio de la Artillería de Campaña*. México: Talleres del Departamento de Estado Mayor, 1908.

———. *Reglamento para las Maniobras y Combate de la Infantería*. México: Tipografía Literaria de Filomeno Mata, 1887.

———. *Reglamento Provisional para el Servicio y Maniobras de la Artillería de Campaña*. México: Librería de la Vda. De Ch. Bouret, 190?

Sun Tzu Wu, *The Art of War: A Business Strategy Handbook*. Austin, TX: Hale Fred Press, 1991.

Troncoso, Francisco de Paula. *Campaña de 1910 a 1911, estudio en general de las operaciones que han tenido lugar del 18 de noviembre de 1910 al 25 de mayo de 1911, en la parte que corresponde a la 2a zona militar*. México: Secretaría de Guerra y Marina, 1913.

Books and Pamphlets

Aguilar Camín, Héctor. *La Frontera Nómada: Sonora y la Revolución Mexicana*. México: León y Cal Editores, S.A. de C.V., 1999.

Aguirre, Amado. *Mis Memorias de Campaña: Apuntes para la Historia*. México: Estampas de la Revolución Mexicana, 1953.

Aguirre Benavides, Luis. *De Francisco I. Madero a Francisco Villa: Memorias de un Revolucionario*. México: A. del Bosque Impresor, 1966.

Aguirre Benavides, Luis and Adrián. *Las Grandes Batallas de la División del Norte al Mando de Pancho Villa*. México: Editorial Diana, S.A., 1964.

Alarcón Amézquita, Saúl Armando. *En la Línea de Fuego, Juan M. Banderas en la Revolución*. Sinaloa: H. Ayuntamiento de Culiacán, 2013.

Alessio Robles, Miguel. *Obregón como Militar*. México: Editorial Cultura, 1935.

Almada, Francisco R. *La Revolución en el Estado de Chihuahua*. 2 vols. Chihuahua: Instituto Nacional de Estudios Históricos de la Revolución Mexicana, 1964.

Ángeles, Felipe. *La Batalla de Zacatecas*. Zacatecas, 1998.

Aragón, Alfredo. *El Desarme del Ejército Federal por la Revolución de 1913*. Paris: Soc. an. des Imprimeries Wellhoff et Roche, 1915.

Ankerson, Dudley. *Agrarian Warlord: Saturnino Cedillo and the Mexican Revolution in San Luis Potosí*. DeKalb: Northern Illinois University Press, 1984.

Barragán Rodríguez, Juan. *Historia del Ejército y de la Revolución Constitucionalista*. 2 Vols. México: Instituto de Estudios Históricos de la Revolución Mexicana, 1985.

Beezley, William H. *Insurgent Governor: Abraham González and the Mexican Revolution in Chihuahua*. Lincoln: University of Nebraska Press, 1973.

Beltrán, Joaquín. *Toma de la Plaza H. Veracruz El 23 de Octubre y la Intromisión*. México: Herrero Hermanos Sucesores, 1930.

Bravo, José (pseudonym for Jorge Useta). *Impresiones de Guerra: Breve relato de los acontecimientos políticos Mexicanos comprendidos entre el mes de agosto y el noviembre de 1914*. Laredo, TX: Edición del Autor, 1915.

Brondo Whitt, Encarnación. *La División del Norte (1914) Por un Testigo Presencial*. Chihuahua: Ayuntamiento de Chihuahua, 1994 [1940].

Bustamante, Luis F. *De El Ébano a Torreón*. Monterrey, NL: Tip. El Constitucional, 1915.

Calles, Plutarco Elías. *Informe Relativo al Sitio de Naco, 1914-1915*. México: Talleres Gráficas de la Nación, 1932.

———. *Partes Oficiales de la Campaña de Sonora rendidos por el General P. Elías Calles al C. General Álvaro Obregón*. México: Talleres Gráficas de la Nación, 1932.

Calzadíaz Barrera, Alberto. *Hechos Reales de la Revolución*. 8 vols. México: Editorial Patria, S.A., 1959-65.

Caraveo, Marcelo. *Crónica de la Revolución (1910-1929)*. México: Editorial Trillas, 1992.

Cerutti, Mario. *Economía de Guerra y Poder Regional en el Siglo XIX: Gastos Militares, Aduanas y Comerciantes en años de Vidaurri (1855-1864)*. Monterrey: Editorial El Sol, S. A., 1983.

Cervantes, Federico. *Felipe Ángeles y la Revolución de 1913: Biografía (1869-1919)*. México: 1942.

———. *Francisco Villa y la Revolución*. México: Ediciones Alonso, 1960.

Corral, Luz. *Pancho Villa, an Intimacy*. Chihuahua: Centro Librero La Prensa, S.A. de C.V., 1981 [1949].

Crónica Ilustrada Revolución Mexicana. 6 vols. México: Publex, S. A., 1966-1972.

Cumberland, Charles C. *Mexican Revolution: The Constitutionalist Years*. Austin: University of Texas Press, 1972.

Davis, Will B. *Experiences and Observations of an American Consular Officer during the Recent Mexican Revolutions*. Chula Vista, CA, 1920.

De Palo, William A., Jr. *The Mexican National Army, 1822-1852*. College Station: Texas A&M University Press, 1997.

Dorador, Silvestre. *Mi Prisión: La Defensa Social y la Verdad del Caso*. México: Departamento de Tallares Gráficos de la Secretaría de Fomento, 1916.

Gavira, Gabriel. *General de brigada Gabriel Gavira, su Actuación Político-Militar Revolucionaria*. México: Talleres, Tipográficos de A. del Bosque, 1933.

González, Manuel W. *Con Carranza Episodios de la Revolución Constitucionalista, 1913-1914*. 2 Vols. México: Instituto Nacional de Estudios Históricas de la Revolución Mexicana, 1933.

———. *Contra Villa: Relato de la Campaña 1914-15*. México: Ediciones Botas, 1935.

González Garza, R.; Ramos Romero, P.; Pérez Rul, J. *La Batalla de Torreón*. México: Secretaría de Educación Pública, [1914], 1964.

González Marín, Silvia. *Heriberto Jara, luchador obrero en la Revolución Mexicana*. México: El Día en libros, 1984.

Guzmán, Martín Luis. *Memorias de Pancho Villa*. México: Compañía General de Ediciones, S. A., 1951.

Grimaldo, Isaac. *Apuntes para la Historia*. San Luis Potosí: Talleres de Imprenta de la Escuela Industrial Militar, 1916.

Sources Consulted

Gracia García, Guadalupe. *El Servicio Médico durante la Revolución Mexicana*. México: Editores Mexicanos, S. A., 1982.

Herrera, Celia. *Francisco Villa ante la Historia*. México: Editorial Libros de México, S. A., 1964 [1939].

Janssens, Joe Lee. *Maneuver and Battle in the Mexican Revolution: A Revolution in Military Affairs*. 2 vols. Houston, TX: Revolution Publishing II LLC, 2016.

———. *Maneuver and Battle in the Mexican Revolution: Rise of the Praetorians*. Houston, TX: Revolution Publishing LLC, 2015.

———. *Maneuver and Battle in the Mexican Revolution: The Military-Agricultural Complex*. 2 vols. Houston, TX: Revolution Publishing III LLC, 2017.

Juvenal (pseudonym for Enrique Pérez Rul). *Quien es Francisco Villa?* Dallas: Impr. Poliglota, 1916.

Katz, Friedrich. *La Guerra Secreta en México: Europa, Estados Unidos, y la Revolución Mexicana*. México: Ediciones Era, 1981.

———. *Pancho Villa*. 2 vols. México: Ediciones Era, S. A. de C. V., 1998.

London, Jack. *México Intervenido: Reportajes desde Veracruz y Tampico, 1914*. Translated by Elisa Ramírez Casteñeda. México: Ediciones Toledo, 1990.

López de Nava Camarena, Rodolfo. *Mis Hechos de Campaña: Testimonios del General de División Rodolfo López de Nava Baltierra, 1911-1952*. México: Instituto Nacional de Estudios Históricos de la Revolución Mexicana, 1995.

Los Verdaderos Acontecimientos en las Campañas del Norte: Los Combates en Torreón, San Pedro de las Colonias, y Monterrey. México: Talleres "La Tribuna," 1914.

Loyo Camacho, Martha Beatriz. *Joaquín Amaro y el Proceso de Institucionalización del Ejército Mexicano, 1917-1931*. México: Universidad Nacional Autónoma de México Instituto de Investigaciones Históricas; Fidecomiso Archivos Plutarco Elías Calles y Fernando Torreblanca; Instituto Nacional de Estudios Históricos de la Revolución Mexicana; Fondo de Cultura Económica, 2003.

Mantecón Pérez, Adán. *Recuerdos de un Villista: Mi Campaña en la Revolución*. México, 1967.

Manzur Ocaña, Justo. *La Revolución Permanente: Vida y Obra de Cándido Aguilar*. México: B. Costa-Amic Editor, 1972.

Marquez, J. M. *El Veintiuno: Hombres de la Revolución y Sus Hechos: Apuntes Sobre el General de División Jesús Agustín Castro, Jefe de la División "Veintiuno," 1a del Cuerpo de Ejército del Sureste, y demás Ciudádanos que lo han acompañado desde 1910*. México, 1916.

McLynn, Frank. *Villa and Zapata: A History of the Mexican Revolution*. New York: Carroll & Graf Publishers, 2000.

Mellado, Guillermo. *Tres Etapas Políticas de Don Venustiano Carranza: Campañas del Cuerpo de Ejército de Oriente*. México, 1916.

Meyer, Michael C. *Huerta: A Political Portrait*. Lincoln: University of Nebraska Press, 1972.

———. *Mexican Rebel: Pascual Orozco and the Mexican Revolution, 1910–1915*. Lincoln: University of Nebraska Press, 1967.

Meyers, William K. *Forge of Progress, Crucible of Revolt: The Origins of the Mexican Revolution in La Comarca Lagunera, 1880-1911*. Albuquerque: University of New Mexico Press, 1994.

Mora-Torres, Juan. *The Making of the Mexican Border: The State, capitalism, and society in Nuevo León, 1848-1910*. Austin: The University of Texas Press, 2001.

Morales Hesse, José. *El General Pablo González: Datos para la Historia, 1910-1916*. México, 1916.

Moreno, Manuel M. *Historia de la Revolución en Guanajuato*. México: Biblioteca del Instituto Nacional de Estudios Históricos de la Revolución Mexicana, 1977.

Muñoz, Ignacio. *Verdad y Mito de la Revolución Mexicana (Relatada por un Protagonista): Gestación, Estallido y Consecuencias*. 4 vols. México: Ediciones Populares, S.A. México, 1960–1965.

Nugent, Daniel. *Spent Cartridges of Revolution: An Anthropological History of Namiquipa, Chihuahua*. Chicago and London: The University of Chicago Press, 1993.

Obregón, Álvaro. *Ocho Mil Kilómetros en Campaña*. México: Fondo de Cultura Económica, 1959.

Ontiveros, Francisco de P. *Toribio Ortega y la Revolución en la Región de Ojinaga*. Chihuahua: Gobierno de Chihuahua, 2003 [1914].

O'Shaughnessy, Edith. *A Diplomat's Wife in Mexico*. New York: Arno Press & the New York Times, 1970 [1916].

Pani, Alberto J. *Apuntes Autobiográficos*. 2 vols. México: Senado de la República, 1945 [2003].

Pazuengo, Matías. *Historia de la Revolución en Durango: de Junio de 1910 a Octubre de 1914*. Cuernavaca, Mor.: Tip. Del Gobierno del Estado, 1915.

Puente, Ramón. *Pascual Orozco y la Revuelta de Chihuahua*. México: Gómez de la Puente, 1912.

Perez, Juan. *Un Haz De Verdades: La Columna "Navarro" en La Campaña del Ébano*. México: Tip. Guerrero Hnos., 1916.

Palomares, Justino N., and Múzquiz, Francisco. *Las Campañas del Norte (Sangre Y Héroes)*. México: Andrés Botas Editor, 1914.

Perez Peña, Aurelio. *Averiguación previa, instruida a solicitud del señor General Juan J. Navarro con motivo de la rendición de la Plaza de Juárez*. México: El Republicano, 1916.

Pasquel, Leonardo. *La Revolución en el Estado de Veracruz*. 2 vols. México, D.F.: Biblioteca del Instituto Nacional de Estudios Históricos de la Revolución Mexicana, 1972.

Ramos, Miguel S. *Un Soldado: José Refugio Velasco*. México: Ediciones Oasis, S.A., 1960.

Reyes Estrada, José. *Sangre y Metralla: Cinco Días de Lucha a las Puertas de Chihuahua*. Chihuahua: El Libro Rojo, 1913.

Ribot, Héctor. *(La) Revolución de 1912: Pascual Orozco en el norte. Zapata en el sur. Los jefes federales y los combates en Rellano, Conejos y Bachimba. Resumen de la situacion politica*. México City: Impenta 1a Calle de Humboldt Num. 5, 1912.

Rivera de la Torre, Antonio. *El Ébano: Los 72 Días de su Heroica Defensa*. México: Imprenta del Departamento de Estado Mayor de la Secretaría de Guerra y Marina, 1915.

Rivera, Francisco. *Hechos Históricos de la Revolución Constitucionalista hasta ahora desconocidos*. México, DF, 1959.

Robles, Rodolfo G. *Sinaloenses en campaña: Labor de la Columna Expedicionaria de Sinaloa en su Campaña contra la Infidencia*. Culiacán: Imp. De F. Díaz, 1959 [1916].

Rodríguez, J., ed. *La Decena Trágica en México: Datos verídicos tomados en el mismo teatro de sucesos por un escritor metropolitano*. León, Gto: Imp. De "El Obrero," 1913.

Rojas, Beatriz. *La Pequeña Guerra: Los Carrera Torres y los Cedillo*. Zamora, Mich.: El Colegio de Michoacán, 1983.

Rubio de De Ita, Amparo (compil.). *La Revolución Triunfante: Memorias del General de División Guillermo Rubio Navarrete*. México: Libros en Red, 2006.

Ruiz, Eduardo. *Un convoy militar de Veracruz a Encarnación: Toma de Aguascalientes*. Guadalajara, 1915.

Sánchez Lamego, Miguel A. *Generales de la Revolución*. 2 vols. México: Biblioteca del Instituto Nacional de Estudios Históricos de la Revolución Mexicana, 1979, 1981.

———. *Historia Militar de la Revolución Constitucionalista*. 5 Vols. México: Talleres Gráficos de la Nación, 1956-1960.

———. *Historia Militar de la Revolución en la Época de la Convención*. México: Talleres Gráficos de la Nación, 1983.

———. *Historia Militar de la Revolución Mexicana en la Época Maderista*. 3 vols. México: Talleres Gráficas de la Nación, 1976–1977.

Serrano, Tomás. *Episodios de la Revolución en el Estado de Chihuahua*. Chihuahua: Universidad Autónoma de Ciudad Juárez; Secretaría de Educación, Cultura y Deporte, Gobierno del Estado de Chihuahua, 2011 [1911].

Simpson, Leslie Byrd. *Many Mexicos*. Berkeley, Los Angeles, New York: University of California Press, 1971 [1941].

Siurob Ramírez, José. *Labor Revolucionaria e Idealista*. México, 1958.

Smith, Jr., Cornelius C. *Emilio Kosterlitzky: Eagle of Sonora and the Southwest Border*. Glendale, CA: The Arthur H. Clark Company, 1970.

Tablada, José Juan. *Historia de la Campaña de la División del Norte*. México: Imprenta del Gobierno Mexicano, 1913.

Taibo, Paco Ignacio, II. *Pancho Villa: Una Biografía Narrativa*. México: Editorial Planeta Mexicana, S.A. de C.V., 2006.

Terrazas, Joaquín. *Memorias: La Guerra Contra Los Apaches*. Chihuahua, Chih.: Centro Librero La Prensa, S. A. de C. V., 1994.

Thord-Gray, I. *Gringo Rebel: Mexico 1913-1914*. Coral Gables, FL: University of Miami Press, 1960.

Treviño, Jacinto B. *Memorias (de) Jacinto B. Treviño*. México: Editorial Orion, 1961.

———. *Parte Oficial rendido al C. Venustiano Carranza, Primer Jefe del Ejército Constitucionalista, con motivo de las operaciones llevadas a cabo por la 3a División del Cuerpo de Ejército del Noreste, del 21 de Marzo al 31 de Mayo de 1915, en El Ébano, S.L.P.* Monterrey, NL: Tip. El Constitucional, 1915.

Ugalde, R. A. *Vida de Pascual Orozco 1882–1915: Orozco, Guerrillero*. El Paso, TX, 1915.

United States Congress, Senate Committee on Foreign Relations. *Revolutions in Mexico: hearing before a subcommittee of the Committee on Foreign Relations, United States Senate, sixty-second Congress, second session, pursuant to S. Res. 335, a resolution authorizing the Committee on Foreign Relations to investigate whether any interests in the United States have been or are now engaged in inciting rebellion in Cuba and Mexico*. Washington, D.C.: Government Printing Office, 1913.

Urquizo, Francisco L. *Memorias de Campaña*. México: Fondo de Cultura Económica, S. A. de C. V., [1971], 1985.

———. *Origen del Ejército Constitucionalista*. México: Biblioteca del Instituto Nacional de Estudios Históricos de la Revolución Mexicana, 1964.

Valadés, José C. *Rafael Buelna: Las Caballerías de la Revolución*. México: Leega-Júcar, 1984.

Vanderwood, Paul J. *Los Rurales Mexicanos*. México: Fondo de Cultura Económica, 1981.

Vargas Arreola, Juan Bautista. *A sangre y fuego con Pancho Villa*. México: Fondo de Cultura Económica, 2010 [1988].

Vasconcelos, José, and Villa, Francisco. *The Sovereign Revolutionary Convention of Mexico and the Attitude of General Francisco Villa*. Washington: Confidential Agency of the Provisional Government of Mexico, 1915.

Vázquez, Ricardo L. *Poncho Vázquez*. Mexico: Ediciones Botas, 1940.

Velasco Ceballos, Rómulo. *La Infamia Yanqui*. México: Edición Privada, 1914.

Villa, Francisco. *Pancho Villa: Retrato autobiográfico, 1894-1914*. Edited by Guadalupe and Rosa Helia Villa. México: Santillana Ediciones Generales, S.A. de C.V.: Universidad Nacional Autónoma de México, 2003.

Voss, Stuart F. *On the Periphery of Nineteenth-Century Mexico: Sonora and Sinaloa 1810-1877*. Tucson: The University of Arizona Press, 1982.

Wasserman, Mark. *Capitalistas, Caciques y Revolución: La Familia Terrazas de Chihuahua, 1854-1911*. Translated by Beatriz Guiza from the 1984 edition by The University of North Carolina Press, 1984. Mexico: Editorial Grijalbo, S.A., 1987.

Other

Bou, Jean. "Cavalry, Firepower, and Swords: The Australian Light Horse and the Tactical Lessons of Cavalry Operations in Palestine." *The Journal of Military History* 71, 1 (Jan. 2007), pp. 99-125.

Echeverría Adame Marquina, Javier. "¡Viva Carranza!: Mis Recuerdos de la Revolución." *El Legionario* 49 5 (Mar. 1955), pp. 17-28.

Holcombe, Harold Eugene. "United States Arms Control and the Mexican Revolution, 1910-1924." Unpublished Ph.D. dissertation, University of Alabama, 1968.

Vanderwood, Paul J. "Response to Revolt: The Counter-Guerrilla Strategy of Porfirio Díaz." *The Hispanic American Historical Review* 56 4 (Nov. 1976): 551-79.

NOTES

Introduction

Weapon Systems and the Fin de Siècle Battlefield

A. S. Martynov, *Enseñanzas de la Guerra Ruso-Japonesa*, transl. by Alfredo Escontria (México: Talleres del Departamento de Estado Mayor, 1908), 46-48. Hans Limmenkohl, *Vom Einzelshuss Zur Feuerwalze: Der Wettlauf Zwischen Technik und Taktik im Ersten Weltkrieg* (Bonn, Germany: Bernard & Graefe Verlag, 1996), 11-12, 18, 36-41. Secretaría de Guerra y Marina. *Reglamento Provisional para el Servicio y Maniobras de la Artillería de Campaña*. México: Librería de la Vda. De Ch. Bouret, 190?, 138, 154, 221-22.

Federal Army Doctrine

Secretaría de Guerra y Marina, Departamento del Estado Mayor, *Reglamento para las Maniobras y Combate de la Infantería*, (México: Tipografía Literaria de Filomeno Mata, 1887), 7-17, 83-84, 202-5, 238, 252-55, 257, 266-67, 271, 316, 331-41, 347, 352, 354, 359-66, 373-75, 378, 383-85, 404-20. Secretaría de Guerra y Marina, *Instrucción Práctica Para el Servicio de Infantería en Campaña*, (México: Imprenta del Gobierno en el Ex-Arzobispado, 1898), 77-79. Eduardo Paz, *Reseña Histórica del Estado Mayor Mexicano*, 2 vols. (México: Talleres del Departamento de Estado Mayor, 1907), vol. 1: *1821-1860*, 269. Miguel A. Sánchez Lamego, *Historia Militar de la Revolución Constitucionalista*, 5 vols. (México: Talleres Gráficas de la Nación, 1956-1960), 1: 23-25, 29. *Crónica Ilustrada Revolución Mexicana*, 6 vols. (México: Publex, S. A., 1966-1972), 5: 234-35. Secretaría de Guerra y Marina, *Reglamento para El Servicio de la Artillería en Campaña*, (México: Talleres Del Departamento de Estado Mayor, 1908), 18, 24-27, 30-31. Secretaría de Guerra y Marina, *Reglamento Provisional para el Servicio y Maniobras de la Artillería*, 104, 154, 163, 202-3, 208-11, 217-22. Comisión para La Formación del Reglamento para el Servicio de las Piezas de 75 milímetros S. Schneider-Canet, Tipo Ligero, *Reglamento Provisional para el servicio de los Cañones de 75mm S. Schneider-Canet Tipo Ligero, Aprobado por la Secretaría de Guerra*, (México: Tipografía del Departamento de Estado Mayor, 1904), 147-49. Secretaría de Estado y del Despacho de Guerra y Marina, *Reglamento para El Ejercicio y Maniobras de la Caballería*, 2 vols. (México: Librería de la Vda. De Ch. Bouret, 1921), 2: 5-24, 187-88, 278, 300. Francisco Urquizo, *Origen del Ejército Constitucionalista* (México: Biblioteca del Instituto Nacional de Estudios Históricos de la Revolución Mexicana), 1964, 23. Jean Bou, "Cavalry, Firepower, and Swords: The Australian Light Horse and the Tactical Lessons of Cavalry Operations in Palestine," *The Journal of Military History*, Vol. 71, No. 1 (Jan., 2007), 104.

1. Madero's Rebellion, November 1910 – May 1911

Contending Forces

Eduardo Paz, *A Donde Debemos Llegar: Estudio Sociológico Militar* (México: Tipografía Mercantil, 5a Ayuntamiento 100, 1910), 87. Joaquín Beltrán, *Toma de la Plaza H. Veracruz el 23 de Octubre y la Intromisión* (México: Herrero Hermanos Sucesores, 1930), 121. Francisco de Paula Troncoso, *Campaña de 1910 a 1911, estudio en general de las operaciones que han tenido lugar del 18 de noviembre de 1910 al 25 de mayo de 1911, en la parte que corresponde a la 2a zona militar* (Mexico: Secretaría de Guerra y Marina, 1913), 1. NARA-USDS 812.22/1: report from Captain Girard Sturtevant. Harold Eugene Holcombe, "United States Arms Control and the Mexican Revolution, 1910-1924" (unpublished PhD dissertation, University of Alabama, 1968), 15. Paul J. Vanderwood, "Response to Revolt: The Counter-Guerrilla Strategy of Porfirio Díaz," *Hispanic American Historical Review* 56, 4 (November 1976): 558.

Initial Operations, 1910

Notes

Francisco de P. Ontiveros, *Toribio Ortega y la Revolución en la Región de Ojinaga* (Chihuahua: Gobierno de Chihuahua, 2003 [1914]), 32–35. Tomás Serrano, *Episodios de la Revolución en el Estado de Chihuahua* (Chihuahua: Universidad Autónoma de Ciudad Juárez; Secretaría de Educación, Cultura y Deporte, Gobierno del Estado de Chihuahua, 2011 [1911]), 26, 189–90, 195–96, 212, 216, 221–23, 225–26, 234–37, 239–46, 251, 259, 262, 265–66, 271–77. Troncoso, *Campaña de 1910 a 1911*, 1–6, 9, 13–16, 19–26, 28–29, 31–35, 41–42, 44–51, 54–56, 58–59. Miguel A. Sánchez Lamego, *Historia Militar de la Revolución Mexicana en la Época Maderista*, 3 vols. (México: Talleres Gráficas de la Nación, 1976), 1: 46, 48–49, 53–60. Marcelo Caraveo, *Crónica de la Revolución (1910–1929)* (México: Editorial Trillas, 1992), 38–45. R. A. Ugalde, *Vida de Pascual Orozco 1882–1915: Orozco, Guerrillero* (El Paso, TX, 1915), 7–8, 10–11. Francisco Villa, *Pancho Villa: Retrato autobiográfico, 1894–1914*, eds. Guadalupe and Rosa Helia Villa (México: Santillana Ediciones Generales, S. A. de C. V.: Universidad Nacional Autónoma de México, 2003), 386, 389–90, 392–99. Martín Luis Guzmán, *Memorias de Pancho Villa* (México: Compañía General de Ediciones, S. A., 1951), 49–52. Alberto Calzadíaz, *Hechos Reales de la Revolución*, 8 vols. (México: Editorial Patria, S.A., 1959–65), 1: 22, 38. "Another Victory for the Rebels," *The New York Times*, December 27, 1910.

Madero Enters Chihuahua

Serrano, *Episodios de la Revolución*, 275–77, 282–83, 290–91, 293–96, 303. Troncoso, *Campaña de 1910 a 1911*, 66–69, 77–99, 109–10, 106–7, 119–120, 133–35, 138–42, 153, 162–65, 170, 172–73, 175–205, 211, 216–17, 221–23, 238–39, 251. Caraveo, *Crónica*, 45–46. United States Congress, Senate Committee on Foreign Relations, *Revolutions in Mexico: hearing before a subcommittee of the Committee on Foreign Relations, United States Senate, sixty-second Congress, second session, pursuant to S. Res. 335, a resolution authorizing the Committee on Foreign Relations to investigate whether any interests in the United States have been or are now engaged in inciting rebellion in Cuba and Mexico* (Washington, D.C.: Government Printing Office, 1913), 287, 556. Ugalde, *Vida de Pascual Orozco*, 12–14. Sánchez Lamego, *Época Maderista*, 1: 62–63, 65, 67–68, 71–72, 93–94, 98–99. Calzadíaz, *Hechos Reales*, 1: 46.

Revolutionaries Wheel North

Serrano, *Episodios de la Revolución*, 32, 249, 303–4, 309, 320. Troncoso, *Campaña de 1910 a 1911*, 245–59, 261–65, 269–70, 288–294, 317, 320. Aurelio Perez Peña, *Averiguación previa, instruida a solicitud del señor General Juan J. Navarro con motivo de la rendición de la Plaza de Juárez* (México: El Republicano, 1916), 8–9, 12–13, 25–26, 43, 60, 64, 72–74, 76–77, 81–83, 105, 113–14, 117, 119, 127–29, 137–38, 141–42, 149–52. Caraveo, *Crónica*, 49–51. Ugalde, *Vida de Pascual Orozco*, 15. Villa, *Pancho Villa*, 436–38. Federico Cervantes, *Francisco Villa y la Revolución* (México: Ediciones Alonso, 1960), 659–60. Guzmán, *Memorias*, 79, 83–84, 91–99. Subcommittee on Foreign Relations, *Revolutions in Mexico*, 550. Sánchez Lamego, *Época Maderista*, 1: 100–101–9. Calzadíaz, *Hechos Reales*, 1: 64–68.

2. Orozco and Zapata, February – October 1912

AHSDN XI/481.5/68.03, February 15, 1912, Major Adolfo Ramírez to War. Sánchez Lamego, *Época Maderista*, 3: 16–17. William H. Beezley, *Insurgent Governor: Abraham González and the Mexican Revolution in Chihuahua* (Lincoln: University of Nebraska Press, 1973), 128.

Contending Forces

Stuart F. Voss, *On the Periphery of Nineteenth-Century Mexico: Sonora and Sinaloa, 1810–1877* (Tucson, AZ: The University of Arizona Press, 1982), 14, 20–22, 31–34, 48, 54, 67–68, 70–72, 82, 87–88, 95–96, 98–99, 111, 149–57, 206–7, 209–12, 220–21. Daniel Nugent, *Spent Cartridges of Revolution: An Anthropological History of Namiquipa, Chihuahua* (Chicago: The University of Chicago Press, 1993), 55. William De Palo, Jr., *The Mexican National Army, 1822–1852* (College Station: Texas A&M University Press), 154–55. Héctor Aguilar Camín, *La Frontera Nómada: Sonora y la Revolución Mexicana* (México: León y Cal Editores, S. A. de C. V., 1999), 43. William K. Meyers, *Forge of Progress, Crucible of Revolt: The Origins of the Mexican Revolution in La Comarca Lagunera, 1880-1911*, (Albuquerque: University of New Mexico Press, 1994), 4, 7, 28, 32–34, 78, 116–7, 119, 121, 125, 128, 136–7, 139, 145, 187, 230.

The Road to Rellano

AHSDN XI/481.5/68.07, March 5, 1912, Ramírez to War. AHSDN XI/481.5/68.08, March 6, 1912, Ramírez to War. AHSDN XI/481.5/68.09, March 7, 1912, A. G. Peña to Ramírez. AHSDN XI/481.5/68.18, March 17, 1912, report

Strategy and Tactics of the Mexican Revolution

on Battle of Santa Cruz de la Neyra, Chihuahua. AHSDN XI/481.5/106.10-15, April 5, 1912, Joaquín Téllez's report. AHSDN XI/481.5/68.492–495, April 12, 1912, First Captain Morales y Buzo's report. AHSDN XI/481.5/68.496–97, April 10, 1912, Aurelio Blanquet's report. AHSDN XI/481.5/68.498, April 2, 1912, Ignacio M. Corona's report. AHSDN XI/481.5/68.505–508, March 24, 1912, Trucy Aubert's report. AREM 735/14/13, March 4, 1912, report on developments. BHN "Se Necesitan 10,000 Hombres para Defender a Torreón," *El Imparcial* (Mexico City). March 28, 1912, 1, 2, 6. Caraveo, *Crónica*, 67. Phil McLaughlin, special correspondent for the El Paso Herald, "Rebels Beaten Back in Attack on Parral," *The Abilene Daily Reporter*, Abilene, Texas, April 5, 1912, 4. Jacinto B. Treviño, *Memorias (de) Jacinto B. Treviño* (México: Editorial Orion, 1961), 14. Villa, *Pancho Villa*, 467–70. Ramón Puente, *Pascual Orozco y la Revuelta de Chihuahua* (México: Gómez de la Puente, 1912), 110–12. Héctor Ribot, *(La) Revolución de 1912: Pascual Orozco en el norte. Zapata en el sur. Los jefes federales y los combates en Rellano, Conejos y Bachimba. Resumen de la situación política* (México City: Impenta la Calle de Humboldt Num. 5, 1912), 33–35, 40. Subcommittee on Foreign Relations, *Revolutions in Mexico*, 508-9, 562. Guzmán, *Memorias*, 119–23, 125–31. Sánchez Lamego, *Época Maderista*, 3: 44–45, 49, 51–54, 56–57. Michael Meyer, *Mexican Rebel: Pascual Orozco and the Mexican Revolution, 1910–1915* (Lincoln: University of Nebraska Press, 1967), 68. Joe Lee Janssens, *Maneuver and Battle in the Mexican Revolution: Rise of the Praetorians* (Houston: Revolution Publishing LLC, 2015), 231-33.

Multi-State Maneuver Warfare

AGN–CDLR 263/1/31/2–5, April 18, 1912, Huerta to President Madero. AHSDN XI/481.5/29.57–58, 72, April 30, 1912, Trucy Aubert's report. AHSDN XI/481.5/29.85–86, May 21, 1912, Colonel Jiménez Riveroll's report. AHSDN XI/481.5/68.12, March 15, 1912, General Sanginés' report. AHSDN XI/481.5/68.27, March 18, 1912, General Sanginés' report. AHSDN XI/481.5/68.71, May 23, 1912, Huerta to War. AHSDN XI/481.5/68.78, June 16, 1912, Huerta's telegram reporting Rancho de la Cruz. AHSDN XI/481.5/106.36–40, May 29, 1912, Federal Major José Gómez's report. AHSDN XI/481.5/106.58–60, May 17, 1912, Ricardo Peña's report. AHSDN XI/481.5/106.103, June 24, 1912, Jiménez Riveroll's actions. AHSDN XI/481.5/106.106, May 21, 1912, Peña's report. AHSDN XI/481.5/106.119–130, 147, Peña's inquiry. AHSDN XI/481.5/106.140–143, 147, 165–166, May 21, 1912, Avilés action report. AHSDN XI/481.5/106.151–152, May 27, 1912, Avilés action telegram. AHSDN XI/481.5/106.665, May 14, 1912, roster of casualties. AREM 681/1/15–16, May 10, 1912, Interior's bulletin. AREM 681/1/27, May 11, 1912, Ignacio Morelos Zaragoza to Madero. AREM 681/1/29–30, 33–34, May 11, 1912, War bulletin. AREM 681/1/51–54, May 13, 1912, War bulletin. AREM 681/1/90–91, May 20, 1912, War bulletin. AREM 681/1/96–98, 101–103, May 21, 1912, update on Huerta's campaign. AREM 681/1/104, May 23, 1912, Huerta's press release. AREM 681/1/114, n.d., Blanquet and Jiménez Riveroll fighting. AREM 681/1/117, May 23, 1912, Huerta to Madero. AREM 681/1/141, 146, May 27, 1912, from Madero's chief of staff. AREM 681/1/161–63, May 29, 1912, Interior circular. AREM 735/9/4, July 3, 1912, Huerta to President Madero. AREM 741/3/1–2, February 24, 1912, consul in Clifton, AZ, to Secretary of Foreign Relations. AREM 743/11/52–53, August 5, 1912, report Alfredo N. Acosta (consul, Los Angeles, CA). AREM 816/7/1, May 22, 1912, Llorente to President Madero. AREM 816/7/2, May 20, 1912, Candelario Cervantes to Llorente. AREM 816/7/7, May 10, 1912, Municipal President of Namiquipa, Adolfo Delgado, to Enrique Llorente. AREM 816/7/11–13, June 13, 1912, consular reports. AREM 867/2/218, June 13, 1912, from Consul Enrique Llorente. BHN "Llega a Consuelo el Grueso de los Federales," *El Imparcial* (Mexico City), July 1, 1912, 2. BHN "Huerta Está a 10 ktros. de Bachimba," *El Tiempo Diario Católica de la Mañana* (Mexico City), July 2, 1912, 1, 6. BHN "Sólo dos kilómetros separan a los federales de los rebeldes," *El Tiempo* (Mexico City), July 3, 1912, 1, 3. BHN "La Sangrienta Batalla de Bachimba Ha Comenzado," *El Imparcial* (Mexico City), July 4, 1912, 1, 2, 5. BHN "Fracasó el Último Esfuerzo de los Revolucionarios Orozquistas," *El Tiempo* (Mexico City), July 4, 1912, 1. BLAC-CPGMR, Reel 2— manuscript by Anonymous, "Heroicos Cuerpos de Ejército del Noreste y Oriente del General Pablo González," 21– 41. CEHM-GRN 1.18.1, map by Rubio. CEHM-GRN 1.14.1, various dates, Carranza to President Madero. CEHM-GRN 1.15.1, May 13, 1912, Carranza to President Madero. CEHM-GRN 1.15.2, diagram by Rubio. CEHM-GRN 2.55.01, n.d., Rubio's transcript of fighting. CEHM-MWG 1.2.44.1–2, Pablo González to Carranza. NARA-GSS, Entry A1-65, Box 1888, 5761, 5761-1091/31, May 14, 1914, Edwin Emerson Jr.'s report. NARA-GSS, Entry A1 65, Box 1888, 5761, dated July 1914, Captain William Mitchell's "Notes on Mexican Constitutionalists Or The Northern Mexican Insurgents." Caraveo, *Crónica*, 7, 70–72. Matías Pazuengo, *Historia de la Revolución en Durango: de Junio de 1910 a Octubre de 1914* (Cuernavaca, Mor.: Tip. Del Gobierno del Estado, 1915), 16–17. Ontiveros, *Toribio Ortega*, 56–57. Treviño, *Memorias*, 17, 19, 21. José Juan Tablada, *Historia de la Campaña de la Divisón del Norte* (México: Imprenta del Gobierno Mexicano, 1913), 18–20, 25–26, 31–33, 37–40, 43–52, 56–57, 61–62, 65-67, 95-96. "Decisive Battle Is On," *The Washington Post*, July 4, 1912, 1. Ribot, *(La) Revolución de 1912*, 44, 47, 54–58, 63, 65, 70–71, 74, 102–3.

Notes

Subcommittee on Foreign Relations, *Revolutions in Mexico*, 286, 296, 400, 483, 486. Guzmán, *Memorias*, 131–43. Sánchez Lamego, *Época Maderista*, 3: 58–79, 84–86. Sánchez Lamego, *Generales de la Revolución*, 1: 15–16. Holcombe, "Arms Control," 22–26, 30–31. Manzur Ocaña, *Cándido Aguilar*, 55. Meyer, *Mexican Rebel*, 73–74, 81.

Orozco's Small War

AHSDN XI/481.5/29.126, October 9, 1912, Major Guajardo's report. AHSDN XI/481.5/68.86–87, July 31, 1912, Sanginés' initial telegram concerning the Battle of Ojitos. AHSDN XI/481.5/68.98–100, 109–110, July 31, 1912, Sanginés' battle report. AHSDN XI/481.5/68.153–154, August 25, 1912, rosters and reports filed by Toribio Ortega. AHSDN XI/481.5/68.163–165, September 16, 1912, and September 17, 1912, General Trucy Aubert's, Colonel Landa's reports. AHSDN XI/481.5/269.385, July 20, 1912, from Interior to War regarding planned Orozquista invasion of Sonora. AHSDN XI/481.5/269.602, 609, September 17, 1912, Federal Major Eleazar C. Muñoz's report of engagement at El Tigre. AHSDN XI/481.5/269.623–624, September 20, 1912, Obregón's report. AREM 681/1/173, July 11, 1912, telegram from Calero with news that Sanginés was arguing with Garbaldi [sic] and José de la Luz Blanco. AREM 738/2/295, August 2, 1912, message from Sanginés passed on by Governor José M. Maytorena. AREM 745/1/28, August 16, 1912, Llorente's telegram to Ministry of Foreign Relations. AREM 744/3/93–101, September 19, 1912, Consul del Toro's report to Foreign Relations. AREM 816/7/19, July 10, 1912, from J. Sterling to Abraham Molina; AREM 816/7/20–21, July 6, 1912, from Manuel Gonzáles to Abraham Molina; planned Orozquista invasion of Sonora and alleged split in rebel camp. AREM 816/7/24, July 6, 1912, report from Sanginés in Colonia Morelos to his "dear friend" Consul Llorente. AREM 816/7/39–41, June 27, 1912, Sanginés confirming his mission from Huerta, clarifying the route he should take, and requesting money, especially small bills, to recruit men as he passed through the sierra into Chihuahua. AREM 825/1/43, 47, September 6, 1912, Consul Cuesta to Foreign Relations. Confirmation of information received by the Under Secretary of War Plata, AREM 825/1/175, September 9, 1912. AREM 825/1/280, July 22, 1912, Ferrer MacGregor ordered Interior and War informed by Consul Miguel López Torres in Naco, Arizona. AREM 825/1/283, July 20, 1912, planned Orozquista invasion of Sonora per Consul López Torres in Naco, Arizona. BHN "Lo Que Harán Los Rebeldes Después de su Derrota," *El Tiempo* (Mexico City), July 6, 1912, 1, 5. "Buried Bombs Await Huerta at Chihuahua," *The Salt Lake Tribune*, July 5, 1912, 1, 4. Caraveo, *Crónica*, 72–74. Obregón, *Campaña*, 10–26. Ontiveros, *Toribio Ortega*, 58–60. Ribot, *(La) Revolución de 1912*, 10, 75, 83. Sánchez Lamego, *Época Maderista*, 3: 88–89, 91–95.

The Citadel

CEHM-GRN 1.6.1, n.d., Rectifies and Clarifies events around La Ciudadela. CEHM-GRN 1.22.2, n.d., orders to Colonel Rubio Navarrete and his response. CEHM-GRN 2.65.02, n.d., Rubio's response to magazine article. Guerra y Marina, *Artillería de Campaña*, 104. Beltrán, *Veracruz*, xii, 126, 129, 134, 146–47, 150, 159-60, 177–78. J. Rodríguez, ed., *La Decena Trágica en México: Datos verídicos tomados en el mismo teatro de sucesos por un escritor metropolitano* (León, Gto: Imp. De "El Obrero," 1913), 4.

3. The Constitutionalist Revolution, February 1913 – January 1914

AHSDN XI/481.5/88.129-130, April 1, 1913, Decree 435 "suppressed" the military zones, converting each into operational divisions. Juan Barragán Rodríguez, *Historia del Ejercito y de la Revolución Constitucionalista*, 2 vols. (México: Instituto de Estudios Históricos de la Revolución Mexicana, 1985), 1: 182-85. Treviño, *Memorias*, 39. Mario Cerutti, *Economía de Guerra y Poder Regional en el Siglo XIX: Gastos Militares, Aduanas y Comerciantes en Años de Vidaurri (1855-1864)* (Monterrey: Editorial El Sol, S. A., 1983), 27. William K. Meyers, *Forge of Progress, Crucible of Revolt: The Origins of the Mexican Revolution in La Comarca Lagunera, 1880-1911* (Albuquerque: University of New Mexico Press, 1994) 25, 57, 205. Juan Mora-Torres, *The Making of the Mexican Border: The State, Capitalism, and Society in Nuevo León, 1848-1910*, (Austin: The University of Texas Press, 2001), 12, 14–16, 18–19, 31-33, 36–42, 44-50, 53, 70, 87, 94, 101-3, 106, 130-32, 155-57, 159, 222-23, 228.

Northern Sonora Campaign

AREM 830/1/4-7, n.d., report by Ángel Aguilar signed by Kosterlitzky and Reyes. AREM 830/1/17, newspaper clipping, "Battle of Nogales," *The Border Vidette* (Nogales, AZ), March 15, 1913. Barragán, *Revolución Constitucionalista*, 1: 132, 135. Ceballos, *La Infamia Yanqui*, 26–27, 31-32, 33-34, 39, 43-44, 49-52, 56-57, 64, 78-79, 88-95. Obregón, *Campaña*, 28-35, 36-39, 41-42, 44-45, 47-50, 53-54. Rubio, *La Revolución Triunfante*, 44,

Strategy and Tactics of the Mexican Revolution

Rubio Navarrete bemoaned Obregón's superior strategic position and numbers. Sánchez Lamego, *Constitucionalista*, 1: 109-10, 112, 115-16, 119-22, 125-32, 137-38. Héctor Aguilar Camín, *La Frontera Nómada: Sonora y la Revolución Mexicana*, (México: León y Cal Editores, S.A. de C.V., 1999), 379-81. Cornelius C. Smith, Jr., *Emilio Kosterlitzky: Eagle of Sonora and the Southwest Border*, (Glendale, CA: The Arthur H. Clark Company, 1970), 195, 209.

Southern Sonora Campaign

Obregón, *Campaña*, 54-56. Ceballos, *La Infamia Yanqui*, 10, 100-5, 115-16, 118-20, 127-34, 161. Sánchez Lamego, *Constitucionalista*, 1: 141-45, 147-49, 150-54, 164.

Southern Chihuahua Campaign

Barragán, *Revolución Constitucionalista*, 1: 240, 242. Ontiveros, *Toribio Ortega*, 10, 71-88. 134. Joaquin Terrazas, *Memorias: La Guerra Contra los Apaches* (Chihuahua, Chih.: Centro Librero La Prensa, S. A. de C. V., 1994), 154-55. Villa, *Pancho Villa*, 534-56. Cervantes, *Francisco Villa*, 50-53, 56. Guzmán, *Memorias*, 193-94, 196, 201. Calzadíaz, *Hecho Reales*, 1: 22-24, 102-104, 116. Sánchez Lamego, *Constitucionalista*, 1: 197-216, 218, 223; 3: 225. William A. De Palo, Jr. *The Mexican National Army, 1822-1852* (College Station: Texas A&M University Press, 1997), 154-55. Friedrich Katz, *Pancho Villa*, 2 vols. (México: Ediciones Era, S. A. de C. V., 1998), 1: 79, 241-45, 244-48. Daniel Nugent, *Spent Cartridges of Revolution: An Anthropological History of Namiquipa, Chihuahua* (Chicago: The University of Chicago Press, 1993), 5, 44, 46-47, 55, 68. Paco Ignacio Taibo, II, *Pancho Villa: Una Biografía Narrativa* (México: Editorial Planeta Mexicana, 2006), 181-82. Mark Wasserman, *Capitalistas, Caciques y Revolución: La Familia Terrazas de Chihuahua, 1854-1911*, transl. Beatriz Guiza from the 1984 edition by The University of North Carolina Press, 1984, (México: Editorial Grijalbo, S.A., 1987), 213-29, 276-78.

The Guaymas Valley Campaign

Ceballos, *La Infamia Yanqui*, 18, 136-61, 167-94, 204-5. Obregón, *Campaña*, 56-71, 73, 76-78. Rubio, *La Revolución Triunfante*, 67. Miguel Alessio Robles, *Obregón como Militar*, (México: Editorial Cultura, 1935), 65. Sánchez Lamego, *Constitucionalista*, 1: 156-86. *Crónica Ilustrada*, 3: 251.

A Plan for the Northeast

AHSDN XI/481.5/95.02, March 13, 1913, Medina Barrón's column. AHSDN XI/481.5/88.292, Decree 438, May 30, 1913. CEHM-GRN 2.54.02, manuscript that is essentially a Federal Army postmortem. Alfredo Aragón, *El Desarme del Ejército Federal por la Revolución de 1913* (Paris: Soc. an. des Imprimeries Wellhoff et Roche, 1915), 54, 58, 60, 64, 66, 68. Barragán, *Revolución Constitucionalista*, 1: 116-17, 120-22, 126-28, 154-55, 189, 631-37. Manuel W. González, *Con Carranza Episodios de la Revolución Constitucionalista, 1913-1914*, 2 vols. (México: Instituto Nacional de Estudios Históricas de la Revolución Mexicana, 1933), 1: 10. J. M. Marquez, *El Veintiuno: Hombres de la Revolución y Sus Hechos: Apuntes Sobre el General de División Jesús Agustín Castro, Jefe de la División "Veintiuno," 1a del Cuerpo de Ejército del Sureste, y demás Ciudádanos que lo han acompañado desde 1910* (México, 1916), 36. Rubio, *La Revolución Triunfante*, 55, 59. Treviño, *Memorias*, 37-39. Sánchez Lamego, *Constitucionalista*, 1: 59-65, 69-72-74, 91-96, 248-59.

Durango State

AGN-CR 135.488.14, "1912 Estado de Fuerzas-Cuerpos Rurales," December 15, 1912. Barragán, *Revolución Constitucionalista*, 1: 150-1, 156-58, 207, 209-12. Silvestre Dorador, *Mi Prisión: La Defensa Social y la Verdad del Caso* (México: Departamento de Tallares Gráficos de la Secretaría de Fomento, 1916), 207. Justino N. Palomares and Francisco Múzquiz, *Las Campañas del Norte (Sangre Y Héroes)* (México: Andrés Botas Editor, 1914), 13-26. Pazuengo, *Historia de la Revolución en Durango*, 7-9, 36-42, 44-47, 49-53, 65-69-72, 74-75. Sánchez Lamego, *Constitucionalista*, 1: 206, 222-46; 3: 55-76, 80-81, 86; 4: 67, 82. Miguel A. Sánchez Lamego, *Generales de la Revolución*, 2 vols. (México: Biblioteca del Instituto Nacional de Estudios Históricos de la Revolución Mexicana, 1979, 1981), 1: 17. Manzur Ocaña, *Cándido Aguilar*, 73-74.

The Federal Coahuila Campaign

Notes

AHSDN XI/481.5/88.426, August 23, 1913, General Blanquet to Secretary of Foreign Relations. CEHM-GRN 1.31.15, September 8, 1913, report from Téllez to Rubio. CEHM-GRN 1.31.28, September 26, 1913, General Rubio to Huerta. CEHM-GRN 2.55.03, June 22, 1913, transcribed telegram from Huerta to Rubio; July 03, 1913, transcribed telegram from Rubio to Minister of War. CEHM-GRN 2.64.02, November 19, 1943, is copy of a letter from General Rubio Navarrete to a magazine correcting some errors of fact in an article. Barragán, *Revolución Constitucionalista*, 1: 190, 196-98, 206-7, 248-50. González, *Con Carranza*, 1: 10, 13-17, 20, 37-42, 56. Rubio, *La Revolución Triunfante*, 60-61. Treviño, *Memorias*, 42. Sánchez Lamego, *Constitucionalista*, 1: 68-69; 3: 13-14, 18, 25-29, 33-41, 45, 48, 50-1.

Villa's Chihuahua Campaign

AHSDN XI/481.5/88.445, August 26, 1913, orders for Colonel Delgadillo. CEHM-HS 1.1.30.1 Cándido Aguilar's Report. Barragán, *Revolución Constitucionalista*, 1: 243-45, 264, 269-70, 684-91. Luis and Adrián Aguirre Benavides, *Las Grandes Batallas de la División del Norte al Mando de Pancho Villa* (México: Editorial Diana, S.A., 1964), 30-31, 47-52, 71-72. Calzadíaz, *Hecho Reales*, 1: 134, 136-37, 140, 143-44, 147-59. Caraveo, *Crónica*, 78-82. Cervantes, *Francisco Villa*, 54-59, 63-65, 70-71. Luz Corral, *Pancho Villa, an Intimacy* (Chihuahua: Centro Librero La Prensa, S.A. de C.V., 1981 [1949]), 47. Guzmán, *Memorias*, 204-12, 215-32, 236-238, 252, 240-1, 240-50, 254, 384. Celia Herrera, *Francisco Villa ante la Historia* (México: Editorial Libros de México, S. A., 1964 [1939]), 104-5. Juvenal (pseudonym for Enrique Pérez Rul), *Quien es Francisco Villa?* (Dallas: Impr. Poliglota, 1916), 65. Ontiveros, *Toribio Ortega*, 95-124. Pazuengo, *Historia de la Revolución en Durango*, 71. Edith O'Shaughnessy, *A Diplomat's Wife in Mexico* (New York: Arno Press & the New York Times, 1970 [1916]), 157. José Reyes Estrada, *Sangre y Metralla: Cinco Días de Lucha a las Puertas de Chihuahua* (Chihuahua?: El Libro Rojo, 1913), 4-18. Rubio, *La Revolución Triunfante*, 115, 117. Sánchez Lamego, *Constitucionalista*, 1: 219; 3: 65, 79-94, 205-11, 213-17, 223-25, 228-33, 241-42, 246-50; 4: 82-84, 94-98, 103-8, 274-75, 277, 285, 295-304. Villa, *Pancho Villa*, 538-40.

Northern Sinaloa

Barragán, *Revolución Constitucionalista*, 1: 212-13. Obregón, *Campaña*, 79-84, 86-102, 105. Treviño, *Memorias*, 47. Sánchez Lamego, *Constitucionalista*, 3: 266-71, 276-91, 300-5, 320-25.

Tamaulipas

AHSDN XI/481.5/88.34, April 4, 1913, Undersecretary of Foreign Relations to Secretary of War. CEHM-MWG 1.6.779.1, November 2, 1913, Villarreal to González. CEHM-MWG 1.6.780.1, November 3, 1913, from Artigas to González. CEHM-MWG 1.6.781.1, November 3, 1913, Castro's report to González. CEHM-MWG 1.6.782.1, November 4, 1913, Major A. de la Vega to González. CEHM-MWG 1.6.783.1, November 4, 1913, Major A. de la Vega to González. CEHM-MWG 1.6.784.1, November 4, 1913, Castro's report to González. CEHM-MWG 1.6.785.1, November 6, 1913, from E. Meade Fierro to Pablo González regarding Blanco. CEHM-MWG 1.6.794.1, November 11, 1913, Castro responding to González. CEHM-MWG 1.6.795.1, November 12, 1913, from E. Meade Fierro to Pablo González. CEHM-MWG 1.6.797.1, November 13, 1913, Cesáreo Castro to Pablo González. CEHM-MWG 1.6.798.1, November 14, 1913, González to his comrades in arms. CEHM-MWG 1.6.800.1, November 14, 1913, from Arzamendi (?) to Division Chief (Téllez). CEHM-MWG 1.6.803.1, November 15, 1913, Constitutionalist Generals José Isabel Robles and Juan E. Gómez at the "Seguín Military Camp" propose campaign to take Saltillo to Lucio Blanco, Pablo González, and Jesús Carranza. CEHM-MWG 1.7.811.1, November 16, 1913, J. Agustín Castro to González. CEHM-MWG 1.7.813.1, November 16, 1913, J. Agustín Castro to González. CEHM-MWG 1.7.814.1, November 17, 1913, J. Agustín Castro to González. CEHM-MWG 1.7.815.1, November 17, 1913, J. Agustín Castro to González. CEHM-MWG 1.7.816.1, November 17, 1913, from Cesario Castro and Lazo de la Vega. CEHM-MWG 1.7.817.1, November 17, 1913, General Cesarío [in early reports, Don Cesáreo signed his name "Cesarío"] Castro to Pablo González. CEHM-MWG 1.7.818.1; CEHM-MWG 1.7.819.1, November 17, 1913, Carranza to González. CEHM-MWG 1.7.824.1, November 18, 1913, J. A. Castro to González. CEHM-MWG 1.7.826.1, November 19, 1913 from González to C. Undersecretary of War and Navy, Hermosillo, Sonora. CEHM-MWG 1.7.829.1, November 20, 1913, Murguía's strength report. CEHM-MWG 1.7.832.1, November 20, 1913, from Cesario Castro to González. CEHM-MWG 1.7.834.1. CEHM-MWG 1.7.836.1, November 23, 1913, from General F. Murguía to González. CEHM-MWG 1.7.837.1, November 24, 1913, from Caballero to González. CEHM-MWG 1.7.840.1, November 25, 1913, from Castro to González. CEHM-MWG 1.7.841.1, November 25, 1913, from Cesáreo Castro to Generals Villarreal, Castro,

Murguía, and Elizondo. CEHM-MWG 1.7.844.1, November 27, 1913, from Caballero to González. CEHM-MWG 1.7.846.1, November 27, 1913, from V. Carranza to González. CEHM-MWG 1.7.850.1, December 10, 1913, from González to Colonel Andrés Saucedo. CEHM-MWG 1.7.866.1, December 6, 1913, from González to V. Carranza. CEHM-MWG 1.7.872.1, December 12, 1913, from Colonel Andrés Saucedo to González. CEHM-MWG 1.7.887.1, December 21, 1913, orders from González to Sr. Col. Andres Saucedo. CEHM-MWG 1.7.908.1, January 5, 1914, from Brigadiers Jesús Dávila Sánchez and Ernesto Santos Coy, to Pablo González. CEHM-GRN 1.7.1, written manuscript. CEHM-GRN 1.32.04, October 1913, General Téllez to General Rubio. CEHM-GRN 1.32.20, November 20, 1913, from Téllez to Rubio. CEHM-GRN 1.32.31, November 25, 1913, from Rubio to Téllez. CEHM-GRN 2.35.03, October 16, 1913, Lieutenant Colonel Enrique Luebbert to General Rubio. CEHM-GRN 2.36.02, typewritten manuscript of November events, "martillo defensive." CEHM-GRN 2.39.01 unofficial report of Santa Engracia. CEHM-GRN 2.39.02, manuscript and notes by Rubio. CEHM-GRN 2.42.01, January 26, 1914, Rubio's report with map. CEHM-GRN 2.62.01 1913, November 22, 1913, Rubio to Téllez. CEHM-GRN 2.64.02, November 19, 1943, letter from General Rubio Navarrete to a magazine correcting some errors of fact in an article. Barragán, *Revolución Constitucionalista*, 1: 58-62, 249-50, 253-58, 260, 289-90, 292-301, 697-98, 700-2. González, *Con Carranza*, 1: 77, 91-92, 94-96, 98-106, 108-17, 119-25, 127-34, 137-52, 155-60, 170-74, 177-81. Marquez, *El Veintiuno*, 11-12, 48, 51, 54, 56-57. Ontiveros, *Toribio Ortega*, 119. Rubio, *La Revolución Triunfante*, 55, 109-13. Urquizo, *Memorias*, 116. Jorge Useta, *Impresiones de Guerra: Breve relato de los acontecimientos políticos Mexicanos comprendidos entre el mes de agosto y el noviembre de 1914* (Laredo, TX: Edición del Autor, 1915), 14-18. Sánchez Lamego, *Constitucionalista*, 3: 87, 93, 120-22, 124, 127-59, 163-71, 178-79, 214-21; 4: 136-49. Guzmán, *Memorias*, 199-201. Meyer, *Huerta*, 146-54. *Crónica Ilustrada*, 3: 139.

4. The Constitutionalist Revolution, January– August 1914

AHSDN XI/481.5/95.29-36, December 1913, General Eduardo Paz's RESERVED [security clearance] study submitted to General of Division Aurelio Blanquet, Minister of War. AHSDN XI/481.5/95.37-43, January 15, 1914, General of Brigade Eduardo Camargo's strategy counterproposal. CEHM-MWG 1.8.945.1, January 12, 1914, González to General Villa. CEHM-MWG 1.8.947.1, January 12, 1914, González to the First Chief. CEHM-MWG 1.8.954.1, January 15, 1914, Villa's reply to González. CEHM-MWG 1.8.968.1, January 14, 1914, V. Carranza to Pablo González. Barragán, *Revolución Constitucionalista*, 1: 316-17. González, *Con Carranza*, 2: 18. Obregón, *Campaña*, 102-4. Rubio, *La Revolución Triunfante*, 124-25. Guzmán, *Memorias*, 284-86. Holcombe, "Arms Control," 72-73. Katz, *La Guerra Secreta*, 184, 195-96, 214, 216, 270-71.

Southern Coahuila

AHSDN XI/481.5/31.115-6, April 21, 1914, report by General Carlos García Hidalgo. AHSDN XI/481.5/31.103, May 7, 1914, de Moure's report. AHSDN XI/481.5/31.104, May 1, 1914, First Captain Julio García's report. AHSDN XI/481.5/31.107, June 9, 1914, Agustín Valdés' report on artillery assets destroyed. AHSDN XI/481.5/31.114-15, April 21, 1914, report by General Carlos García Hidalgo. AHSDN XI/481.5/31.118-9, April 27, 1914, affidavit written by a Federal Lieutenant Franco. AHSDN XI/481.5/88.790, December 29, 1913, letter to Secretary of War from Treasury. AHSDN XI/481.5/96.315, 323, 335, March 26-29, 1914, regards the new Federal Division of the North. AHSDN XI/481.5/108.12, March 4, 1914, Velasco to War. AHSDN XI/481.5/198.139-148, March, 1914, Federal General Jacinto Guerra's report on the Monterrey defenses. AHSDN XI/481.5/252.498-502, April, 25, 1914, from General Antonio G. Olea to War. AHSDN XI/481.5/252.526-528, May 4, 1914, General Alberto T. Rasgado's report. BLAC-CPGMR, Reel 2—manuscript by Anonymous, "Heróicos Cuerpos de Ejército del Noreste y Oriente del General Pablo González," 140. BLAC-CPGMR, Reel 48, April 18, 1914, message from F. Neira B to Chief of Staff, Alberto Fuentes. BLAC-CPGMR, Reel 48, April 12, 1914, messages to González from First Chief. BLAC-CPGMR, Reel 48, April 1914 message from Pablo González to Tampico. BLAC-CPGMR, Reel 48, April 22, 1914, message from Caballero to General González. BLAC-CPGMR, Reel 48, April 22, 1914, Jesús Carranza to Pablo González. BLAC-CPGMR, Reel 48, April 21, 1914, message from J. Agustín Castro to González. BLAC-CPGMR, Reel 48, April 20, 1914, from E. Reyna to Pablo González. BLAC-CPGMR, Reel 48, correspondence dated April 14, 15 and 16, 1914 related to General Coss. BLAC-CPGMR, Reel 48, n.d. from General Teodoro Elizondo to González. BLAC-LGC IX/A/9, March 12, 1914. CEHM-HS 1.1.30.1 Cándido Aguilar's Report. CEHM-MWG 1.8.972.1, January 14, 1914, Venustiano Carranza to Pablo González. CEHM-MWG 1.8.1081.1, January 22, 1914, Orders of the Day. CEHM-MWG 1.9.1206.1, January 31, 1914, González's orders to Murguía. CEHM-MWG 1.9.1209.1, January 31, 1914, González's

Notes

orders to Coss. CEHM-MWG 1.9.1220.1, January 31, 1914, González's orders to Cesáreo Castro. CEHM-MWG 1.9.1231.1, February 1, 1914, Cándido Aguilar to Pablo González. CEHM-MWG 1.10.1322.1, February 8, 1914, report from Pablo González to the First Chief. CEHM-MWG 1.10.1366.1, February 10, 1914, Teodoro Elizondo to Pablo González. CEHM-MWG 1.10.1367.1, February 11, 1914, report from Elizondo to Pablo González. CEHM-MWG 1.10.1414.1, February 14, 1914, Pablo A. de la Garza to González. CEHM-MWG 1.11.1441.1, February 16, 1914, González to Captain Victoriano Sarmiento. CEHM-MWG 1.11.1589.1, February 25, 1914, from Colonel Emilio Salinas to General González. CEHM-MWG 1.11.1600.1, February 25, 1914, Jesús Carranza to Pablo González. CEHM-MWG 1.12.1625.1, February 26, 1914, Pablo González to Jesús Carranza. CEHM-MWG 1.12.1759.1, March 5, 1914, from González to Fortunato Zuazua. CEHM-MWG 1.13.1819.1, May 8, 1914, Pablo González to Blas Corral. CEHM-MWG 1.13.1824.1, March 8, 1914, González to Castro. CEHM-MWG 1.13.1869.1, March 11, 1914, A. C. Ravelo to Pablo González. CEHM-MWG 1.13.1885.1, March 11, 1914, González confirms Nicéforo Zambrano's report. CEHM-MWG 1.13.1912.1, March 13, 1914, González's report to First Chief. CEHM-MWG 1.13.1913.1, March 13, 1914, Cesáreo Castro's report to Pablo González. CEHM-MWG 1.13.1927.1, March 14, 1914, Nicéforo Zambrano reports on Guardiola. CEHM-MWG 1.14.1941.1, March 15, 1914, from Juan Guimbardo to Nicéforo Zambrano. CEHM-MWG 1.14.1946.1, March 16, 1914, Nicéforo Zambrano to Pablo González. CEHM-MWG 1.14.1949.1, March 16, 1914, from Guimbardo to Zambrano. CEHM-MWG 1.14.1955.1, March 17, 1914, Zambrano's report to González. CEHM-MWG 1.14.1963.1, March 17, 1914, Pablo González to Zambrano. CEHM-MWG 1.14.1968.1, March 18, 1914, Zambrano to Pablo González. CEHM-MWG 1.14.1990.1, March 20, 1914, Castro's after-action report to González. CEHM-MWG 1.14.1991.1, March 20, 1914, Murguía's report to Pablo González. CEHM-MWG 1.14.1995.1 and CEHM-MWG 1.14.1998.1, March 20, 1914, González's orders to Castro and Murguía, respectively. CEHM-MWG 1.14.2035.1, March 24, 1914, report from Major Carlos Prieto. CEHM-MWG 1.14.2069.1, March 27, 1914, Nicéforo Zambrano to Pablo González. CEHM-MWG 1.14.2085.1, March 29, 1914, Cesáreo Castro to Pablo González. CEHM-MWG 1.15.2177.1, April 23, 1914, Maass to Massieu. CEHM-MWG 1.15.2190.1, April 29, 1914, Quintana to Pablo González. CEHM-MWG 1.15.2197.1, May 2, 1914, Pablo González to General A. I. Villarreal. CEHM-MWG 1.15.2232.1, May 13, 1914, Castro reports. CEHM-MWG 1.16.2332.1, May 20, 1914, First Chief's orders to shoot any Federal envoys. CEHM-MWG 1.16.2341.1, May 20, 1914, Colonel Manuel C. Lárraga's men. CEHM-GRN 2.41.18, March 30, 1914, from Rubio to Huerta, Rubio has repaired rails to Gutierrez, thirty kilometers north of Fresnillo. CEHM-GRN 2.64.02, November 19, 1943, General Rubio Navarrete's letter to a magazine. NARA-AGOMI Box 7473, April 16, 1914, Special Plan A for Armed Intervention in Mexico. NARA-AGOMI Box 7473, April 24, 1914, McAdoo to Secretary of War quotes instructions to "Collectors of Customs." NARA-AGOMI Box 7473, September 14, 1914, General Wotherspoon to Winchester Bennett, 1st Vice President of Winchester Repeating Arms Company. NARA-GSS, Entry A1-65, 5768-3, January 21, 1910, Sturtevant No. 306. NARA-GSS, Entry A1-65, 5761-381, February 16, 1912, Sturtevant's terminal report: Conditions in Mexico, January 31, 1912. NARA-GSS, Entry A1-65, 5761-1091/31, May 14, 1914, Edwin Emerson Jr.'s report. Barragán, *Revolución Constitucionalista*, 1: 357, 370-73, 462, 469-70, 472-76, 736-56. Benavides, *Grandes Batallas*, 72. E. Brondo Whitt, *La División del Norte* (Chihuahua: Ayuntamiento de Chihuahua, 1984 [1940]), 43-44, 46, 49-51, 54, 62, 77, 80, 109-10, 124-26, 131-39, 153-54, 159. Calzadíaz, *Hecho Reales*, 1: 81-82, 161, 167, 170, 180-81, 193-94, 198, 200, 244. Federico Cervantes, *Felipe Ángeles y la Revolución de 1913: Biografía (1869-1919)* (México: 1942), 65, 80-83, 85. Cervantes, *Francisco Villa*, 109-11, 139-40. Gabriel Gavira, *General de brigada Gabriel Gavira, su Actuación Político-Militar Revolucionaria* (México: Talleres, Tipográficos de A. del Bosque, 1933), 84-94. González, *Con Carranza*, 2: 24-29, 31, 36-38, 42, 44-47, 56, 59-64, 67-69, 71-85, 93-96. R. González Garza, P. Ramos Romero, J. Pérez Rul, *La Batalla de Torreón*, (México: Secretaría de Educación Pública, [1914], 1964), 5, 7-10, 12-17, 20, 28-43. Guzmán, *Memorias*, 271, 286, 307-12, 314-36, 341-48, 356-60, 362-73, 382, 387-99, 401, 413-14, 418-24, 430-31, 443-45. Joe Lee Janssens, *Maneuver and Battle in the Mexican Revolution: A Revolution in Military Affairs*, 2 vols. (Houston, TX: Revolution Publishing II LLC, 2016), 2: 118, 158. Janssens, *Rise of the Praetorians*, 355. Juvenal, *Quien es Francisco Villa*, 66, 68. *Los Verdaderos Acontecimientos en las Campañas del Norte: Los Combates en Torreón, San Pedro de las Colonias, y Monterrey*, (México: Talleres "La Tribuna, 1914), 6-8, 10-11, 14-18, 20-21, 25-27, 29-31. Manzur Ocaña, *Cándido Aguilar*, 74. Marquez, *El Veintiuno*, 66, 68-69. Ignacio Muñoz, *Verdad y Mito de la Revolución Mexicana (Relatada por un Protagonista): Gestación, Estallido y Consecuencias*, 4 vols. (México: Ediciones Populares, S.A. México, 1961), 3: 119-20, 183, 276-77. Ontiveros, *Toribio Ortega*, 139-40, 143-53. Palomares and Múzquiz, *Las Campañas del Norte*, 34-40, 44-45, 47, 54. Miguel S. Ramos, *Un Soldado: José Refugio Velasco* (México: Ediciones Oasis, S.A., 1960), 139-67. Sánchez Lamego, *Constitucionalista*, 5: 15-26, 30-36, 39-43, 49-51, 56-74, 78-91, 94-97, 103-4, 109, 112-20, 143, 162-68, 170-77, 346. Sánchez

Lamego, *Generales de la Revolución*, 1: 17-21. Sun Tzu Wu, *The Art of War: A Business Strategy Handbook* (Austin, TX: Hale Fred Press, 1991), 68.

The Heartland

AHSDN XI/481.5/31.136-38, July 8, 1914, Villa to Obregón. AHSDN XI/481.5/150.106, July 16, 1914, telegram from General García Hidalgo to Secretary of War. AHSDN XI/481.5/150.108-9, July 20, 1914, Colonel Miguel Lizama's report. AHSDN XI/481.5/150.213-17, August 4, 1914, Doctor José de la Vega's report. AHSDN XI/481.5/334.260-1, June 9, 1914, telegram from Medina Barrón. AHSDN XI/481.5/334.264-265, August 13, 1914, Federal Second Captain Baldomero García's report. AHSDN XI/481.5/334.266, June 11, 1914, Medina Barrón to Secretary of War. AHSDN XI/481.5/334.267, June 11, 1914, Federal report from Aguascalientes. AHSDN XI/481.5/334.271, 275, June 18, 1914, notes regarding Argumedo. AHSDN XI/481.5/334.288-90, July 14, 1914 report filed by officers of the de los Santos Regiment. AREM 787/3/1, May 6, 1914, from R. E. Múzquiz to Isoro Fabela. AREM 789/2/1-2, July 7, 1914, from Inspector of Consuls, Arturo M. Elías, to Foreign Affairs. CEHM-MWG 1.15.2194.1, April 30, 1914, from J. Carranza regarding operations in Center. CEHM-MWG 1.16.2395.1, May 29, 1914, Pablo González to Eulalio Gutiérrez. CEHM-HS 1.1.30.1, Cándido Aguilar's Report. Amado Aguirre, *Mis Memorias de Campaña: Apuntes para la Historia*, (México: Estampas de la Revolución Mexicana, 1953), 47-54. Felipe Ángeles, *La Batalla de Zacatecas* (Zacatecas, 1998), 29-43, 67, 80-81. Anonymous, "Heróicos Cuerpos," 181. Barragán, *Revolución Constitucionalista*, 1: 368-69, 462-63, 485-87, 515-21, 539, 565, 568-69, 576, 585, 599-603, 757-60. Benavides, *Grandes Batallas*, 159-61. Brondo, *La División del Norte*, 123, 168-77, 190-91, 198-99, 206-8, 213, 235-36, 306. Calzadíaz, *Hecho Reales*, 1: 277. Ceballos, *La Infamia Yanqui*, 73-74. Cervantes, *Felipe Ángeles*, 89-106, 109-10, 132-133, 135-40. Cervantes, *Francisco Villa*, 117, 183-84, 719, 722-25, 728, 732-33. Charles C. Cumberland, *Mexican Revolution: The Constitutionalist Years* (Austin: University of Texas Press, 1972), 138. Gavira, *General de Brigada*, 101-3, 108. González, *Con Carranza*, 2: 100, 120, 125-26, 148, 164, 166, 172-73. Silvia González Marín, *Heriberto Jara, luchador obrero en la Revolución Mexicana* (México: El Día en libros, 1984), 156-57. Guzmán, *Memorias*, 452-54, 497-503, 507-13, 515, 518-23, 537. Juvenal, *Quien es Francisco Villa*, 72. B. H. Liddell Hart, *Strategy*, (New York: Meridian, 1954, [1991]), 10, 93, 125, 188, 212, 276-7, 322, the "indirect method." Manuel M. Moreno, *Historia de la Revolución en Guanajuato* (México, D.F.: Biblioteca del Instituto Nacional de Estudios Históricos de la Revolución Mexicana, 1977), 123. Muñoz, *Verdad y Mito*, 2: 102, 184-200, 202-12, 218-19, 258-63-64, 280, 340; 3: 101-2, 137-39. Obregón, *Campaña*, 104-15, 119-31, 135-39, 143-45, 151-52, 158-61, 478. Ontiveros, *Toribio Ortega*, 155-66, 181. Alberto J. Pani, *Apuntes Autobiográficos*, 2 vols. (México: Senado de la República, 1945 [2003]), 1: 183-85. Pazuengo, *Historia de la Revolución en Durango*, 88-96, 101-2, 104-6. Ramos, *Un Soldado*, 58-59, 76, 71-75, 125, 156-61. Francisco Rivera, *Hechos Históricos de la Revolución Constitucionalista hasta ahora desconocidos* (México, D.F., 1959), 2. Beatriz Rojas, *La Pequeña Guerra: Los Carrera Torres y los Cedillo* (Zamora, Mich.: El Colegio de Michoacán, 1983), 41. Sánchez Lamego, *Constitucionalista*, 5: 215-24, 244, 250-53, 257-58, 278-79, 283-84, 287-88, 293-96, 301-3, 307, 312-13. Sánchez Lamego, *Generales de la Revolución*, 1: 21. I. Thord-Gray, *Gringo Rebel: Mexico 1913-1914* (Coral Gables: University of Miami Press, 1960), 228-32, 249, 335-37. Treviño, *Memorias*, 50, 55, 60, 65-68. Urquizo, *Origen*, 21-22, the "Sonoran infantry fought in the Federal system" with "the first sections deployed to form the firing line; the second sections remained as support and the third as reserve." Useta, *Impresiones de Guerra*, 23-4.

5. Civil War, September 1914 – May 1915

Northwest

AGN-CONV 194/1/7/9, October 23, 1914, Carranza to A. I. Villarreal. AGN-CONV 194/1/9/17, October 15, 1914, Convention agrees to cessation of hostilities. AGN-CONV 194/1/9/24, October 16, 1914, Convention asks Governor Maytorena to explain attacks on Naco. AGN-CONV 194/1/9/31-2, October 20, 1914, Secretary of Convention V. Alessio Robles to Villa and Maytorena. AGN-CONV 194/1/9/52, October 22, 1914, A. I. Villarreal to General Ramón V. Sosa. AGN-CONV 194/1/9/65-68, October 23, 1914, from Parral, Maclovio Herrera attacks plaza. AGN-CONV 194/1/9/69-72, October 24, 1914, Francisco Villa to Antonio I. Villarreal. AGN-CONV 194/3/4/105, January 22, 1915. AHSDN XI/481.5/14.80, n.d., from R. V. Pesqueira to V. Carranza. AHSDN XI/481.5/14.83, October 7, 1914, regarding Maytorenistas leaving Santa Rosalía for Guaymas. AHSDN XI/481.5/41.14-15, December 17, 1914, Jesús Carranza to V. Carranza. AHSDN XI/481.5/96.1082, n.d., Mariano Arrieta to First Chief. AHSDN XI/481.5/96.1089-90, October 10, 1914, Carranza to Mariano Arrieta. AHSDN XI/481.5/96.1097, October 16, 1914, Villa takes control

Notes

of line between Aguascalientes and Zacatecas. AHSDN XI/481.5/96.1115, October 22, 1914, Yaquis from Ensenada reach Nogales. AHSDN XI/481.5/96.1146, October 29, 1914, Hill to Carranza. AHSDN XI/481.5/96.1296, n.d., González asks for rearguard attack. AHSDN XI/481.5/96.1306-11, n.d., initial Operations Orders from Carranza. AHSDN XI/481.5/96.1347-1348, Obregón appoints Hill as interim governor of Sonora. AHSDN XI/481.5/96.1435, November 21, 1914, Salazar escapes from prison. AHSDN XI/481.5/96.1529-31, November 30, 1914, Hill to V. Carranza. AHSDN XI/481.5/96.1578-9, December 12-13, 1914, from Calles to Carranza. AHSDN XI/481.5/97.64, January 12, 1915, from J. M. Amador to Carranza. AHSDN XI/481.5/108.18-20, October 5, 1914, from Domingo Arrieta to Carranza. AHSDN XI/481.5/128.374-75; Jesús Carranza on Guaymas. AHSDN XI/481.5/262.117, October 9, 1914, from Mariano Arrieta to Carranza. AHSDN XI/481.5/262.125-27, 143, October 2, 1914, from Maclovio Herrera to the Arrieta brothers. AHSDN XI/481.5/262.145-46, October 17, 1914, Maytorenistas in Baja California. AHSDN XI/481.5/262.149-50, October 22, 1914, M. Arrieta to the First Chief. AHSDN XI/481.5/262.172, November 11, 1914, Herreras arrive in Mazatlán. AHSDN XI/481.5/262.191-94, November 17, 1914, Herrera declines Sonora campaign. AHSDN XI/481.5/271.288-320, Ortigoza's report. AHSDN XI/481.5/271.325, October 15, 1914, Calles to Carranza. AHSDN XI/481.5/271.326-7, October 24, 1914, Calles to Carranza. AHSDN XI/481.5/271.332, November 30, 1914, R.P. Denegri and A.R. Campillo to Carranza. AHSDN XI/481.5/271.338, September 29, 1914, from Hill to Jacinto B. Treviño. AREM 784/6/1, July 2, 1914, from U.S. Consulate, Douglas, Arizona.AREM 789/9/1, October 20, 1914, *Mazatlán* leaves Ensenada, B. C., for Guaymas. AREM 789/10/1 and AREM 789/14/1-2, October 19, 1914, Maytorenistas confiscate steamer *Mazatlán*. AREM 789/11/7-8, September 21, 1914. AREM 836/1/35, February 5, 1915, Cabral and Calles were friends. AREM 836/1/58, January 8, 1915, Pan American News Service reports on Cabral. AREM 836/1/72, January 31, 1915, Pan American News Service report. AREM 836/1/101, February 24, 1915, Cabral quits fight, blames Villa. AREM 836/1/105, February 24, 1915, Pan American News Service report. CEHM-VC 23/2285/1, December 29, 1914, Jesús Carranza to Venustiano Carranza. NARA-USDS 812.2311/120, August 18, 1914, General Hill requests permission to travel through U.S. NARA-USDS 812.2311/129, August 23, 1914, Hill leaves (Ciudad Porfirio) Díaz at noon. NARA-USDS 812.2311/159, from Zachary Cobb, El Paso Collector of Customs and Intelligence. NARA-USDS 812.2311/163, August 30, 1914, Villa leaves Sonora abruptly. NARA-USDS 812.2311/177, September 19, 1914, from Nogales, Arizona, to Secretary of State. Anonymous, "Heróicos Cuerpos," 188. Barragán, *Revolución Constitucionalista*, 1: 497-503, 586. Benavides, *Grandes Batallas*, 201. José Bravo (pseudonym for Jorge Useta), *Impresiones de Guerra: Breve relato de los acontecimientos políticos mexicanos comprendidos entre el mes de agosto y el de noviembre de 1914* (Laredo, TX: 1915), 12-13. "Cabral Leaves Mexico," *The Salt Lake Tribune*, February 25, 1915, 3. Plutarco Elías Calles, *Informe Relativo al Sitio de Naco, 1914-1915* (México: Talleres Gráficas de la Nación, 1932), 3-23. Manuel W. González, *Contra Villa: Relato de la Campaña 1914-15* (México: Ediciones Botas, 1935), 137. Isaac Grimaldo, *Apuntes para la Historia* (San Luis Potosí: Talleres de Imprenta de la Escuela Industrial Militar, 1916), 59. Juvenal, *Quien es Francisco Villa*, 47, 57. "Maytorena Reoccupies Naco," *The Galveston Daily News*, February 22, 1915, 2. Muñoz, *Verdad y Mito*, 2: 223-25, 226-29, 293-95. Obregón, *Campaña*, 153-54, 156-58, 167-69, 172-80, 183-87, 199-214, 246. Pani, *Apuntes Autobiográficos* , 1: 199. Rodolfo G. Robles, *Sinaloenses en campaña: Labor de la Columna Expedicionaria de Sinaloa en su Campaña contra la Infidencia*, (Culiacán: Imp. De F. Díaz, 1959 [1916]), 10-12, 27-29. Vargas, *A Sangre y Fuego*, 193. Guzmán, *Memorias*, 646. Herrera, *Francisco Villa*, 62-63, 66-67, 97, 99-100, 102. Sánchez Lamego, *Convención*, 57-61, 81-82. "Two Thousand Conventionists Force out Maytorena Troops," *The Galveston Daily News*, February 16, 1915, 2.

Mexico City-Puebla

AGN-CONV 194/3/4/94, January 21, 1915, some of Robles' units defect. AGN-CONV 194/4/2/48-9, February 22, 1915, Banderas against Barona. AGN-EZ 256/2/5/34-5, Colonel Flores Alatorre's men disarmed. AGN-EZ 256/2/5/117-24, December 20, 1914, Colonel José Flores Alatorre's complaint. AGN-EZ 256/2/6/21-22, December 22, 1914, parade in homage to General Morelos. AGN-EZ 256/2/6/25, December 23, 1914, Palafox to Juan Martínez Carrasco. AGN-EZ 256/2/6/26, December 14, 1914, the Robles Brigade was assigned to the Engineer Park. AGN-EZ 256/2/6/34, Alfredo Serrratos to Gen. Rafael Cal y Mayor. AGN-EZ 256/2/7/70-1, December 29, 1914. AGN-EZ 256/2/7/108-27, December 30, 1914, Palafox's circulars to jefes, officers, and troops. AGN-EZ 256/3/1/50, December 31, 1914, Palafox to Robles. AGN-EZ 256/3/1/52, December, 31, 1914, to Cotero. AGN-EZ 256/3/1/55, December, 31, 1914, to Chavarría. AGN-EZ 256/3/1/65, December 31, 1914, Astrolabio F. Guerra confirms receipt of circular; AGN-EZ 256/3/2/38, January 1, 1915. AGN-EZ 256/3/2/2, General Marcelino Rodríguez's report to Zapata. AGN-EZ 256/3/2/39-40, January 1, 1915, fighting at Amozoc Station. AGN-EZ 256/3/2/48, January 2, 1915, Palafox to General Francisco V. Pacheco. AGN-EZ 256/3/2/66-7, January 2, 1915, General Aurelio Bonilla's report. AGN-EZ 256/3/2/70-1, January 2, 1915, General Gilberto Camacho to Zapata. AGN-EZ 256/3/2/141, January 5, 1915, Palafox to Colonel Moisés Camacho. AGN-EZ 256/3/3/45-47, January 6, 1915, from Colonel Gustavo Pérez Figueroa. AGN-EZ 256/3/3/49, January 6, 1915, Marcelino Rodríguez's report to Zapata. AGN-EZ 256/3/3/63, 78, January 7, 1915. AGN-EZ 256/3/3/64, January 7, 1915, from General Ricardo Reyes Márquez. AGN-EZ 256/3/3/110, January 7, 1915, Palafox to General Pedro M. Morales. AGN-EZ 256/3/3/122, January 8, 1915, from Colonel Pedro D. Torres. AGN-EZ 256/4/1/20-1, January 11, 1915, General Camacho on Puebla's fall. AGN-EZ 256/4/1/51, January 13, 1915, from Colonel Manuel Herrera y Herrera, in Almazán's column. AGN-EZ 256/4/1/96, January 16, 1915, Benjamín Argumedo wounded. AGN-EZ 256/4/2/68, January 23, 1915, Generals Marcelino Rodríguez, Fortino Ayaquica, and others form junta to agree on battles. AGN-EZ 256/4/2/165, January 28, 1915, Palafox to Colonel Trinidad Sánchez. AGN-EZ 256/4/2/172, January 28, 1915, Palafox to Bonilla. AGN-EZ 256/4/3/39, February 1, 1915, Manuel Palafox to General Domingo Arenas. AGN-EZ 256/4/3/52, 55, February 2, 1915, Palafox to Almazán. AGN-EZ 256/4/3/51, February 2, 1915, Palafox to Colonel Gil R. Serra. AGN-EZ 256/4/3/123, 131, February 3, 1915, Palafox's request. AGN-EZ 256/4/3/206, February 5, 1915, Cal y Mayor's orders from Zapata; AGN-EZ 256/5/1/12. AGN-EZ 256/5/1/33, February 6, 1915, Palafox to Pacheco, Cal y Mayor. AGN-EZ 256/5/1/39, February 6, 1915, Palafox to General Jesús Cázares. AGN-EZ 256/5/1/110-11, February 7, 1915, Pacheco to Zapata. AGN-EZ 256/5/1/117, February 9, 1915, General Chavarría to Licenciado Mauricio L. Chirino. AGN-EZ 256/5/1/126, February 9, 1915, Palafox to General Pacheco. AGN-EZ 256/5/1/131, February 9, 1915, General Pacheco to Palafox. AGN-EZ 256/5/1/141, February 9, 1915, General Inocencio Quintanilla to Zapata. AGN-EZ 256/5/1/147, February 9, 1915, General Pacheco to undersecretary Santiago Orozco. AGN-EZ 256/5/1/161, February 10, 1915. AGN-EZ 256/5/1/171, February 10, 1915, Santiago Orozco's confirmation. AGN-EZ 256/5/3/13, February 16, 1915, Headquarters tells Palafox to order men to stop wasting ammunition by shooting in the air, see also AGN-EZ 256/5/3/22, February 16, 1915; AGN-EZ 256/5/3/58, February 17, 1915; and AGN-EZ 256/5/3/97, February 18, 1915. AGN-EZ 256/5/3/56, February 17, 1915, General of Brigade Francisco Pacheco to Zapata. AGN-EZ 256/5/3/124, February 19, 1915, Colonel Margarito Aguas' encounters with the Bonilla Brigade. AGN-EZ 256/6/1/18-19, February 22, 1915, Pacheco to Zapata. AGN-EZ 256/6/1/41, February 22, 1915, Palafox to Aureliano Azar. AGN-EZ 256/6/1/94, February 24, 1915, Gildardo Magaña to Iriarte. AGN-EZ 256/6/1/125, 135, February 25, 1915, from Pacheco to Zapata. AGN-EZ 256/6/2/55, February 27, 1915, from Pacheco to Zapata. AGN-EZ 256/6/2/76, February 27, 1915, General Santiago Orozco "recommended" that Porfirio Bonilla and Amador collaborate in the operations. AGN-EZ 256/6/3/21, March 1, 1915, from Ayaquica to Zapata. AGN-EZ 256/6/3/44, March 3, 1915, from General Guerra to Zapata. AGN-EZ 256/6/3/78, March 4, 1915, from General Antonio Barona to Zapata. AGN-EZ 256/6/4/32, March 6, 1915, General of Brigade Irineo Albarrán to Zapata. AGN-EZ 256/6/4/44, March 7, 1915, Colonel Federico Córdoba to Zapata. AGN-EZ 256/6/4/49-51, March 7, 1915, Justino Cotero to Zapata. AGN-EZ 256/6/4/68, March 8, 1915, Porfirio Bonilla to Zapata. AGN-EZ 256/6/4/69, March 8, 1915, General Antonio Barona to Zapata. AGN-EZ 256/6/4/73-80, March 8, 1915, Emiliano Zapata to all commanders in the field. AGN-EZ 256/6/4/115, March 9, 1915, Manuel F. Vega to Zapata. AGN-EZ 256/6/4/136, March 10, 1915, Manuel Bonilla to Zapata. AGN-EZ 256/6/4/148-50, March 10, 1915, General Inocencio Quintanilla to Gildardo Magaña. AGN-EZ 256/7/1/20, March 11, 1915, Liberating Army headquarters on soldiers selling ammunition. AGN-EZ 256/8/1/91, May 5, 1915, disagreements between de la O's men and General Antonio Barona. AHSDN XI/481.5/96.1096, October 16, 1914, Pablo González to V. Carranza. AHSDN XI/481.5/96.1595, December 17, 1914, Arredondo to Carranza, regarding General Scott. AHSDN XI/481.5/97.178-9, New York Times describing Constitutionalists as better disciplined, armed, and marksmen. AHSDN XI/481.5/97.186, February 10, 1915, from battalion commander to Military Commander of Mexico City. AHSDN XI/481.5/97.189-90, February 15, 1915, Colonel Tomás Estrada's report. AHSDN XI/481.5/97.191, February 15, 1915, General Porfirio González's report. AHSDN XI/481.5/97.192, February 28, 1915, Alejo González's report. AHSDN XI/481.5/97.222, March 6, 1915, from Obregón to Carranza. AHSDN XI/481.5/220.632-33, October 11, 1914, Pablo González to V. Carranza, regarding precariousness of situation in Puebla. AHSDN XI/481.5/220.700, December 8, 1914, Coss to V. Carranza, fighting around Puebla, requests artillery. AHSDN XI/481.5/220.703, December 8, 1914, Coss to V. Carranza, abandons Atilxco. AHSDN XI/481.5/220.707, December 9, 1914, Coss to V. Carranza. AHSDN XI/481.5/220.709, December 12, 1914, Francisco Coss to Carranza. AHSDN XI/481.5/220.712, December 13, 1914, Zapatista Generals Herminio Chavarría and Agustín Cázares to Conventionist Secretary of War. AHSDN XI/481.5/220.714, December 12, 1914, Zapatista 2nd Regiment, "A. Barona" Brigade reports to Conventionist General Muñoz Santarriaga. AHSDN XI/481.5/220.715, December 13, 1914, from Zapatista Colonel José Hernández to General Muñoz Santarreaga. Other reports of fighting at San Antonio Chautla: AHSDN XI/481.5/220.716, December 13, 1914; AHSDN XI/481.5/220.718,

Notes

December 12, 1914; AHSDN XI/481.5/220.731, December 13, 1914; AHSDN XI/481.5/220.733, December 17, 1914. AHSDN XI/481.5/220.722-23, December 12, 1914, report from the Muñoz Santarreaga and Díaz Brigades. AHSDN XI/481.5/220.748, December 29, 1914, Argumedo's report. AHSDN XI/481.5/220.749, December 30, 1914, from J. I. Robles to General Rafael Espinosa. AHSDN XI/481.5/253.21, January 11, 1915, from General Matías Ramos to J. I. Robles. AHSDN XI/481.5/303.126-28, December 31, 1914, from Obregón to Carranza. AHSDN XI/481.5/304.11-13, January 1, 1915, Obregón to Carranza. AHSDN XI/481.5/304.28-29, January 4, 1915, from Obregón to Carranza. AHSDN XI/481.5/316.5-7, January 1, 1915. AHSDN XI/481.5/315.692, December 13, 1914, General Salvador Alvarado to V. Carranza. AJBT 1/4/221, March 17, 1915, Obregón on arrival of Amaro and Elizondo. AREM 836/1/49, January 9, 1915, Pan American New Service report. AREM 836/1/53-5, January 7, 1915, Benjamín Argumedo wounded. AREM 836/1/63-4, January 20, 1915, Benavides in San Luis Potosí with 4,000 men. AREM 836/1/103, February 24, 1915, Pan American News Service report. CEHM-JA 3/179/1, February 25, 1915, Zapata to all jefes besieging Mexico City. CEHM-MWG 1.11.1562.1, February 23, 1914, General Pablo González congratulates Colonel Porfirio González. CEHM-VC 25/2476/1, January 19, 1915, Gutiérrez's flight. For the meager Constitutionalist ammunition issues to commanders fighting Zapatistas see: CEHM-VC 26/2677/1, February 4, 1915; CEHM-VC 27/2755/1, February 8, 1915; CEHM-VC 27/2793/1, February 10, 1915; CEHM-VC 27/2878/1, February 15, 1915; CEHM-VC 30/3219/1, March 9, 1915. Archivo General de la Nación, *Documentos inéditos sobre Emiliano Zapata y el Cuartel General, seleccionados del archivo de Genovevo de la O, que conserva el Archivo General de la Nación* (México: Comisión para la Conmemoración del Centenario del Natalicio del General Emiliano Zapata, 1979), 76-77. Barragán, *Revolución Constitucionalista*, 2: 170-71, 228-31, 233-37. Calles, *Informe Relativo al Sitio*, 27-28. Cervantes, *Felipe Ángeles*, 103, 183. Cervantes, *Francisco Villa*, 345, 381, 411, 475-76. Javier Echeverría Adame Marquina, "¡Viva Carranza!: Mis Recuerdos de la Revolución," *El Legionario* 49, 5 (March, 1955), 19, 21, 26. Gavira, *General de Brigada*, 107-8, 113-15, 118-19. González, *Contra Villa*, 149. Juvenal, *Quien es Francisco Villa*, 82. Muñoz, *Verdad y Mito*, 2: 335-37. Obregón, *Campaña*, 232-33, 239-40, 245, 249-54, 257-59, 262-65, 269-70, 273-4, 279-81, 287-89. Treviño, *Memorias*, 11-12. Vargas, *A Sangre y Fuego*, 69-70, 197-201. Vasconcelos and Villa, *The Sovereign Revolutionary Convention*, 24-25 decree by General Villa, Assuming Political Power. Guzmán, *Memorias*, 850-51, 857, 865-66. Sánchez Lamego, *Convención*, 42-43. Leslie Byrd Simpson, *Many Mexicos* (Berkeley: University of California Press, 1971 [1941]), 217-23, 226, 241. Friedrich Katz, *La Guerra Secreta en México: Europa, Estados Unidos, y la Revolución Mexicana* (Mexico: Ediciones Era, 1981), 148-49. Loyo Camacho, *Joaquín Amaro*, 32. Paul J. Vanderwood, *Los Rurales Mexicanos* (México: Fondo de Cultura Económica), 39, 53. Paul J. Vanderwood, "Response to Revolt: The Counter-Guerrilla Strategy of Porfirio Díaz," *The Hispanic American Historical Review* 56, 4 (November, 1976), 558.

East-Northeast

AGN-CONV 194/2/7/18, January 6, 1915, General Carrera Torres to Lieutenant Colonel Lamberto C. Chávez. AHSDN XI/481.5/31.234, December 30, 1914, Herrera in command improves morale. AHSDN XI/481.5/32.02, January 2, 1915, and AHSDN XI/481.5/294.18, January 5, 1915, Villarreal requests rearguard attack. AHSDN XI/481.5/32.03, January 5, 1915, González to First Chief. AHSDN XI/481.5/32.12-13, March 5, 1915, from Governor of Coahuila Jesús Acuña to Carranza. AHSDN XI/481.5/32.21-22, March 8, 1915, from F.A. Pereyra, Jesús Acuña, S. Aguirre to Carranza. AHSDN XI/481.5/32.24, May 25, 1915, and AHSDN XI/481.5/32.35, May 26, 1915, Constitutionalists take Monclova. AHSDN XI/481.5/32.28, March 12, 1915, Rosalío Hernández occupies Piedras Negras. AHSDN XI/481.5/32.30, April 23, 1915, Hernández and Pereyra enter Piedras Negras. AHSDN XI/481.5/41.14-15, December 17, 1914, Jesús Carranza in Manzanillo to V. Carranza, regarding the Northwest. AHSDN XI/481.5/96.1044-46, September 23, 1914, Gutiérrez to Carranza. AHSDN XI/481.5/96.1068, October 6, 1914, and AHSDN XI/481.5/198.107, n.d., regarding ex-Federals. AHSDN XI/481.5/97.523, May 8, 1915, Carrera Torres reinforces Raúl Madero at Monterrey. AHSDN XI/481.5/198.89, December 9, 1914, Villarreal to Carranza. AHSDN XI/481.5/198.101, October 5, 1914, A. I. Villarreal to Carranza. AHSDN XI/481.5/198.103, October 5, 1914, Carranza to Villarreal. AHSDN XI/481.5/198.111, October 7, 1914, A. I. Villarreal to Carranza. AHSDN XI/481.5/198.105-6, October 5, 1914, Carranza to Villarreal. AHSDN XI/481.5/199.102, October 5, 1914, Carranza to Antonio I. Villarreal. AHSDN XI/481.5/252.583-84, December 15, 1914, from General Alberto Carrera Torres to Secretary of War. AHSDN XI/481.5/252.595-96, December 19, 1914, Carrera Torres to Robles. AHSDN XI/481.5/252.598, December 19, 1914, Saturnino Cedillo moving to El Ébano. AHSDN XI/481.5/252.606-7, December 23, 1914, Conventionist movement to El Ébano. AHSDN XI/481.5/252.608, December 22, 1914, Cedillo at El Ébano. AHSDN XI/481.5/252.612, December 22, 1914, Robles to Cedillo. AHSDN XI/481.5/252.631-32, December 22, 1914, Conrado Hernandez's men fight at El Higo. AHSDN XI/481.5/252.636-38, December 28, 1914, M.

Cedillo's battle report. AHSDN XI/481.5/252.644, December 25, 1914, Military Commander of Cerritos to War. AHSDN XI/481.5/253.34-36, May 4, 1915, Colonel Apolonio Treviño's report. AHSDN XI/481.5/253.39, May 31, 1915, General Gabriel González Cuéllar's report. AHSDN XI/481.5/253.41-42, Major Fernando Vázquez's report. AHSDN XI/481.5/253.47-49, Benigno's report. AHSDN XI/481.5/293.266, December 18, Carrera Torres to War. AHSDN XI/481.5/293.270, December 19, 1914, Caballero to Carranza. AHSDN XI/481.5/293.274, January 9, 1915, González to First Chief. AHSDN XI/481.5/294.12-13, January 3, 1915, General Caballero to Carranza. AHSDN XI/481.5/294.17, January 5, 1915, Villarreal to V. Carranza. AHSDN XI/481.5/294.20, January 5, 1915, Villarreal to First Chief. AHSDN XI/481.5/294.24, January 7, 1915, González to First Chief. AHSDN XI/481.5/294.31, January 9, 1915, attack on Ciudad Victoria. AHSDN XI/481.5/294.55, March 26, 1915, González to I. Pesqueira. AHSDN XI/481.5/294.66, April 28, 1915, from Pintos to General M. Mendez. AHSDN XI/481.5/294.67, May 4, 1915, from Pablo González to Carranza. AHSDN XI/481.5/316.28, January 5, 1915, González to Carranza. AHSDN XI/481.5/316.39, January 5, 1915, Carranza to González. AHSDN XI/481.5/316.87-88, March 14, 1915, the taking of Dzitbalché. AJBT 1/2/75-78, December 22, 1914, Pablo González to Treviño. AJBT 1/3/134-35, January 6, 1915, Alberto Carrera Torres to M. Cedillo. AJBT 1/3/142-44, January 11, 1915, González Cuéllar in Hidalgo. AJBT 1/3/148-49, planned attack on San Luis Potosí. AJBT 1/3/163-64, January 19, 1915, López de Lara relieves González Cuéllar at Hidalgo. AJBT 1/3/165, January 19, 1915, Treviño to General González Cuéllar. AJBT 1/3/167-69, January 21, 1915, and AJBT 1/3/172-73, January 26, 1915, Cosío Robelo's troops. AJBT 1/3/177-79, January 28, 1915, González Cuéllar to march on Linares. AJBT 1/3/187-88, February 8, 1915, González Cuéllar to General Treviño. AJBT 1/4/202, March 16, 1915, General Alvarado's victories in Yucatán. AJBT 1/4/206, March 17, 1915, from Francisco del Arco to Jacinto B. Treviño. AJBT 1/4/219-20, March 17, 1915, Allende on March 16. AJBT 1/4/231, March 19, 1915, Colorado to join Army Corps of the Northeast. AJBT 1/4/256, April 28, 1915, Constitutionalist air force bombs Villista trains. AJBT 1/4/268-9, May 19, 1915, González to Treviño. AJBT 1/4/271, May 24, 1915, Carranza to Treviño. AREM 836/1/30, January 23, 1915, Pan American News Service report. AREM 836/1/45, January 30, 1915; AREM 836/1/66, January 14, 1915; AREM 836/1/82, January 27, 1915; AREM 836/1/91, January 24, 1915; and AREM 836/1/94, January 30, 1915; Pan American News Service reports about Monterrey. AREM 836/1/47-48, Valles taken January 12, 1915. AREM 836/1/71, January 31, 1915, Pan American News Service report. AREM 836/1/85-86, January 27, 1915, Pan American News Service report. AREM 836/1/103, February 24, 1915, Pan American News Service report. CEHM-MWG 1.19.2755.1, Emilio Salinas' apologia. CEHM-MWG 1.19.2761.1, Lárraga's manuscript. CEHM-MWG 1.19.2769.1, May 15, 1915, Villarreal's letter. CEHM-VC 23/2285/1, December 29, 1914, Jesús Carranza to V. Carranza. CEHM-VC 24/2423/1, January 14, 1915, Villarreal's report. CEHM-VC 26/2703/1, February 4, 1915, from Emilio Salinas to Venustiano Carranza. CEHM-VC 27/2762/1, February 8, 1915, González to V. Carranza. CEHM-VC 29/3114/1, León J. Aguirre reporting on the Northeast. CEHM-VC 31/3311/2, March 15, 1915, González to Carranza. CEHM-VC 32/3380/1, March 19, 1915, from González to Carranza. CEHM-VC 33/3556/1, March 30, 1915, Francisco Aguirre León's report to First Chief Carranza. CEHM-VC 36/3962/1, April 24, 1915, González to Carranza regarding Caballero. CEHM-VC 40/4359/1, May 23, 1915, Carranza to Luis Gutiérrez. NARA-AGOVR, box 7643, September 10, 1915, the spreadsheet titled "Report of Exportations of Arms, Ammunition and Munitions of War through Border Ports into Mexico, March 15, 1915, to September 4, 1915." Anonymous, "Heróicos Cuerpos," 214-16, 207, 218, 224-25. Barragán, *Revolución Constitucionalista*, 2: 135-36, 162, 164-65, 177-80, 231, 242-54, 256, 258-59, 305, 309-10, 312-17, 363, 459-60, 494, 562-63, 602-5. Luis F. Bustamante, *De El Ébano a Torreón*, (Monterrey, NL: Tip. El Constitucional, 1915), 9, 11, 51, 64, 80-83, 90, 93-94, 107. Calzadíaz, *Hechos Reales*, 2: 146-7. Cervantes, *Felipe Ángeles*, 179-88. Cervantes, *Francisco Villa*, 371, 450-51. González, *Con Carranza*, 2: 67-68. González, *Contra Villa*, 95, 123, 126, 130, 143, 145-46, 151, 153-57, 162-63, 168, 172-73, 181, 185-87, 190-97, 201-3, 208, 211, 213-14, 219-21, 226-29, 234-35, 237-38, 241, 243-44, 246, 249-50, 252, 255-56, 263, 268, 270, 277-78. Grimaldo, *Apuntes para la Historia*, 36, 54-55. Guzmán, *Memorias*, 891-95, 904-5, 927-28, 954-55. Juvenal, *Quien es Francisco Villa*, 84-85. Mellado, *Cuerpo de Ejército de Oriente*, 138-39, 215-16, 224-29, 236-37. José Morales Hesse, *El General Pablo González: Datos para la Historia, 1910-1916* (México, 1916), 70-73, 78, 84-85. Muñoz, *Verdad y Mito*, 2: 303-4, 307-8, 327-28. Juan Perez, *Un Haz De Verdades: La Columna "Navarro" en La Campaña del Ébano* (México: Tip. Guerrero Hnos., 1916), 7-8, 11-17, 20-23, 25, 31-32, 34-38, 50, 54-55, 59, 64-68. Antonio Rivera de la Torre, *El Ébano: Los 72 Días de su Heroica Defensa*, (México: Imprenta del Departamento de Estado Mayor de la Secretaría de Guerra y Marina, 1915), 8-9, 10, 12-16, 20-21, 25, 36, 49. Treviño, *Memorias*, 74, 80-84, 86-95. Jacinto B. Treviño, *Parte Oficial rendido al C. Venustiano Carranza, Primer Jefe del Ejército Constitucionalista, con motivo de las operaciones llevadas a cabo por la 3ª División del Cuerpo de Ejército del Noreste, del 21 de Marzo al 31 de Mayo de 1915, en El Ébano, S.L.P.* (Monterrey, N.L.: Tip. El Constitucional, 1915), 4-5.

Notes

Treviño, *Parte Oficial*, 6-24, 26, 29-34, 36-48, 52-61. Vargas, *A Sangre y Fuego*, 149-52, 214-15, 303. Ricardo L. Vázquez, *Poncho Vázquez* (Mexico: Ediciones Botas, 1940), 141-42. "Brownsville Faces Three-Inch Shells," *The Galveston Daily News*, March 29, 1915, 1-2. "Villaistas Delay Matamoros Attack," *The Galveston Daily News*, April 1, 1915, 1. "Carrancistas Claim Victory at Celaya," *The Galveston Daily News*, April 9, 1915, 2. "Defeat Overtakes Villaista Troops," *The Galveston Daily News*, April 13, 1915, 2. "Matamoros Sortie Punishes Besiegers," *The Galveston Daily News*, April 14, 1915, 1. "Zapata Ready to Invade Mexico City," *The Salt Lake Tribune*, April 14, 1915, 4. "Armies Mark Time in Matamoros Lines," *The Galveston Daily News*, April 15, 1915, 1. "Obregon Whips Villa, Takes 8000 Prisoners," *The San Antonio Light*, April 16, 1915, 1. "Matamoros Siege Raised, Is Report," *The Galveston Daily News*, April 17, 1915, 1. "May Announce Firmer Policy as to Mexico," *The Salt Lake Tribune*, April 21, 1915, 4. "Villa, Disorganized Is Not Defeated," *The Galveston Daily News*, April 23, 1915, 1. "U.S. Not Ready to Recognize any Faction," *The San Antonio Light*, April 29, 1915, 5. "President Receives Evidence on Mexico," *The Galveston Daily News*, May 25, 1915, 2. "La Plaza de Ciudad Victoria, Tamaulipas, Quedó Bajo el Control Constitucionalista," *El Pueblo*, May 30, 1915, 1. "Se Confirma La Toma de Monclova Coahuila," *El Pueblo*, May 30, 1915, 1. Dudley Ankerson, *Agrarian Warlord: Saturnino Cedillo and the Mexican Revolution in San Luis Potosí*, (DeKalb, IL: Northern Illinois University Press, 1984), 75. González, *Heriberto Jara*, 177. Janssens, *A Revolution in Military Affairs*, 2: 158-59. Sánchez Lamego, *Convención*, 51-54, 90, 92-93, 96-102, 105-6, 109-10, 112-13.

Bajío

AHSDN XI/481.5/96.1040-43, September 23, 1914, Carranza to Eulalio Gutiérrez. AHSDN XI/481.5/96.1097, October 16, 1914, Villa takes control of line between Aguascalientes and Zacatecas. AHSDN XI/481.5/97.152, January 25, 1915, Urbina's campaign for Tampico forced by a lack of fuel. AHSDN XI/481.5/118.22-25, April 6, 1915, General Hill's report. AHSDN XI/481.5/118.27, April 15, 1915, Obregón's Second Celaya report. AHSDN XI/481.5/118.33-34, April 6, 1915, Obregón to the First Chief. AHSDN XI/481.5/118.48, April 20, 1915, Obregón to V. Carranza. AHSDN XI/481.5/118.50, April 20, 1915, Obregón to Maycotte. AHSDN XI/481.5/118.53-5, 59-61, April 22, 1915, Obregón to V. Carranza. AHSDN XI/481.5/118.64-6, April 22, 1915, Obregón to V. Carranza. AHSDN XI/481.5/118.91, June 22, 1915, Manzo's Report on Battle of Trinidad. AHSDN XI/481.5/161.03, González's report. AHSDN XI/481.5/293.290-2, November 9, 1914, Caballero to V. Carranza. AHSDN XI/481.5/316.103, n.d., Obregón to Military Commander, Jalapa. AJBT 1/2/70-1, November 16, 1914, Pablo A. de la Garza to Treviño. AJBT 1/2/72, November 17, 1914, Vizcayno to Treviño. AJBT 1/2/119-22, December 26, 1914, Treviño to Venustiano Carranza. AJBT 1/3/154, January 17, 1915. AJBT 1/3/155-6, January 18, 1915, Treviño to González. AREM 836/1/60, January 20, 1915, Alvarez defects. CEHM-VC 41/4485/1, June 6, 1915, Obregón's report. Aguirre, *Mis Memorias de Campaña*, 203-4. Luis Aguirre Benavides, *De Francisco I. Madero a Francisco Villa: Memorias de un Revolucionario* (México: A. del Bosque Impresor, 1966), 238. Saúl Armando Alarcón Amézquita, *En la Línea de Fuego, Juan M. Banderas en la Revolución* (Sinaloa: H. Ayuntamiento de Culiacán, 2013), 395-97. Francisco R. Almada. *La Revolución en el Estado de Chihuahua*. 2 vols. (Chihuahua: Instituto Nacional de Estudios Históricos de la Revolución Mexicana, 1964), 2: 276. Barragán, *Revolución Constitucionalista*, 2: 203, 221-25, 229-31, 260-64, 267-70, 273-98, 324-55. Calzadíaz, *Hechos Reales*, 2: 126-27, 182-84, 186-87, 226-27, 234-35, 239-41, 243. Cervantes, *Francisco Villa*, 384-85, 387-88, 395, 415-16, 424, 451-53, 458, 460-65. Gavira, *General de Brigada*, 117-25. González, *Contra Villa*, 64-65, 162. González, "El Centinela Fiel del Constitucionalismo," 320. Gracia García, *El Servicio Médico*, 226, 228. Guzmán, *Memorias*, 851-4, 857, 865-68, 886, 903, 908-27, 933, 939-40, 942-45, 948-52, 955, 961-63. Janssens, *A Revolution in Military Affairs*, 2: 430-32. Baron Antoine Henri de Jomini, *The Art of War*, introduction by Charles Messenger, first published 1838, this translation published by J.B. Lippincott, Co., Philadelphia, 1862, (Novato, CA: Presidio Press, 1992), 296. Juvenal, *Quien es Francisco Villa*, 84, 86-87. Mantecón, *Recuerdos de un Villista*, 84, 87-89. Muñoz, *Verdad y Mito*, 2: 338-39, 357-59, 362-66, 374-77, 380, 382-83; 3: 72-82, 129-30, 135-36, 152-55. Obregón, *Campaña*, 263, 274, 279-80, 293-95, 298-303, 305-11, 314-15, 320, 324, 327, 330-40, 342-45, 347-50, 352-55, 357, 362, 365-76. Perez, *La Campaña del Ébano*, 59-60. Rojas, *La Pequeña Guerra*, 45-46, 48-50. Ruiz, *Un convoy militar*, 20-21, 53. Sánchez Lamego, *Convención*, 42-45, 98, 104. Secretaría de Guerra y Marina, *Maniobras y Combate de la Infantería*, 407-8, see #357 "Inhabited Areas" and #356 "Combat." José Siurob Ramírez, *Labor Revolucionaria e Idealista* (México, 1958), 83-90. Treviño, *Memorias*, 76, 93. Vargas, *A Sangre y Fuego*, 202-4, 214, 217-23, 226-31, 233-49, 257, 310. "Villa Receiving Needed Ammunition," *The Galveston Daily News*, April 19, 1915, 3. "Villa, Disorganized Is Not Defeated," *The Galveston Daily News*, April 23, 1915, 1. "Both Mexican Factions are Claiming Victory," *The Atlanta Constitution*, May 3, 1915, 3. "Yaquis Surround Fifty Americans," *The San Antonio Light*, May 15, 1915, 1. "Hard Battle at Leon," *The Washington Post*, June 5, 1915, 2.

Strategy and Tactics of the Mexican Revolution

West

AHSDN XI/481.5/150.133-4, December 18, 1914, Diéguez's report. AHSDN XI/481.5/151.48-53, January 14, 1915, official report of Estrada's preparatory movement. AREM 836/1/79, January 28, 1915, Villistas admit loss of Guadalajara. Aguirre, *Mis Memorias de Campaña*, 90-92, 113-28, 143-44, 147-52, 167, 175, 177-78. Barragán, *Revolución Constitucionalista*, 2: 212-20. Calzadíaz, *Hechos Reales*, 2: 126-28, 134-37. Will B. Davis, *Experiences and Observations of an American Consular Officer During the Recent Mexican Revolutions* (Chula Vista, CA, 1920), 94-95, 123-29. González, *Contra Villa*, 163, 189. Guzmán, *Memorias*, 853-54, 857, 864-65, 869, 875-85, 890. Juvenal, *Quien es Francisco Villa*, 83-84. Martha Beatriz Loyo Camacho, *Joaquín Amaro y el Proceso de Institucionalización del Ejército Mexicano, 1917-1931*, (México: Universidad Nacional Autónoma de México Instituto de Investigaciones Históricas; Fidecomiso Archivos Plutarco Elías Calles y Fernando Torreblanca; Instituto Nacional de Estudios Históricos de la Revolución Mexicana; Fondo de Cultura Económica, 2003), 32-33. Muñoz, *Verdad y Mito*, 2: 335-37. Obregón, *Campaña*, 259-62. Valadés, *Rafael Buelna*, 70, 74-5. Vargas, *A Sangre y Fuego*, 100-101, 203-8, 212-13.

Strategy

BLAC-CPGMR, Reel 48. Salvador Ruiz Celis, "Campaña Constitucionalista Dirigida por el Señor General de División Don Pablo González." AHSDN XI/481.5/96.1443, November 21, 1914, Obregón's requests to Carranza. AHSDN XI/481.5/96.1445-6, n.d., Obregón to Carranza. AHSDN XI/481.5/129.31, November 5, 1914, General M. C. Zarraga [Lárraga] loosely secures El Ébano. AHSDN XI/481.5/139.338, November 22, 1914, González to Carranza. AHSDN XI/481.5/150.118-21, November 13, 1914, Diéguez to V. Carranza. AHSDN XI/481.5/220. 681, November 21, 1914, Coss to V. Carranza. AHSDN XI/481.5/252.586-87, December 15, 1914, to the President of the Republic from General A. Carrera Torres. AHSDN XI/481.5/293.247, November 22, 1914, Caballero needs López de Lara. AHSDN XI/481.5/293.252, November 23, 1914, Caballero warns: send López de Lara. AHSDN XI/481.5/293.257, November 26, 1914, Caballero sends López de Lara to Tampico. AHSDN XI/481.5/294.11, January 2, 1915, González to V. Carranza on his plans. AHSDN XI/481.5/315.594, November, 1914, J. Carranza to V. Carranza. AHSDN XI/481.5/315.632-3, November 16, 1914, Obregón to Carranza regarding his strategy idea. AJBT 1/2/32-3, November 3, 1914, "Project of Operations," written on the letterhead of the 1st Hidalgo Brigade of the 2nd Division of the Center. AJBT 1/2/126-129, December 28, 1914, Gabriel González Cuéllar to Treviño. AJBT 1/2/131, December 31, 1914, González Cuéllar reports. AJBT 1/5/415-16, August 15, 1915, from Lieutenant Colonel Canuto Villarreal to General Jacinto Treviño. CEHM-VC 21/2144/1, November 24, 1914, Jesús Carranza to Pascual Morales y Molina. CEHM-VC 23/2275/1, December 28, 1914, Caballero to V. Carranza. Anonymous, "Heróicos Cuerpos," 174-5, 199-200. Barragán, *Revolución Constitucionalista*, 2: 132-33, 185, 221-25, 259, 263-64, 559-61. Brondo, *La División del Norte*, 329-30. Cervantes, *Francisco Villa*, 369, 452. Cervantes, *Felipe Ángeles*, 176. González, *Contra Villa*, 70-71, 75-76, 81, 93, 149-50, 193. González, "El Centinela Fiel del Constitucionalismo," 320-21. Guzmán, *Memorias*, 786, 797-99, 884-85, 890, 955-56, 969-71. Katz, *Pancho Villa*, 2: 59. López de Nava, *Mis Hechos de Campaña*, 48. Frank McLynn, *Villa and Zapata: A History of the Mexican Revolution*, (New York: Carroll, Graf Publishers, 2000), 273-85. Mellado, *Cuerpo de Ejército de Oriente*, 136-37. Obregón, *Campaña*, 227, 231-33. Leonardo Pasquel, *La Revolución en el Estado de Veracruz*, 2 vols. (México, D.F.: Biblioteca del Instituto Nacional de Estudios Históricos de la Revolución Mexicana, 1972), 2: 184-86. Rojas, *La Pequeña Guerra*, 46. Treviño, *Memorias*, 77, 103-4. Vargas, *A Sangre y Fuego*, 192, 212-13. "U.S. Troops to Leave Vera Cruz on Nov. 23," *The Atlanta Constitution*, November 14, 1914, 1, 4. "Mexican Flag Files over Vera Cruz," *The Atlanta Constitution*, November 24, 1914, 1.

6. Civil War, June – December 1915

Northeast

AHSDN XI/481.5/32.27, June 12, 1915, and CEHM-VC 42/4570/1, June 14, 1915, from Director General of Telegraphs to the First Chief. AHSDN XI/481.5/97.687, May 27, 1915, Madero retakes Las Vacas and Monclova. AHSDN XI/481.5/294.76, June 11, 1915, from A. Garza González to the First Chief. AHSDN XI/481.5/316.232, June 6, 1915, Veracruz to Eliseo Arredondo in Washington, D.C. AJBT 1/5/341-342, June 13, 1915, Carranza to Treviño regarding materiel. AJBT 2/6/433-4, September 1, 1915, Icamole battle plan . AJBT 2/6/444-5, September 3, 1915. AJBT 2/6/454, September 4, 1915. AJBT 2/6/490, September 24, 1915, Obregón to Treviño. AJBT 2/6/511, September 24, 1915, orders to General Vicente Dávila. AJBT 2/6/537-40, September 25, 1915, General Fortunato

Notes

Zuazua to Treviño. AJBT 2/6/541, September 25, 1915, Treviño to González Cuéllar. AJBT 2/6/559, September 25, 1915, Fran(cisco) González in Hipólito to General Caballero. AJBT 2/6/561, September 25, 1915, Treviño to Matías Ramos. AJBT 2/6/572, September 25, 1915, General Obregón to Matías Ramos. AJBT 2/6/591, September 27, 1915, Treviño to General José Santos. AJBT 2/6/613, September 28, 1915, Treviño to General José Santos. AJBT 2/6/615, September 30, 1915, Major José Riojas in Boquillas reports. AREM 730/2/1, September 16, 1915, Gen. Luis Gutiérrez regarding Seguín Station and Las Vacas, Coah. AREM 796/1/21, September 26, 1915, Lieutenant Colonel F. del Arco, Office of Information, to M. Davalos, Secretary of Foreign Affairs. BLAC-LGC VI/G/8 and 12, October 4, 1915, R. E. Navarro to Lázaro de la Garza. CEHM-VC 40/4399/1, May 28, 1915, E. Gutiérrez to Robles. CEHM-VC 53/5842/1, September 25, 1915, from Treviño to V. Carranza. CEHM-VC 57/6388/1-2, September 22, 1915, from Cayetano González Pérez to Treviño. Almada, *Estado de Chihuahua*, 2: 236. Barragán, *Revolución Constitucionalista*, 2: 373-75, 459-60, 466, 468. Bustamante, *De El Ébano a Torreón*, 11, 13-15, 18, 39, 49-51, 53-74, 77, 115-16. González, *Contra Villa*, 281-83. Juvenal, *Quien es Francisco Villa*, 88. Mellado, *Cuerpo de Ejército de Oriente*, 141, 231-35, 237. Obregón, *Campaña*, 434-45. Sánchez Lamego, *Convención*, 98. Treviño, *Memorias*, 95-102. José C. Valadés, *Rafael Buelna: Las Caballerías de la Revolución* (México: Leega-Júcar, 1984), 87-92. Vargas, *A Sangre y Fuego*, 273, 279, 284-85. "Monclova Recaptured by Villistas Under Madero," *Laredo Weekly Times*, May 30, 1915, 11. "Fall of Mexico City Probable," *The Galveston Daily News*, June 14, 1915, 2. "Final Appeal to Mexico," *The Galveston Daily News*, August 11, 1915, 2. "Siguendo el Plan de Campaña Convenido, El Gral. Treviño Capturó la Importante Plaza de Icamole," *El Demócrata*, September 6, 1915, 1. "A Última Hora Los Reaccionarios Raúl Madero San Román y Ramírez Fueron Aniquilados en Icamole, *El Demócrata*, September 12, 1915, 1-2. "Torreon is Evacuated by Villa," *The San Antonio Light*, September 19, 1915, 1.

El Bajío

AHSDN XI/481.5/06.1-2, June 30, 1915, report from General Martín Triana. CEHM-VC 46/5110/1, July 30, 1915, from Chief of the Office of Information and Propaganda. AHSDN XI/481.5/118.30, July 20, 1915, from Hacienda Santiaguillo to Chief of Operations. AHSDN XI/481.5/118.102-4, June 29, 1915, detailed description of Sánchez's action. AHSDN XI/481.5/118.127-29, July 31, 1915, Obregón to Amaro. AHSDN XI/481.5/118.131-32, trying to locate Fierro and Reyes. AHSDN XI/481.5/118.138, and AHSDN XI/481.5/118.141-2, July 23, 1915, from Amaro to Obregón. AHSDN XI/481.5/118.142, July 22, 1915, from Amaro to Obregón. AHSDN XI/481.5/118.150, July 20[?] from Amaro in Irapuato to General Cecilio García. AHSDN XI/481.5/118.151, July 20, 1915, from Amaro to Obregón. AHSDN XI/481.5/118.153, n.d., from General Amaro to Obregón. AHSDN XI/481.5/118.158, June 22, 1915, from Celaya to Amaro. AHSDN XI/481.5/118.164-65, June 16, 1915, to Amaro. AHSDN XI/481.5/118.166-67, July 16, 1915, from Espinosa Córdoba to Amaro. AHSDN XI/481.5/118.172-173, June 19, 1915, to Amaro. AHSDN XI/481.5/118.175-176, June 20, 1915, from Espinosa Córdoba to Amaro. AHSDN XI/481.5/118.183-4, June 23, 1915, from Espinosa Córdoba to Amaro. AHSDN XI/481.5/122.90, June 27, 1915, report on Sánchez's attack on San Felipe. AHSDN XI/481.5/151.58, June 30, 1915, initial action report. AHSDN XI/481.5/231.19, July 23, 1915, Amaro to Elizondo. CEHM-VC 47/5166/1, August 3, 1915, from Obregón to V. Carranza. Aguirre, *Mis Memorias de Campaña*, 236-37. Alarcón, *En la Línea de Fuego*, 414, 420-21. Anonymous, "Heróicos Cuerpos," 270. Barragán, *Revolución Constitucionalista*, 2: 371, 376-7, 389-90, 392-97, 459-60. Gavira, *General de Brigada*, 126-31. Gracia García, *El Servicio Médico*, 232, 234. Joe Janssens, *Maneuver and Battle in the Mexican Revolution: The Military-Agricultural Complex*, 2 vols. (Houston, TX: Revolution Publishing, 2017), 2: 182. Mantecón, *Recuerdos de un Villista*, 89. Obregón, *Campaña*, 388-98, 400-9, 412-27. Ruiz, *Un convoy militar*, 11, 27-28, 30-32, 36-37, 44, 46, 58-60, 81, 85, 92-95, 99-102, 105-6. Vargas, *A Sangre y Fuego*, 250, 252-53, 260-62, 274-77. "Villa Troops Abandon San Luis Potosi, Too," *The San Antonio Light*, July 15, 1915, 1. "Carranza Soldiers Scatter Zapatistas," *The Galveston Daily News*, July 17, 1915, 1. "Villa Plans Campaign in Two States," *The San Antonio Light*, July 17, 1915, 2.

Strategy

Barragán, *Revolución Constitucionalista*, 2: 372-76. González, *Contra Villa*, 244-6, 263-5, 268.

Mexico City-Puebla

AGN-EZ 256/8/4/51, June 3, 1915, Constitutionalists attacked and took Calpulalpam. AGN-EZ 256/8/6/63, June 28, 1915, General Mejía's report. AGN-EZ 256/9/1/18-19, July 1, 1915 Agustín Cortés reports. AGN-EZ 256/9/1/37, July 3, 1915, Manuel Bonilla reports. AGN-EZ 256/9/2/32, July 17, 1915, General Cal y Mayor's report.

Strategy and Tactics of the Mexican Revolution

Various communications related to Zapata's Puebla campaign: AGN-EZ 256/9/2/20-1, July 15, 1915; AGN-EZ 256/9/2/25, July 16, 1915; AGN-EZ 256/9/2/26, July 16, 1915; AGN-EZ 256/9/2/27, July 16, 1915; AGN-EZ 256/9/2/33, July 17, 1915; AGN-EZ 256/9/2/34, July 17, 1915; AGN-EZ 256/9/2/35, July 17, 1915; AGN-EZ 256/9/2/36, July 17, 1915; AGN-EZ 256/9/2/37, July 17, 1915; AGN-EZ 256/9/2/45, July 18, 1915; AGN-EZ 256/9/2/57-58, July 18, 1915; AGN-EZ 256/9/2/60, July 18, 1915; AGN-EZ 256/9/2/63, July 19, 1915; AGN-EZ 256/9/2/65, July 19, 1915; AGN-EZ 256/9/2/67, July 19, 1915; AGN-EZ 256/9/2/77, July 20, 1915; AGN-EZ 256/9/2/78, July 20, 1915; AGN-EZ 256/9/3/3, July 21, 1915; AGN-EZ 256/9/3/4, July 21, 1915; AGN-EZ 256/9/3/6, July 21, 1915; AGN-EZ 256/9/3/12, July 23, 1915; AGN-EZ 256/9/3/15, July 23, 1915; AGN-EZ 256/9/3/22, July 26, 1915; AHSDN XI/481.5/97.842-43, June 19, 1915, Zapatista report. AHSDN XI/481.5/97.864, June 22, 1915, 1st Regiment of Cepeda's Brigade. AHSDN XI/481.5/97.869, June 23, 1915, from Colonel Secretary A. García to General Francisco V. Pacheco. AHSDN XI/481.5/97.920, June 30, 1915, Zapatista report. AHSDN XI/481.5/161.07, June 14, 1915, Argumedo to Secretary of War. AHSDN XI/481.5/161.08, June 13, 1915, General Cepeda to Coss. AHSDN XI/481.5/161.09, June 18, 1915, transcription of Azar's report to Secretary of War. AHSDN XI/481.5/161.10-12, June 25, 1915, report from Major of 3rd Battalion of Veracruz to General Odilón V. Moreno. AHSDN XI/481.5/161.26, August 1, 1915, General Cepeda to González. AHSDN XI/481.5/161.27, August 3, 1915, Cepeda to Coss. AHSDN XI/481.5/221.233, July 22, 1915, report to Cepeda. AHSDN XI/481.5/304.87, June 13, 1915, Manuel Bonilla's report. AHSDN XI/481.5/304.89, June 13, 1915, Zapatista report. AHSDN XI/481.5/316.259, June 21, 1915, Carranza to Obregón. AHSDN XI/481.5/316.260, June 27, 1915, Carranza to Arredondo about the decisive blow. AREM 836/2/37-39, July 13, 1915, Burns to Denegri. CEHM-VC 46/5084/1, 2, July 28, 1915, from the Constitutionalist Office of Information and Propaganda. CEHM-VC 46/5110/1, July 30, 1915, from the Constitutionalist Office of Information and Propaganda. CEHM-VC 47/5149/1, August 1, 1915, report from Chief of the Office of Information and Propaganda. Alarcón, *En la Línea de Fuego*, 411-17. Almada, *Estado de Chihuahua*, 2: 197. Anonymous, "Heróicos Cuerpos," 248-51, 254-57, 259, 270-72, 277-78. Archivo General de la Nación, *Genovevo de la O*, 78. Barragán, *Revolución Constitucionalista*, 2: 358-61, 390, 608-9, 612-17. Cervantes, *Francisco Villa*, 450. González, *Contra Villa*, 113, 196-7, 273-4, 279-80, 286, 291-88, 293, 296-300, 303, 307, 309-10, 313, 321-25, 342-43, 345, 349, 351-52, 354-60. López de Nava, *Mis Hechos de Campaña*, 53-56. Mellado, *Cuerpo de Ejército de Oriente*, 85-86, 89-90, 93-94, 128, 132-34, 274. Morales Hesse, *El General Pablo González*, 95-98, 101-3, 108-9, 114. Sánchez Lamego, *Convención*, 206. "Mexico Needs Shipments of Food," *Nevada State Journal*, May 23, 1915, 3. "City of Mexico Quiet with Its Capture," *The Galveston Daily News*, July 14, 1915, 1.

Northwest

Calles and Obregón on the Villista invasion force: AREM 836/3/195-8, October 23, 1915; AREM 836/3/199-200, October 24, 1915; AREM 836/3/10, October 27, 1915; AREM 836/3/12, October 28, 1915. Progress of Constitutionalist troop trains: AREM 809/2/44, October 29, 1915; AREM 809/2/45, 47, October 30, 1915; AREM 809/2/48-51, October 30, 1915; AREM 810/2/158; AREM 810/2/159-60, October 30, 1915; NARA-USDS 812.2311/191, October 25, 1915; NARA-USDS 812.2311/194, October 28, 1915; NARA-USDS 812.2311/195, October 30, 1915; NARA-USDS 812.2311/198, October 30, 1915. Approval for Constitutionalists to pass through U.S. territory: NARA-USDS 812.2311/185, October 19, 1915; NARA-USDS 812.2311/188, October 20, 1915; and, NARA-USDS 812.2311/190, October 23, 1915. AHSDN XI/481.5/272.23-26, September 29, 1915, to Diéguez from Navojoa. AHSDN XI/481.5/272.34, November 4, 1915, General Flores' report. AHSDN XI/481.5/32.38, November 5, 1915, Obregón to the governor of Durango. AHSDN XI/481.5/272.42-45, November 18, 1915, Colonel Melitón Albáñez's report. AHSDN XI/481.5/272.54, December 1, 1915, Flores' report. AHSDN XI/481.5/272.56, December 3, 1915, Diéguez reports. AHSDN XI/481.5/272.58, December 12, 1915, Carranza to Undersecretary of War. AHSDN XI/481.5/272.60, November 21, 1915, Diéguez's report. AJBT 1/5/420, August 22, 1915, Treviño to Pablo González. AJBT 2/6/435, September 1, 1915, from Treviño to Enrique W. Paniagua. AJBT 2/8/848, November 23, 1915, Arturo Prieto in El Paso reports. AJBT 2/8/853, November 30, 1915, Cruz Domínguez (Agustín Estrada Brigade) to Villa. AJBT 2/8/861, December 3, 1915, General Genaro Baca to Governor General Fidel Ávila. AJBT 2/8/868, December 5, 1915, General Roberto Limón to General Baca. AJBT 2/8/872, December 6, 1915, General Baca to General Roberto Limón. AJBT 2/8/873, December 6, 1915, Colonel Valentín Vázquez to General Cruz Domínguez. AJBT 2/8/893, December 21, 1915, the order of march into Chihuahua City. AREM 729/4/4-5, October 14, 1915, Urbina's men rebel against Villistas. AREM 730/2/1, September 16, 1915, regarding Herrera. AREM 808/8/21, 24-27, December 20, 1915, surrender and amnesty agreements. AREM 808/8/38, December 21, 1915, Consul García to Major José Riojas. AREM 808/8/49, December 21, 1915, Villistas fleeing to mountain. AREM 810/2/5, October 5, 1915, regarding Villa

Notes

surrender. AREM 810/1/10, November 28, 1915, Vice-consul Soriano Bravo to Carranza. AREM 810/1/35-6, December 9, 1915, from Andrés G. García to First Chief. AREM 810/1/60, December 16, 1915, Andrés García to the First Chief. AREM 819/2/8-11, June 16, 1915, Domingo Arrieta's 1915 campaigns. AREM 819/2/31, June 13, 1915, Calles to Denegri. AREM 836/3/78, November 15, 1915, from Obregón to Diéguez. AREM 836/3/89, November 13, 1915, regarding Diéguez. AREM 836/3/91, October 25, 1915, from General Calles for Diéguez. AREM 836/3/101, December 4, 1915, Obregón's staff reports. AREM 836/3/176, October 19, 1915, reports regarding Diéguez. AREM 836/3/186-87, October 23, 1915, from General M. M. Diéguez to Consul Ramón P. Denegri. AREM 836/3/214-15, November 26, 1915, Obregón to Diéguez. BLAC-LGC VI/G/13, October 5, 1915, Urbina's men may have joined Rosalío Hernández's command. CEHM-VC 47/5166/1, August 3, 1915, from the Chief of the Office of Information and Propaganda. CEHM-VC 51/5606/1, September 3, 1915. CEHM-VC 53/5877/1, November 5, 1915, Carlos Carranza's service. CEHM-VC 61/6760/1, November 17, 1915, from Mariano Arrieta to Carranza. Aguirre, *Mis Memorias de Campaña*, 240, 242-49. Alarcón, *En la Línea de Fuego*, 426-39. Almada, *Estado de Chihuahua*, 2: 219, 238-41, 257, 290-91, 299-300. Barragán, *Revolución Constitucionalista*, 2: 460-65, 487, 492, 512-14, 516-21, 525-30. Plutarco Elías Calles, *Partes Oficiales de la Campaña de Sonora rendidos por el General P. Elías Calles al C. General Álvaro Obregón*, (México: Talleres Gráficas de la Nación, 1932), 82-92. Cervantes, *Francisco Villa*, 527-30, 532. Gavira, *General de Brigada*, 149-50, 154-66. Herrera, *Francisco Villa*, 68. Juvenal, *Quien es Francisco Villa*, 87-89. Mantecón, *Recuerdos de un Villista*, 97-103, 108. Mellado, *Cuerpo de Ejército de Oriente*, 347-49. Muñoz, *Verdad y Mito*, 2: 404-5; 3: 16-19, 91. Obregón, *Campaña*, 428-34, 449-52, 456-70, 476, 478-83. Robles, *Sinaloenses en campaña*, 49-65, 69-73. Sánchez Lamego, *Convención*, 65-66, 73, 81-82. Treviño, *Memorias*, 104-6. Vargas, *A Sangre y Fuego*, 247-48, 264-71, 278, 286-93, 295-304, 310-11, 315. "Intention of Villa Puzzle to Officers," *The Galveston Daily News*, November 4, 1915, 1. "Obregon Will Change Plans Against Villa," *The San Antonio Light*, November 24, 1915, 7.

Conclusion

AHSDN XI/481.5/88.262, Decree 436, May 1, 1913, per Article 1, "the School of the Soldier of the Army will be rudimentary." June 1, 1913, the Law of Obligatory Military Service took effect, AHSDN XI/481.5/88.265-266. AHSDN XI/481.5/198.139-148, Federal General Jacinto Guerra's report on the defenses at Monterrey in March, 1914. AHSDN XI/481.5/271.317, Ortigoza's report. NARA-GSS, Entry A1-65, 5761-1091/31, Edwin Emerson Jr.'s report of May 14, 1914. NARA-GSS, 7141/1, July 30, 1912, Burnside No. 756: Artillery. Calles, *Partes Oficiales*, 39-50, 92. Calles, *Informe Relativo al Sitio*, 20. Calzadíaz, *Hechos Reales*, 1: 43, 137, 140; 2: 211-12. Gavira, *General de Brigada*, 113. Guzmán, *Memorias*, 245. Muñoz, *Verdad y Mito*, 2: 277; 3: 72-73. Sánchez Lamego, *Constitucionalista*, 1: 376-78. Treviño, *Memorias*, 84. Urquizo, *Origen*, 19-20 (quotes about the qualities of Federal Army personnel). "Horse is Shot Under Villa in a Charge," *The San Antonio Light*, April 26, 1915, 3, railborne Villista assault block quote. "General Murguía Was Not Burned to Death," *The San Antonio Light*, May 3, 1915, 2, (quote).

Appendix B: Weapon Systems

CEHM-GRN 1.9.1, n.d., Spring 1913 Federal arsenal. NARA-GSS 6934-3, January 2, 1912, Girard Sturtevant. NARA-GSS 6158-6, December 5, 1910, Girard Sturtevant. Comisión para La Formación del Reglamento para el Servicio de las Piezas de 75 milímetros S. Schneider-Canet, Tipo Ligero, *Reglamento Provisional para el Servicio de los Cañones de 75mm S. Schneider-Canet Tipo Ligero, Aprobado por La Secretaría de Guerra*, (México: Tipografía del Departamento de Estado Mayor, 1904), 191. *Crónica Ilustrada*, 3: 251; 4:91; 5: 234-35; 6: 170-1. Eduardo Paz, *Temas Tácticos Aplicados a Pequeñas Unidades de Infantería*, (México: Talleres del Departamento de Estado Mayor, 1908), 66. Secretaría de Guerra y Marina, *Reglamento de Maniobras de Infantería*, (México: Imprenta y Fototipia de la Secretaría de Industria y Comercio, 1914), 106-7. Barragán, *Revolución Constitucionalista*, 1: 58. Committee on Foreign Relations, *Revolutions in Mexico*, 121-22. James B. Hughes, Jr., *Mexican Military Arms: The Cartridge Period, 1866-1967*, (Houston, TX: Deep River Armory, Inc., 1968), 63-79, 100. Jack London, *México Intervenido: Reportajes desde Veracruz y Tampico, 1914*, trans. Elisa Ramírez Casteñeda (México: Ediciones Toledo, 1990), 104-5. Obregón, *Campaña*, 348. https://en.wikipedia.org/wiki/No_2_grenade Treviño, *Memorias*, 22. Sánchez Lamego, *Constitucionalista*, 5: 18-19. Urquizo, *Origen*, 20.

INDEX

Index

Index

Index

Index

Index

Index

Index

Index

Printed in Great Britain
by Amazon